Manufacturing Processes of Metallic Materials

Manufacturing Processes of Metallic Materials

Guest Editors

Hong Xiao
Chao Yu

MDPI

Basel • Beijing • Wuhan • Barcelona • Belgrade • Novi Sad • Cluj • Manchester

Guest Editors
Hong Xiao
National Engineering
Research Center for
Equipment and Technology of
Cold Strip Rolling
Yanshan University
Qinhuangdao
China

Chao Yu
National Engineering
Research Center for
Equipment and Technology of
Cold Strip Rolling
Yanshan University
Qinhuangdao
China

Editorial Office
MDPI AG
Grosspeteranlage 5
4052 Basel, Switzerland

This is a reprint of the Special Issue, published open access by the journal *Metals* (ISSN 2075-4701), freely accessible at: https://www.mdpi.com/journal/metals/special_issues/U128A5GDVC.

For citation purposes, cite each article independently as indicated on the article page online and as indicated below:

Lastname, A.A.; Lastname, B.B. Article Title. *Journal Name* **Year**, *Volume Number*, Page Range.

ISBN 978-3-7258-6402-7 (Hbk)
ISBN 978-3-7258-6403-4 (PDF)
https://doi.org/10.3390/books978-3-7258-6403-4

Contents

metals

MDPI

Editorial

Manufacturing Processes for Metallic Materials

Chao Yu and Hong Xiao *

National Engineering Research Center for Equipment and Technology of Cold Strip Rolling, Yanshan University, Qinhuangdao 066004, China; yuc@ysu.edu.cn

* Correspondence: xhh@ysu.edu.cn

1. Introduction and Scope

As the global manufacturing industry transitions toward high-value-added and low-carbon development, metal material manufacturing processes are undergoing systemic transformations. The explosive growth of new energy vehicles and the pursuit of superior performances in the aerospace field [1,2] have fostered the growth of a huge market for lightweight and high-strength metal components (such as aluminum alloys and magnesium alloys). These components often require a combination of complex topological structures, gradient functional characteristics, and ultra-high precision, which traditional metal manufacturing processes have struggled to achieve. In addition, as manufacturing systems impose strict requirements for the sustainable development of the whole life cycle of metal materials, resource efficiency and environmental pollution control have become prerequisites for process innovation.

Against this background, traditional processes have revealed their inherent limitations [3,4]. For instance, although the typical melt casting rolling process is mature and stable, it faces the dilemma of high energy consumption and low material utilization and often finds it difficult to form complex cavities. To address these challenges, researchers have integrated numerical simulations into manufacturing processes, enabling multi-physics field coupling simulations (thermal–mechanical–electrical–magnetic–fluidic) of the manufacturing process [5–7]. This makes it possible to predict the evolution of material microstructures (such as dynamic recrystallization and phase transformation kinetics) and key performance indicators before material manufacturing. Meanwhile, the precise control technology of microstructure is used to improve the performance of products from the perspective of material origin. For example, strategies such as grain refinement, interface strengthening, and in situ reaction layer design [8–10] optimize phase composition and defect distribution at the atomic scale, achieving a coordinated improvement in strength/toughness/corrosion resistance. Furthermore, additive manufacturing and hybrid processing technologies that have subverted traditional manufacturing logic are also utilized [11–14]. With layer-by-layer accumulation or in situ material synthesis as their core principles, these technologies break through the constraints of geometric complexity and open up new pathways for lightweight topological structures, functionally graded materials, and multi-material integration. In addition, systematic research on fatigue-resistant design and joining mechanisms has been conducted [15], which provides crucial theoretical support for the service reliability of products. Therefore, the chain of innovation in modern metal manufacturing process has been expanded from single equipment innovation to the full-dimensional collaboration of "basic theory—Numerical Simulation—process development—virtual verification—green assessment".

 https://doi.org/10.3390/met15111203

2. Contributions

Numerical simulation and intelligent monitoring technology are reconfiguring the process design paradigm. This kind of technology not only avoids the high cost of the traditional trial-and-error method, but also lays the foundation for digital twin technology and virtual manufacturing.

Russo Spen (Contribution 1) proposed an efficient calibration-free method for predicting welding deformation. The welding thermal cycle data were obtained In Situ by infrared thermal imaging, and directly input into the forced Imposed Thermal Cycle (ITC) model for finite element simulation. This method avoids the destructive metallographic calibration required by the traditional Moving Heat Source (MHS) model. The experimental and numerical verification show that the predicted deformation error is less than 0.3%. The calculation time is significantly reduced by more than 85% compared with the classical method, which significantly improves the efficiency, although the error estimation is slightly increased by 0.03%. Based on the actual process data-driven simulation process, a virtual representation of this process can be realized, enabling the realization of future digital twin applications.

Dai (contribution 2) established an incremental coupling model, which effectively solved the problems of inaccurate coil temperature and nonlinear interactions between stress and deformation during the coiling process of hot-rolled steel strips. The model breaks through the limitations of the traditional method, and obtains the mechanical properties and radial elastic modulus of the strip steel by coupling the temperature and stress of the coil, and conducting tensile tests and laminated compression tests at different temperatures. Meanwhile, the effects of key process parameters such as strip thickness, coiling tension and initial mandrel temperature on the internal state of the coil core are analyzed. The results of the model were compared with the measured values and analytical solutions, which confirmed the effectiveness of the incremental coupling model, and provided a theoretical basis for optimizing and precisely controlling the hot strip coiling process.

Kendall (Contribution 3) investigated the key influence of surface geometry on the oxidation kinetics of conveying pipes. Based on the Stefan problem, the customized thermochemical database and the numerical solution of the diffusion equation, a prediction model for the oxide thickness of typical curved surface was developed. The critical effect of radial diffusion term on oxidation kinetics was quantitatively revealed for the first time when the radius of the curvature of pipe was $\leq$200 mm, and proved the necessity of the cylindrical coordinate system in the oxidation modeling of curved surfaces. The model provides a theoretical basis for the high-precision prediction of scale layer growth, the optimization of descaling processes and the reduction of tool damage.

Precise control of microstructure and interface engineering is the key to optimizing material performance.

Zou (Contribution 4) significantly improved the comprehensive properties of MIG welded joints of 5xxx series aluminum alloys by adding scandium (Sc) and erbium (Er) to the welding wire: the grain size was refined from 47 μm to 29 μm and 31 μm, respectively, which was attributed to the heterogeneous nucleation effect of submicron-sized coherent Al3Er and Al3Sc phases (L12 structure). The tensile strength, fracture elongation and microhardness of the welded joint were significantly improved due to refining and dispersion strengthening. In addition, the corrosion resistance of the joint was significantly enhanced, which showed that the corrosion current density decreased and the corrosion potential increased, which was attributed to the formation of a denser oxide film and the balance of the potential difference between the precipitates and matrix. The addition of Sc and Er elements have similar and significant effects on improving the properties of welded joints.

Sisodia (Contribution 5) innovatively used local electron beam-post-weld heat treatment (LEB-PWHT) to regulate the electron beam welded joint of 12 mm thick S960QL high-strength steel, and achieved accurate local energy input through a defocusing beam. The results show that LEB-PWHT significantly optimizes the joint performance: the hardness of the weld seam and heat-affected zone (HAZ) was reduced by 23% and 21%, respectively, the tensile strength was increased by 3% (reaching 1082 MPa), and the ductility was improved. However, attention should be paid to the decrease in the impact toughness of the weld metal at −40 °C (from 63 J to 27 J). Microstructure, LEB-PWHT promoted the transformation of martensite in the coarse-grain heat-affected zone (CGHAZ) into tempered martensite + carbide precipitation, and equiaxed grains and dispersed carbides were formed in the fine-grain heat-affected zone (FGHAZ). This reveals the directional control mechanism of local heat treatment on microstructure and properties. Local post weld heat treatment provides technical guidance for the precise control of micro zone properties for high-strength steel thick plate welding.

Wang (Contribution 6) developed a simple one-step method without additives or pre modification to directly prepare ZIF-8 films with strong adhesion on zinc substrates, revealing the dual-role mechanism of the ZnO interlayer. The ZnO interlayer is formed by the decomposition products of solvent and zinc ions. It not only serves as a sacrificial precursor to promote the nucleation and continuous growth of ZIF-8 crystal, but also acts as an anchoring site to significantly enhance the adhesion between the film and the substrate. This work provides a simple direct production process that can be applied to other metal substrates experimentally to further study the complete film formation mechanism, and will provide more comprehensive theoretical support for the preparation of metal substrate films.

At the same time, additive manufacturing (AM) technology has opened up new dimensions for the fabrication of complex metal components.

Cui (Contribution 7) successfully fabricated 22MnCrNiMo mooring chain steel with excellent comprehensive mechanical properties by utilizing the selective laser melting (SLM) additive manufacturing process and determining the combination of key process parameters (laser power: 200 W, scanning speed: 800 mm/s, layer thickness: 30 μm, scanning spacing: 110 μm) through experiments. The SLM-formed parts exhibit a microhardness of 513.2 HV0.5, a tensile strength of 1223 MPa, a yield strength of 1114 MPa, an elongation of 8.5%, and an impact energy of 127 J. The research findings reveal the microstructure evolution law and the strengthening–toughening mechanism of 22MnCrNiMo steel fabricated via SLM technology, providing a new method and technical basis for the direct fabrication of high-performance mooring chains

Gracheva (Contribution 8) systematically reviewed the current research findings on selective laser melting (SLM) technology for biodegradable metals (Mg, Fe, Zn), and quantified the key properties of three types of materials: Mg alloys fabricated via SLM achieve an elastic modulus matching that of bone (40–45 GPa) and a moderate degradation rate (1–3 mm/year); Fe-based metals exhibit excellent strength (400–600 MPa) but a relatively slow degradation rate (0.1–0.5 mm/year); Zn alloys offer moderate overall performance. Meanwhile, the design strategy of porous/lattice structure is proposed to enhance bone integrity and achieve performance gradients, and it is pointed out that the four core challenges in this field include controlling degradation kinetics, optimizing the SLM process for active metals, standardizing testing methods, and coordinating regulatory frameworks. This work provides systematic theoretical support and technical guidance for the development of the next generation of personalized biodegradable implants.

In the preparation process of metal materials, it is necessary to achieve collaborative optimization of design and process parameters to ensure that the materials possess sufficient fatigue resistance and reliable service life. Therefore, it is essential to study the failure mechanisms of materials under fatigue loads.

Meilinger (Contribution 9) addressed the lack of systematic high cycle fatigue (HCF) research on key aluminum/steel hybrid joints for lightweight vehicle structures. Through experiments, aluminum/steel hybrid joints were prepared using 5754-H22, 6082-T6 aluminum alloys and DP600 substrates, respectively, through resistance spot welding (RSW) and conventional clinching (CCL), and HCF testing was conducted. An HCF design curve with a 50% failure probability was established, and the study indicated a significant correlation between interface structure and fatigue life. The relative load bearing capacity of different joints under HCF was quantified (based on the steel/steel joint): the load bearing capacity of the aluminum/steel hybrid joint was significantly reduced (RSW was 48.7% and 73.0%, CCL was 35.0% and 38.7%), while that of the aluminum/aluminum joint was even lower (RSW was 39.9% and 50.4%, CCL was 31.7% and 35.0%). In addition, the study clearly shows that, with one exception, the load bearing capacity of CCL joints is better than that of RSW joints (156.1–108.3%). This study provides an important basis for the selection of lightweight vehicle bodies connection process.

Basak (Contribution 10) systematically reviewed the laws governing the performance of common non thermal mechanical connections (adhesive, bolted, clinched and riveted joints) under fatigue loads, mainly introducing the influencing factors of different connection types. For instance, the fatigue behavior of bonded connections was affected by the bond length, thickness, and the properties of different materials—increasing the bond length can improve its fatigue strength until a certain length is reached, while increasing the thickness of the laminate or adhesive reduces fatigue life unless the surface roughness is increased; the differences in mechanical properties of different laminated materials directly determine their fatigue performance. This review integrates the fatigue behavior of mechanical bonds and the effects of various internal and external parameters, providing a key theoretical basis for the selection of optimal parameters in product design.

3. Conclusions and Outlook

The contributions summarized above provide an overview of recent progress in the research on metal material manufacturing processes. Through theoretical research, numerical simulation and experimental verification, advanced manufacturing technologies are provided for the preparation of high-performance metal materials. Numerical simulation and intelligent detection have overcome the limitations of traditional manufacturing processes, while virtual manufacturing design has replaced the experience driven trial and error model. Studies on the microstructure and macroscopic properties of the materials further show the importance of reforming the manufacturing process. Meanwhile, emerging manufacturing technologies such as additive manufacturing show significant advantages in terms of their material forming limits. It is worth noting that the significance of material fatigue failure in actual service has been highlighted. As we look ahead, the manufacturing processes employed in the fabrication of metallic materials must further embrace intelligence, functional integration, and green development. It is essential to reduce energy consumption and resource usage in manufacturing to meet the higher standards required for sustainable industrial development while ensuring material performance.

Conflicts of Interest: The authors declare no conflicts of interest.

List of Contributions:

1. Spena, P.R.; De Maddis, M.; Razza, V.; Santoro, L.; Mamarayimov, H.; Basile, D. Infrared-Guided Thermal Cycles in FEM Simulation of Laser Welding of Thin Aluminium Alloy Sheets. *Metals* **2025**, *15*, 830. https://doi.org/10.3390/met15080830.

2. Dai, M.; Hu, Y.; Hao, Y.; Qiu, P.; Xiao, H. Analysis of Temperature and Stress Fields in the Process of Hot-Rolled Strip Coiling. *Metals* **2025**, *15*, 111. https://doi.org/10.3390/met15020111.
3. Kendall, M.; Coleman, M.; Cockings, H.; Sackett, E.; Owen, C.; Auinger, M. Computational Thermochemistry for Modelling Oxidation During the Conveyance Tube Manufacturing Process. *Metals* **2024**, *14*, 1402. https://doi.org/10.3390/met14121402.
4. Zou, C.; Wu, R.; Yang, X.; Ma, Z.; Hou, L. Effects of a Welding Wire Containing Er or Sc on the Microstructure, Mechanical Properties, and Corrosion Resistance of the 5xxx Aluminum Alloy MIG Joint. *Metals* **2025**, *15*, 287. https://doi.org/10.3390/met15030287.
5. Sisodia, R.P.S.; Sliwinski, P.; Koncz-Horváth, D.; Węglowski, M.S. Influence of Post-Weld Heat Treatment on S960QL High-Strength Structural Steel Electron-Beam-Welded Joint. *Metals* **2024**, *14*, 1393. https://doi.org/10.3390/met14121393.
6. Wang, H.; Liu, J.; Liu, B.; Zhang, Z.; Ren, X.; Wang, X.; Wu, P.; Zhang, Y. Direct In Situ Fabrication of Strong Bonding ZIF-8 Film on Zinc Substrate and Its Formation Mechanism. *Metals* **2024**, *14*, 1403. https://doi.org/10.3390/met14121403.
7. Cui, X.; Li, X.; Hu, C.; Zhao, D.; Liu, Y.; Wang, S. Microstructure and Properties of Mooring Chain Steel Prepared by Selective Laser Melting. *Metals* **2025**, *15*, 541. https://doi.org/10.3390/met15050541.
8. Gracheva, A.; Polozov, I.; Popovich, A. Additive Manufacturing of Biodegradable Metallic Implants by Selective Laser Melting: Current Research Status and Application Perspectives. *Metals* **2025**, *15*, 754. https://doi.org/10.3390/met15070754.
9. Meilinger, Á.; Kovács, P.Z.; Lukács, J. High-Cycle Fatigue Characteristics of Aluminum/Steel Clinched and Resistance-Spot-Welded Joints Based on Failure Modes. *Metals* **2024**, *14*, 1375. https://doi.org/10.3390/met14121375.
10. Basak, A.K.; Bajwa, D.S.; Pramanik, A. Fatigue Behaviour of Mechanical Joints: A Review. *Metals* **2024**, *15*, 25. https://doi.org/10.3390/met15010025.

References

1. Zhang, W.; Xu, J. Advanced lightweight materials for Automobiles: A review. *Mater. Des.* **2022**, *221*, 110994. [CrossRef]
2. Ding, W.; Wan, N.; Zhao, B.; Fu, Y.; Xu, J. Research status and tendency of advanced manufacturing theory and technology in aerospace. *Acta Aeronaut. Astronaut. Sin.* **2025**, *46*, 531309.
3. Vergara, D.; Fernández-Arias, P.; Ariza-Echeverri, E.A.; del Bosque, A. Residual Stresses in Metal Manufacturing: A Bibliometric Review. *Materials* **2025**, *18*, 3612. [CrossRef] [PubMed]
4. Song, X.; Feih, S.; Zhai, W.; Sun, C.-N.; Li, F.; Maiti, R.; Wei, J.; Yang, Y.; Oancea, V.; Brandt, L.R.; et al. Advances in additive manufacturing process simulation: Residual stresses and distortion predictions in complex metallic components. *Mater. Des.* **2020**, *193*, 108779. [CrossRef]
5. Xu, Z.-J.; Zheng, Z.; Gao, X.-Q. Operation optimization of the steel manufacturing process: A brief review. *Int. J. Miner. Met. Mater.* **2021**, *28*, 1274–1287. [CrossRef]
6. Dogan, A.; Birant, D. Machine learning and data mining in manufacturing. *Expert Syst. Appl.* **2021**, *166*, 114060. [CrossRef]
7. Amaral, J.V.S.D.; Miranda, R.d.C.; Montevechi, J.A.B.; dos Santos, C.H.; Gabriel, G.T. Metamodeling-based simulation optimization in manufacturing problems: A comparative study. *Int. J. Adv. Manuf. Technol.* **2022**, *120*, 5205–5224. [CrossRef]
8. Liu, C.; Rao, J.; Sun, Z.; Lu, W.; Best, J.P.; Li, X.; Xia, W.; Gong, Y.; Wei, Y.; Zhang, B.; et al. Near-theoretical strength and deformation stabilization achieved via grain boundary segregation and nano-clustering of solutes. *Nat. Commun.* **2024**, *15*, 9283. [CrossRef] [PubMed]
9. Chandra, S.; Wang, C.; Tor, S.B.; Ramamurty, U.; Tan, X. Powder-size driven facile microstructure control in powder-fusion metal additive manufacturing processes. *Nat. Commun.* **2024**, *15*, 3094. [CrossRef] [PubMed]
10. Zhang, Y.; Zhao, Y.; Kai, X.; Yang, J.; Zhu, H.; Shan, Y. Study on the Microstructure and Mechanical Properties of 7085 Aluminum Alloy Reinforced by In Situ (ZrB_2 + Al_2O_3) Nanoparticles and Rare Earth Er. *Materials* **2025**, *18*, 2009. [CrossRef] [PubMed]
11. Zhou, H.-R.; Yang, H.; Li, H.-Q.; Ma, Y.-C.; Yu, S.; Shi, J.; Cheng, J.-C.; Gao, P.; Yu, B.; Miao, Z.-Q.; et al. Advancements in machine learning for material design and process optimization in the field of additive manufacturing. *China Foundry* **2024**, *21*, 101–115. [CrossRef]
12. Liu, Z.; Ding, M.; Wang, P.; Cheng, Z.; Yang, C.; Xu, B.; Peng, D.; Liu, C.; Shen, J. The application status and prospects of ma-chine learning in metal additive manufacturing. *Aeronaut. Manufact. Technol.* **2022**, *65*, 14–28. [CrossRef]
13. Chia, H.Y.; Wu, J.; Wang, X.; Yan, W. Process parameter optimization of metal additive manufacturing: A review and outlook. *J. Mater. Inform.* **2022**, *2*, 16. [CrossRef]

14. Blakey-Milner, B.; Gradl, P.; Snedden, G.; Brooks, M.; Pitot, J.; Lopez, E.; Leary, M.; Berto, F.; du Plessis, A. Metal additive manufacturing in aerospace: A review. *Mater. Des.* **2021**, *209*, 110008. [CrossRef]
15. Molaei, R.; Fatemi, A. Fatigue Design with Additive Manufactured Metals: Issues to Consider and Perspective for Future Research. *Procedia Eng.* **2018**, *213*, 5–16. [CrossRef]

metals

MDPI

Article

Infrared-Guided Thermal Cycles in FEM Simulation of Laser Welding of Thin Aluminium Alloy Sheets

Pasquale Russo Spena [1,*,†], Manuela De Maddis [1,†], Valentino Razza [1,†], Luca Santoro [2,†], Husniddin Mamarayimov [1,†] and Dario Basile [1,†]

1 Dipartimento di Ingegneria Gestionale e della Produzione, Politecnico di Torino, Corso Duca degli Abruzzi 24, 10129 Torino, Italy; manuela.demaddis@polito.it (M.D.M.); valentino.razza@polito.it (V.R.); husniddin.mamarayimov@polito.it (H.M.); dario.basile@polito.it (D.B.)

2 Dipartimento di Ingegneria Meccanica e Aerospaziale, Politecnico di Torino, Corso Duca degli Abruzzi 24, 10129 Torino, Italy; luca.santoro@polito.it

* Correspondence: pasquale.russospena@polito.it

† Current address: Advanced Joining Technologies Laboratory J-Tech@PoliTO, Corso Duca degli Abruzzi 24, 10129 Torino, Italy.

Abstract: Climate concerns are driving the automotive industry to adopt advanced manufacturing technologies that aim to improve energy efficiency and reduce vehicle weight. In this context, lightweight structural materials such as aluminium alloys have gained significant attention due to their favorable strength-to-weight ratio. Laser welding plays a crucial role in assembling such materials, offering high flexibility and fast joining capabilities for thin aluminium sheets. However, welding these materials presents specific challenges, particularly in controlling heat input to minimize distortions and ensure consistent weld quality. As a result, numerical simulations based on the Finite Element Method (FEM) are essential for predicting weld-induced phenomena and optimizing process performance. This study investigates welding-induced distortions in laser butt welding of 1.5 mm-thick Al 6061 samples through FEM simulations performed in the SYSWELD 2024.0 environment. The methodology provided by the software is based on the Moving Heat Source (MHS) model, which simulates the physical movement of the heat source and typically requires extensive calibration through destructive metallographic testing. This transient approach enables the detailed prediction of thermal, metallurgical, and mechanical behavior, but it is computationally demanding. To improve efficiency, the Imposed Thermal Cycle (ITC) model is often used. In this technique, a thermal cycle, extracted from an MHS simulation or experimental data, is imposed on predefined subregions of the model, allowing only mechanical behavior to be simulated while reducing computation time. To avoid MHS-based calibration, this work proposes using thermal cycles acquired in-line during welding via infrared thermography as direct input for the ITC model. The method was validated experimentally and numerically, showing good agreement in the prediction of distortions and a significant reduction in workflow time. The distortion values from simulations differ from the real experiment by less than 0.3%. Our method exhibits a slight decrease in performance, resulting in an increase in estimation error of 0.03% compared to classic approaches, but more than 85% saving in computation time. The integration of real process data into the simulation enables a virtual representation of the process, supporting future developments toward Digital Twin applications.

Keywords: laser welding; IR camera; FEM simulation; thermal cycle; distortion

 https://doi.org/10.3390/met15080830

1. Introduction

Laser welding is a leading joining technique in lightweight manufacturing. It is widely recognized for its precision and efficiency in joining aluminium alloys, which are critical to the automotive industry. The increasing demand for lighter, fuel-efficient vehicles has driven the widespread use of thin aluminium sheets, leveraging their strength-to-weight ratio [1,2]. However, welding thin sheets poses a major challenge: improper heat control can lead to distortion and compromise the integrity of the weld [3]. In addition, compared to carbon steels, the large thermal expansion coefficient and the low elastic modulus of aluminium alloys make them particularly prone to significant deformation and residual stresses during welding [4]. Consequently, thorough evaluation and precise control of the welding process are essential to ensure joint quality and performance [5]. Although helpful for small-scale laboratory specimens, traditional trial-and-error methods are inadequate for industrial applications, particularly when dealing with large or intricately shaped components [6]. Finite Element Method (FEM) simulations have become an indispensable tool, enabling detailed predictions of welding outcomes before production and significantly reducing both time and material waste [7].

While FEM simulations provide valuable insights into the welding process, they face a main limitation: calibrating the thermal source requires a repetitive and iterative process [8]. It involves continuous comparison between metallographic cross-sections of actual welds and their simulated counterparts, demanding considerable time and resources to achieve accurate alignment [9–12]. Moreover, because these methods are destructive and time-consuming, they restrict the ability to make real-time adjustments, ultimately hindering the efficient optimization of welding parameters.

Recent advancements have introduced non-destructive testing (NDT) and prediction modeling methods to address these limitations. For instance, [13] explores automated calibration techniques, while [14] proposes inverse modeling approaches to streamline the calibration process. In addition, digital twins have also emerged as a powerful tool, enabling real-time simulation and monitoring of welding processes without the need for physical testing [9].

Thermography, an imaging technique based on infrared radiation, provides a non-contact and efficient method for real-time monitoring of temperature distribution in the welding zone [15–19]. The ability to detect temperature anomalies is crucial for predicting and preventing defects such as deformation, which can compromise the structural integrity of the weld [20–26]. Thermography can be used in combination with a laser system to non-destructively inspect welded joints, such as resistance spot weld [27–29]. Integrating thermography into the laser welding of aluminium not only enhances quality assurance but also enables real-time process control, significantly reducing the likelihood of weld defects and the need for costly rework [15,16,30–34].

This paper proposes a novel thermography-based approach not only for monitoring the laser welding process and ensuring joint quality, but also as a support for a fully non-destructive calibration strategy in FEM-based simulations. In this method, the transient temperature field recorded by an infrared camera is directly imposed as the heat-source boundary condition, dramatically reducing development time.

Unlike conventional methods that rely on the iterative tuning of moving heat-source parameters against metallographic cross-sections, the proposed *infrared-guided imposed thermal cycle approach*, hereafter referred to as IR-ITC, *eliminates* the need for macrographic calibration. It reduces the calibration parameter set to a single LOAD definition and cuts the setup time from roughly ten hours to less than thirty minutes. The simulation demonstrated 1.5 mm Al 6061 butt joints welded with a 2.8 kW diode laser in SYSWELD 2024.0. This strategy is readily transferable to any process where surface temperature data is available.

Beyond the drastic time savings, IR-ITC enables the development of real-time digital twins, facilitating *closed-loop weld quality control* and adaptive manufacturing workflows.

The remainder of this paper is organized as follows. Section 2 reviews conventional industry-standard FEM workflows and the limitations of moving heat-source calibration. Section 3 details the proposed IR-ITC methodology, including thermal-cycle acquisition and model implementation. Section 4 describes the experimental setup and validates the simulations through comparison with thermal maps, melt-pool geometry, and distortion measurements. Finally, Section 5 presents concluding remarks and directions for future research.

2. Industry Standard Approaches

2.1. Numerical Model

Traditional methods to simulate the laser welding process involve a two-step procedure, first computing the thermal behavior, then assessing the mechanical response.

Thermal analysis in laser welding process simulation is typically modelled using the heat diffusion equation [35]

$$\rho C_p \frac{\partial T}{\partial t} = \nabla \cdot (k \nabla T) + \dot{Q}, \tag{1}$$

where ρ ($\mathrm{kg\,m^{-3}}$) is the material density, C_p ($\mathrm{J\,K^{-1}\,kg^{-1}}$) is the specific heat capacity, k ($\mathrm{W\,m^{-1}\,K^{-1}}$) is the thermal conductivity, $\dot{Q}$ ($\mathrm{W\,m^{-3}}$) is the internally generated heat per unit volume, and T (K) is the temperature distribution.

By applying heat conduction and energy conservation laws to an infinitesimally small control volume, it becomes possible to determine the instantaneous temperature at any point within the welded material. Solving the heat diffusion equation provides the transient temperature distribution $T(x, y, z, t)$ as a function of time t and spatial coordinates (x, y, z).

It is worth noting that key thermophysical properties, such as density, specific heat capacity, and thermal conductivity, are temperature-dependent. The primary heat source in the welding process is the external heat input. This heat input drives the thermo-mechanical changes in the material. Thus, defining a proper heat source model (i.e., a laser beam) is critical for ensuring the accuracy of the theoretical model.

A three-dimensional truncated conical heat source model with a Gaussian distribution is commonly used to model highly concentrated energy sources, such as laser beams [36]. This model assumes a Gaussian distribution of heat intensity with the peak located at the apex of the cone and decreasing both radially and axially. Figure 1 illustrates the geometric characteristics of the model, where r_e and r_i are the upper and lower radii of the cone, and z_e and z_i are the distances from the welding trajectory to the top and bottom surfaces of the heat source. The thermal energy delivered to the welding plates by a laser beam with power P and absorption efficiency η is defined in the laser reference system (x_ℓ, y_ℓ, z_ℓ) by [37]

$$\dot{Q}(x_\ell, y_\ell, z_\ell) = \frac{9\eta P \exp\left(3 - \frac{f_{las} r^2}{r_0^2}\right)}{\pi(1 - e^{-3})(z_e - z_i)\left(r_e^2 - r_e r_i + r_i^2\right)} \tag{2}$$

where $(x_\ell, y_\ell, z_\ell) = (0, 0, z_e)$ is the laser spot center, $f_{las} = 3$ is the heat source intensity factor [37], Q is the volumetric heat flux, r represents the radial distance from the heat source center, i.e., $r = \sqrt{x_\ell^2 + y_\ell^2}$, and

$$r_0(z) = r_e - (r_e - r_i)\frac{(z_e - z_\ell)}{(z_e - z_i)}. \tag{3}$$

Assuming the laser moves at a velocity v (mm s^{-1}) along the x direction in the global reference system, this results in

$$\begin{bmatrix} x \\ y \\ z \end{bmatrix} = \begin{bmatrix} x_0 + vt \\ y_0 \\ z_0 \end{bmatrix}, \tag{4}$$

where (x_0, y_0, z_0) is the initial laser spot position.

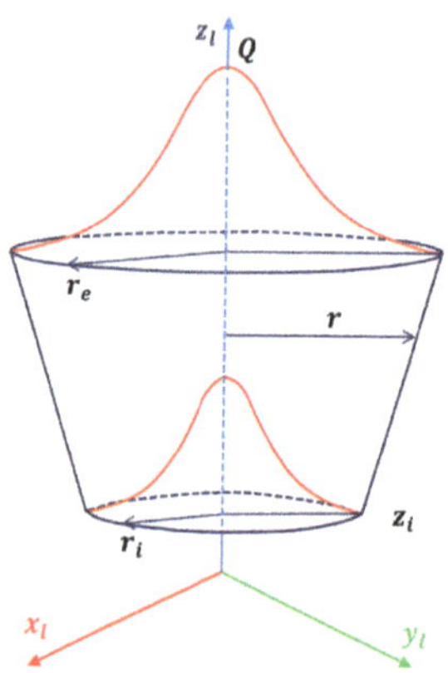

Figure 1. A 3D Gaussian truncated conical heat source model.

Air cooling occurs through the outer surfaces of the plates, resulting in both convective and radiative heat losses. In the numerical model, these surface (skin elements) are coupled to the surrounding environment using Newton's law of cooling for convection and the Stefan–Boltzmann equation for radiation [35]. The initial temperature condition is

$$T(x, y, z, t)\big|_{t=0} = T_0 \tag{5}$$

where T_0 is set at room temperature and assumed to remain constant during the welding process. Convective and radiative heat losses result in the following boundary condition

$$k\nabla T \cdot \vec{n} = h_c(T - T_0) + \sigma\epsilon(T^4 - T_0^4), \quad \forall(x, y, z) \in S \tag{6}$$

where $S \subset \mathbb{R}^3$ is the set of plate surface points, $\vec{n}$ is normal vector to the surface point (x, y, z), h_c is the convection heat transfer coefficient, ϵ is the emissivity, and $\sigma \approx 5.67 \cdot 10^{-8}$ W m^{-2} K^{-4} is the Stefan–Boltzmann constant. Based on the temperature distribution of $T(x, y, z, t)$ within the welded workpiece obtained from the thermal simulation, the following mechanical analysis is conducted. This phase is essential for evaluating the structural integrity of the welded joint, quantifying residual deformations, and the stress state induced by the welding thermal cycle. The thermal effects influence the mechanical response through temperature-dependent material properties, such as Young's modulus and yield strength, as well as thermal expansion or contraction.

The total strain at any node can be conceptually decomposed into four components: elastic strain ε_e, plastic strain ε_p, thermal component ε_{th}, and strain from volumetric changes due to metallurgical phase transformations ε_{phase}

$$\varepsilon = \varepsilon_e + \varepsilon_p + \varepsilon_{th} + \varepsilon_{phase} \tag{7}$$

For the mechanical analysis, only elastic, plastic, and thermal strain components are considered because these components predominantly govern deformation in aluminium alloy sheets under laser welding conditions. Strains due to metallurgical phase transformations are considered negligible, given the absence of significant solid-state phase changes in aluminum alloys during the welding thermal cycle [38,39].

Elastic strain is modeled by accounting for the temperature-dependent Young's modulus and Poisson's ratio. Plastic strain is described using an isotropic hardening model in combination with the Von Mises yield criterion. Thermal strain is approximated as

$$\varepsilon_{th} = \alpha(T)(T - T_{ref})I \tag{8}$$

where T_{ref} is the reference temperature and $\alpha(T)$ is the temperature-dependent average coefficient of thermal expansion

In welding processes, localized high temperatures generate non-uniform thermal expansion. As the material cools, differential contraction between the heated regions and the surrounding cooler regions results in residual stresses. When the resulting deformations are not negligible, mechanical analysis must account for non-linear geometric effects. In this context, the Green–Lagrange strain tensor E plays a crucial role, as it characterizes the deformation relative to the initial configuration of the body [40]

$$E = \frac{1}{2}(F^T F - I) \tag{9}$$

where I is the identity tensor and F is the deformation gradient tensor, mapping a material point from its initial position p_0 to its current position p, according to the relation $p = \chi(p_0, t)$. The components of F are computed as the Jacobian of p with respect to the initial position p_0.

The numerical implementation of such mechanical models, typically through the FEM, allows for the solution of non-linear systems of equations that account for temperature-dependent material properties, mechanical boundary conditions, and the entire deformation history. Post-processing of the simulation results provides a detailed characterization of residual stresses, plastic strains, and distortions introduced by the welding process.

Clamping conditions are simulated in the FEM model using two different kinds of nodes, as shown in Figure 2. Three nodes located at the three bottom corners of the model are assigned as free-clamp conditions to allow rigid body motion. Rigid clamps are used to replicate the physical clamping conditions in the experimental setup. Specifically, six rows of nodes on the top surfaces of each plate are rigidly constrained in all three spatial directions. Meanwhile, six rows of nodes on the bottom surfaces are constrained only in the vertical direction [41]. All clamps are considered active during the melting stage. After the laser fusion process is completed, the rigid clamps are released, while the free clamps remain in place until the end of the welding process.

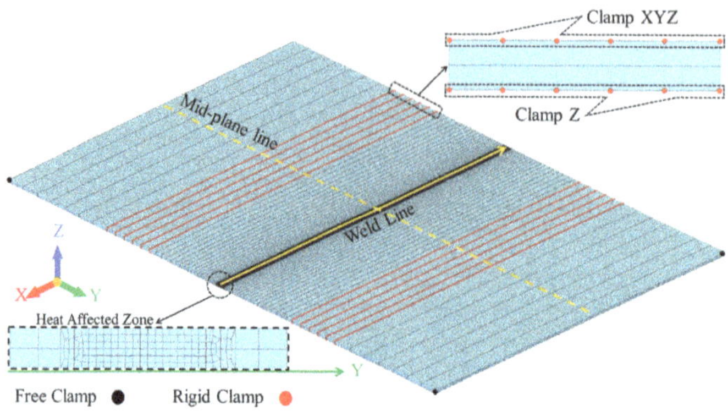

Figure 2. Finite element model schematic and clamping positions.

2.2. Iterative Calibration via Moving Heat Source

Accurate numerical simulation of welding processes using a moving heat source model, implemented in commercial finite element software such as SYSWELD 2024.0, critically relies on an iterative calibration procedure to bridge the gap between idealized computational models and the complexities of real-world welding phenomena.

This calibration process involves systematically tuning the heat source parameters, namely the radii of the upper (r_e) and lower (r_i) cones, the height of the cone ($z_e - z_i$), and efficiency (η) [8]. These parameters serve as key input parameters to the numerical model. Additionally, it is necessary to identify the LOAD, the set of nodes instantaneously affected by the moving heat source along its predefined path (Figure 3). Calibration is considered successful when the geometry of the simulated fusion zone, including width, penetration depth, and asymmetry, closely matches the experimental fusion zone, as observed in cross-sectional metallographic micrographs of the weld bead. Ultimately, this iterative calibration procedure ensures a robust and predictive simulation capable of reliably forecasting temperature fields, phase transformations, residual stresses, and distortion in welded structures [42].

According to the initial assumptions, preliminary welding parameters are introduced to run the thermo-metallurgical FEM simulation. Based on the authors' experience, the upper and lower radii of the conical heat source model mainly influence the width of the melt pool, while the cone height affects penetration depth. Numerical simulations also reveal that the efficiency parameter η impacts both the width and penetration of the melt pool. Thus, η is used to fine-tune the overall dimensions of the melt pool in the final trials. Calibration is performed in quasi-steady state regions of the weld, where the transverse cross-section is separated by the mid-plane line, as illustrated in Figure 2.

It should be noted that the calibration process is time-consuming and subject to human influence. The overall heat source calibration time includes two major tasks: metallographic analysis (i.e., specimen preparation and microscopic examination), and iterative tuning of the input parameters (i.e., inner and outer radii, height, efficiency, and LOAD definition) in SYSWELD 2024.0. In this study, around 4 h were required for metallographic preparation and examination. Each single thermo-metallurgical simulation run, including post-processing and result analysis, took approximately 15 min. We have performed the simulations with a HP 290 G4 Microtower PC equipped with an Intel® Core (TM) i7-10700 CPU and 64 GB of memory RAM. On average, 4.5 h were spent adjusting the input parameters to complete the thermo-metallurgical calibration, and 1.5 h were spent running a full thermo-metallurgical–mechanical simulation. Altogether, a total of approximately 10 h were required to achieve a fully calibrated laser welding FEM model using the traditional moving heat source (MHS) method with a transient approach.

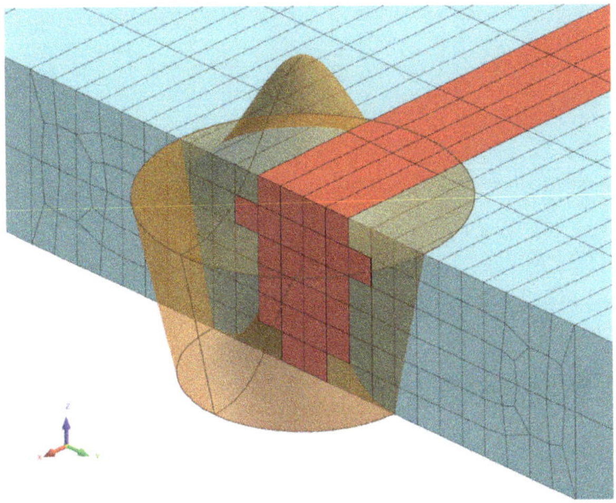

Figure 3. Selected mesh elements (LOAD) heated by MHS.

2.3. Imposed Thermal Cycle

The MHS model, while accurate, demands considerable computational resources, especially for large components with complex geometry. It requires high memory capacity and extended computation times. To overcome these limitations, the Imposed Thermal Cycle (ITC) approach is often preferred due to its simplicity and efficiency. Instead of computing heat transfer from a moving source, this method directly applies a predefined thermal cycle to specific regions of the simulated specimen [43]. Although less accurate than the MHS method, the ITC approach remains a practical and computationally efficient tool for analyzing metallurgical transformations and mechanical deformations. The key challenge of this methodology lies in defining the thermal cycle reliably, as the accuracy of thermo-mechanical and metallurgical predictions depends strongly on the fidelity of the imposed temperature data. Despite the simplification, the method still enables effective analysis of metallurgical transformations and mechanical deformations. Therefore, a well-defined thermal cycle is essential for the effectiveness of this method. In the traditional Imposed Thermal Cycle approach (T-ITC), the thermal cycles are derived from a previously calibrated and validated MHS model [36]. The LOAD region is designed to closely match the shape and dimensions of the melt pool identified in the MHS simulation. According to [44], the discrepancies between experimental and numerical temperature values tend to increase as the thermal data points get close to the welding line, primarily because the ITC model does not account for metal fusion. In the SYSWELD 2024.0 Visual-Viewer module, temperature contour plots are used to define the melt pool boundaries. The required nodes for the thermal cycle are selected from the outer boundary of this molten zone, as presented in Figure 4. The extracted thermal cycles undergo pre-processing to remove any offset data and/or negative values, if any, and to shorten the tail of the temperature curve, typically ending around 400 °C for aluminium parts. In experimental tests, the T-ITC model did not require additional calibration. Interestingly, the T-ITC method provided larger distortion values than the MHS method, with the magnitude of this difference influenced by factors such as geometry, material, and the joining technique [36]. Unlike the MHS approach, the ITC method required only 24 min to complete the full thermo-metallurgical–mechanical simulation for the same FEM model. This efficiency was achieved using the predefined LOAD, based on the theoretical melt pool area derived from the MHS model.

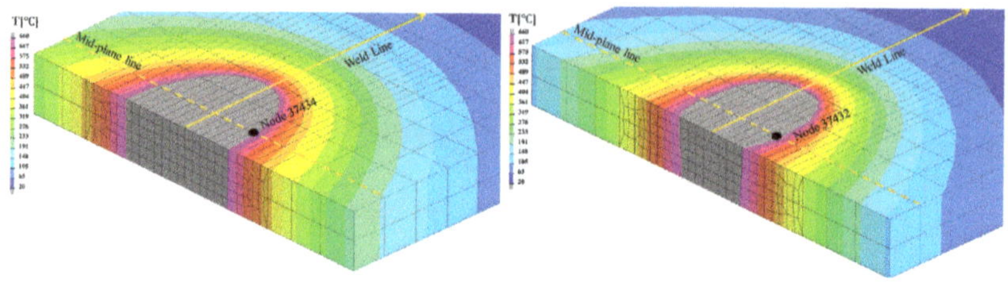

(a) (b)

Figure 4. Thermal cycle extraction node positions from the moving heat source models: (**a**) Sample 1, (**b**) Sample 2.

3. Infrared-Guided Imposed Thermal Cycle Method

The thermal cycle used in the T-ITC model is generally extracted from a calibrated MHS simulation. Nevertheless, it could also be derived directly from experimental data obtained through temperature measurements on the welded sheets. This is commonly performed by using contact thermocouples placed along a line perpendicular to the weld seam, covering the entire width of the heat-affected zone [9,36]. In this study, an innovative alternative is proposed: using an infrared (IR) camera to capture temperature data in real time during the welding process. This approach eliminates the need for MHS simulation or contact thermocouples, enabling continuous acquisition of thermal cycle data and providing real, directly process-related temperature measurements. Unlike the conventional iterative calibration process required by the MHS approach, the proposed infrared-guided imposed thermal cycle (IR-ITC) method integrates infrared thermography directly into the simulation workflow. This allows the experimentally measured thermal cycles to be imposed in the ITC model without the need for extensive iterative parameter tuning. Although the IR-ITC method still requires calibration, it is significantly simpler, due to the considerably smaller number of input parameters, and faster than MHS calibration. The only parameter requiring adjustment is the LOAD—the area to which the thermal cycle is applied—based on its influence on mechanical distortion outputs. Figure 5 presents three different LOAD configurations used for IR-ITC calibration. Consequently, all LOAD elements are subjected to the same thermal cycle, receiving the same thermal input during both the heating and cooling stages. The size of the LOAD area directly influences the extent of distortion in the component, since it defines the effective heating zone. The IR-ITC model is calibrated based on mechanical distortion observed in the experiment; therefore, thermo-metallurgical–mechanical computation is required. Given that a single thermo-metallurgical–mechanical simulation run takes approximately 24 min, the total estimated time to complete the IR-ITC calibration process was around 90 min, significantly faster than traditional approaches.

IR Camera Monitoring

A FLIR A700 microbolometric (Teledyne FLIR LLC; Wilsonville, OR, USA) infrared camera equipped with an 24 × 18 lens was used to monitor the laser welding process. This camera featured a long-wave sensor (7–14 µm), making it compatible with laser welding sources operating near 1 µm wavelength. It was positioned 600 mm from the weld seam at a 40° angle to the horizontal plane, providing an IFOV of 676 mrad. The temperature measurement range is 0–700 °C, and the system was set at a fixed distance of 1.00 m from the specimen to ensure consistent data acquisition. For butt joint welds, the specimens were carefully oriented to provide an acceptable view and spatial resolution. Thermal data was post-processed using FLIR Research Studio 2024.07.1 software.

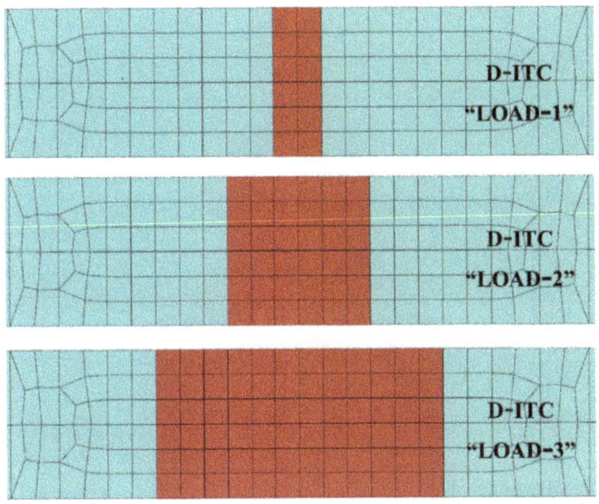

Figure 5. Three different LOAD configurations for the IR-ITC method.

Calibration of the IR system is necessary because the surface thermal emissivity varies strongly with temperature. By identifying an isothermal plateau corresponding to the solid–liquid phase transition (around 650–670 °C) and comparing it with the literature values for aluminium 6061 and 6082 alloys, the surface emissivity was set to 0.35.

To extract a thermal cycle suitable for use in the simulation, which is comparable to that obtained from the MHS model, the thermograms were analysed to identify a region located just outside the molten zone. The thermal history in this region was extracted and used as input for the IR-ITC model.

4. Experimental Results

This section presents the experimental tests and numerical simulations used to validate both the conventional and proposed modeling approaches. Comparisons focus on cross-sectional metallography, thermal distribution, and mechanical distortions.

4.1. Materials and Experimental Methodology

To demonstrate the effectiveness of the proposed approach, we have performed two laser welding experiments to acquire reference data.

The laser beam source is a diode laser (LDF4000-40, Laserline, Mülheim-Kärlich, Germany) with a maximum power of 4 kW and two wavelengths of 1020 and 1060 $\pm$ 10 nm. The welding head (OTS-5, Laserline, Mülheim-Kärlich, Germany) is mounted on a 6-axis industrial robot (IRB2400, ABB, Zurich, Switzerland) Welding is conducted on 1.5 mm-thick Al 6061-T6 plates, each measuring 150 $\times$ 200 mm, in a butt configuration. Prior to welding, the joint area was sanded and polished to remove contaminants (grease and dirt traces) and then cleaned thoroughly with acetone. The two aluminium plates were clamped to ensure a zero-gap joint, as illustrated in Figure 6. The laser beam was oriented perpendicular to the plate surface, and the focal point was adjusted to 3 mm above the plate surface, resulting in a beam spot diameter of 1.2 mm. Neither shielding gas nor filler material was used during welding. We have set the laser power at 2800 W. The two welding experiments differ only in the welding speed: 35 and 50 mm s^{-1} for Sample 1 and Sample 2, respectively. The welding parameters were selected based on preliminary trials aimed explicitly at maximizing weld-induced distortions. The goal is to generate significant mechanical deformation, which is particularly challenging to predict by FEM simulations. Thus, parameter selection prioritizes distortion magnitude over weld joint quality.

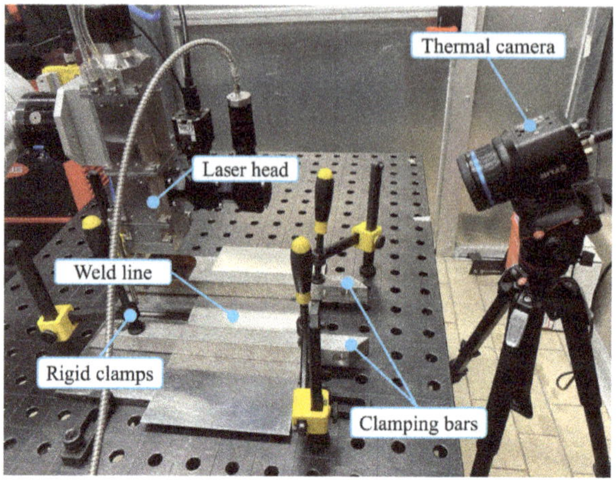

Figure 6. Welding setup and thermal camera.

To examine the joint geometry, a cross-sectional specimen was prepared by cutting through the weld zone. The samples were hot-mounted, polished to achieve a homogeneous surface finish, and then chemically etched using Keller's reagent for metallographic analysis. The cross-sectional images of the welding zone were captured using an (optical microscope (Axiovert a1, Zeiss, Oberkochen, Germany). Subsequently, the molten pool geometry is measured using the software ImageJ 1.54g. Post-weld distortion of the aluminium sheets is assessed using a direct contact profilometer. One side of the sheet is fixed, as shown in Figure 7, and the measurement line is oriented at a 90° angle to the weld seam, positioned at the midpoint of the sheet.

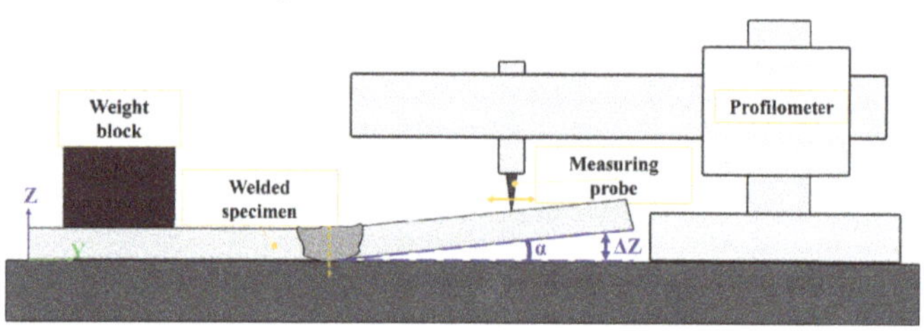

Figure 7. Profilometer measurements set-up.

4.2. Numerical Solver

Considering the complexity of the welding process and focusing on understanding the effect of heat input on mechanical distortion, the models described in Section 2.1 were numerically solved through FEM analysis. A single FEM model was employed for both samples, with varying input parameters, and simulations were carried out using SYSWELD 2024.0 software in two steps: a thermo-metallurgical simulation followed by a mechanical simulation [36,45]. Material properties and model parameters were selected according to the SYSWELD 2024.0 database. The conductive heat transfer is set to $h_c = 25\,\mathrm{W\,m^{-2}\,K^{-1}}$, and surface emissivity as $\epsilon = 0.8$.

Using the Visual-Mesh (SYSWELD 2024.0) module, mesh was defined with 54,400 8-node hexahedral linear 3D prism elements representing the solid domain; 36,544 4-node rectangular linear 2D elements to model heat exchange with the environment; and 200 1D line elements to define both the welding trajectory and reference lines.

Figure 2 provides an overview of the model, including spatial orientation, clamping positions, and Heat Affected Zone (HAZ) mesh appearance. The coordinate system is defined as follows: x-axis corresponds to the welding direction, the y-axis to the width of the weldment, and the z-axis to the direction of weld pool penetration (i.e., thickness). Mesh density was refined in thermally critical regions to ensure accuracy while maintaining reasonable computational time. Specifically, finer mesh grids of $0.25 \times 0.25 \times 1.0$ mm were used in the HAZ, with a transition to coarser center-concentric biased elements beyond the Fusion Zone (FZ) [8]. The Broyden–Fletcher–Goldfarb–Shanno (BFGS) algorithm was used to ensure the convergence and the stability of the numerical model computation [9].

In the same welding conditions, thinner plates are more prone to longitudinal and transverse shrinkage, as well as angular distortions due to thermal volumetric expansion and contraction during welding [46]). In FEM simulations based on small deformation theory, where strain is linearly dependent on displacement, large distortions are often underestimated. However, the interaction of large distortions (longitudinal, transverse, and bidirectional angular) leads to the buckling phenomenon. In large distortion theory, instead, strain is a nonlinear function of displacement. The nonlinear geometry option of SYSWELD 2024.0 accounts for such behavior, though some discrepancy between experimental and FEM numerical results still remains [47]. In this work, the *bending optimized* function of SYSWELD 2024.0 was used to improve the accuracy of distortion computation. Standard hexahedral elements are limited in accurately capturing bending kinematics. This limitation is addressed by using incompatible mode elements, which incorporate additional shape functions to better model bending behavior [48]. As a result, all solid elements, except those in the HAZ, were defined as *type 2* (i.e., *incompatible mode elements*) to enhance the accuracy of distortion computations.

4.3. Cross-Section Comparison and Time Analysis

Figure 8 illustrates the criteria used for measuring the melt pool and presents a comparative analysis between the metallographic cross-sections of the laser-welded aluminium alloys (Sample 1 and Sample 2) and the corresponding results from MHS simulations. Both experimental and simulated results clearly display the Base Metal (BM) and Fusion Zone (FZ). Table 1 summarizes the fusion zone dimensions obtained from the cross-sectional analysis.

The calibrated model exhibits only minor differences between the experimental and FEM results for the weld widths $\mathbf{w_1}$ and $\mathbf{w_2}$. In Sample 2, a smaller fusion zone is observed, which can be attributed to the lower heat input per unit length caused by the higher welding speed.

Table 1. Comparison of FEM and experimental weld pool dimensions.

Sample	FEM (mm)		Experiment (mm)	
	$\mathbf{w_1}$	$\mathbf{w_2}$	$\mathbf{w_1}$	$\mathbf{w_2}$
1	3.22	3.18	3.20	3.20
2	2.65	2.30	2.64	2.29

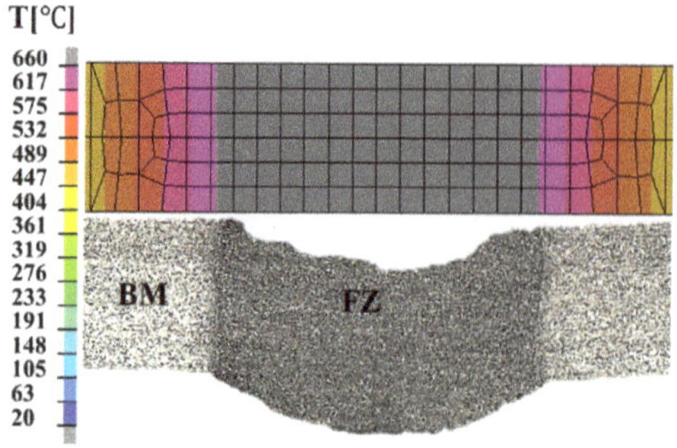

(a)

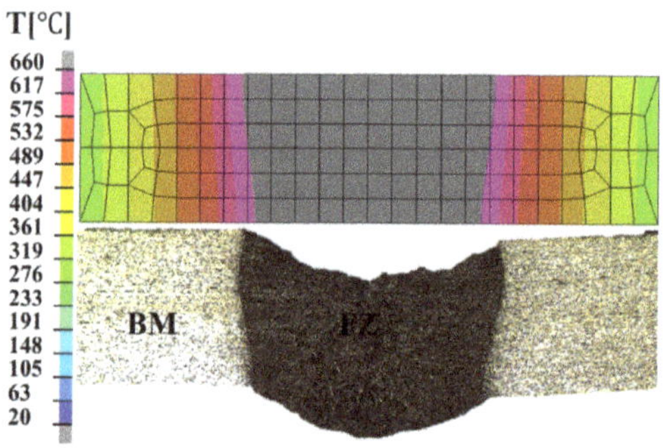

(b)

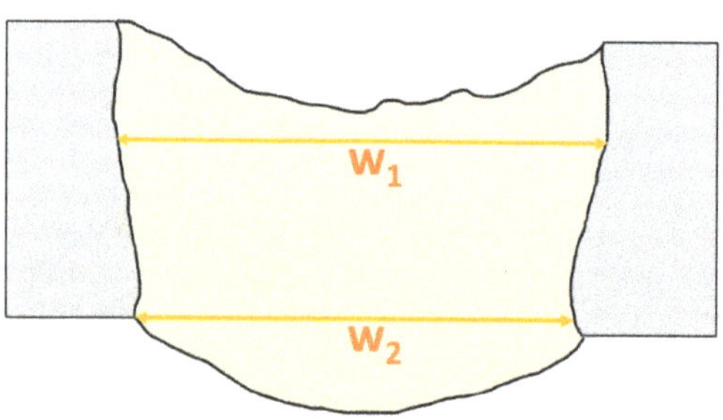

(c)

Figure 8. Comparison of numerical and experimental melt pool morphology: (**a**) Sample 1, (**b**) Sample 2, (**c**) fusion zone measurement criteria.

Within the scope of this study, numerical simulations based on the MHS, T-ITC, and IR-ITC approaches are carried out for Sample 1 and Sample 2. Table 2 outlines the simulation workflow for each model and reports the elapsed time per single run. The calibration of the MHS model is the most time-consuming stage, requiring 10 h. Because the T-ITC model derives its thermal cycle from the calibrated MHS model, its calibration time is identical. In contrast, a single thermo-metallurgical–mechanical run for LOAD calibration in the IR-ITC approach requires only 0.4 h. Three such runs, corresponding to three trial LOAD, were sufficient to calibrate the IR-ITC model. The proposed IR-ITC method have required almost seven times less calibration time than the MHS model, leading to a total time saving of approximately 85% compared to the MHS approach.

Table 2. Breakdown of simulation workflow stages and corresponding time requirements for each modeling approach.

Stage	Task	MHS (h)	T-ITC (h)	IR-ITC (h)
Calibration MHS	Metallographic sample preparation	3.75	3.75	0.00
	Macroscopic analysis	0.25	0.25	0.00
	LOAD adjustment	0.90	0.90	0.00
	Heat Source inner radius adjustment	0.90	0.90	0.00
	Heat Source outer radius adjustment	0.90	0.90	0.00
	Heat Source height adjustment	0.90	0.90	0.00
	Efficiency adjustment	0.90	0.90	0.00
Thermal Cycle Preparation	Thermal cycle extraction and preparation	0.00	0.30	0.30
Calibration IR-ITC	LOAD adjustment	0.00	0.00	1.20
Computation Time	Thermo-metallurgical–mechanical simulation	1.50	0.40	0.00
	Total	**10.00**	**9.20**	**1.50**

4.4. Thermal Analysis Results

Thermal distribution during the welding process was measured through thermographic monitoring. Figure 9 shows a representative frame from the thermal video captured during the experiment. The boundary of the weld pool can be identified from the temperature distribution, with the white region indicating temperatures at or above the melting point of aluminium, approximately 660 °C. It is worth noting that temperature acquisition through an IR thermal camera relies on the knowledge of the material emissivity, which varies with its physical state. In the weld pool, the emissivity of liquid aluminium differs from that of its solid form, introducing a potential error in measuring temperature inside the weld pool. Despite this limitation, the frame in Figure 9 clearly shows the temperature gradient from the outer edges of the metal sheets (cooler, darker regions) toward the weld center (hotter, lighter), where the laser is focused (i.e., about 660 °C). The gray region at the center represents pixel saturation due to the upper temperature limit of the IR camera. However, in this study, the primary aim is on the thermal behavior outside the weld pool, but still inside the welded joint, where emissivity is more stable and temperature measurements are considered reliable. Thermal cycles were extracted at the boundary of the molten zone (refer to Figure 4) from both the MHS simulation and the IR-guided measurements. To validate the proposed methodology, the thermal cycles for T-ITC and IR-ITC are compared with those obtained from the MHS simulation for Sample 1 and Sample 2, labeled as V35 and V50, respectively. The strong correlation between these thermal profiles, as shown in Figure 10, confirms the ability of the simulation framework to capture the thermal dynamics of the welding process accurately.

Following the procedures outlined in Sections 2.3 and 3, the thermal cycles shown in Figure 11 were extracted from the thermal profiles in Figure 10 and subsequently imposed on the T-ITC and IR-ITC numerical models.

Addressing the spatial resolution problem, the main concern might be related to the pixel-to-mm ratio. However, to use the experimental data in the simulation, the best thing is to have a pixel-to-mm ratio at least the minimum mesh element size. This avoids additional work on retrofitting temperature value in unknown areas, but at the same time avoid to have an unnecessary performing thermal camera.

The simulation successfully captures the distinct thermal cycles observed in the two samples, accurately reflecting the different heat inputs applied during welding. These differences are consistent with observations from thermographic analysis. To investigate transverse temperature variations in different models of both samples, five common reference points (P1–P5) were selected along the mid-plane line on the top surface, as shown in Figure 12. These points are located at 1.5 mm, 2.5 mm, 4.73 mm, 8.56 mm, and 13.88 mm from the weld line. Initially, temperature variation was evaluated for the MHS model at P1, P2, and P3 by comparing thermal cycles.

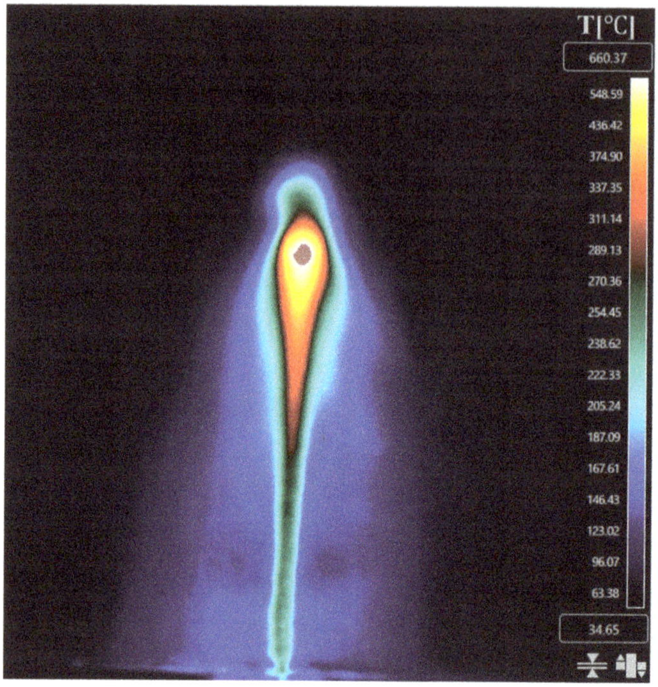

Figure 9. Thermography map highlighting the weld pool.

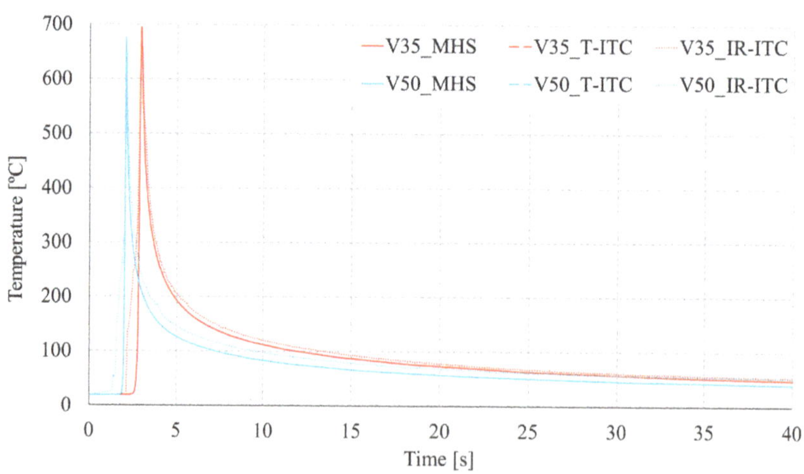

Figure 10. Comparison of experimental and numerical thermal cycles.

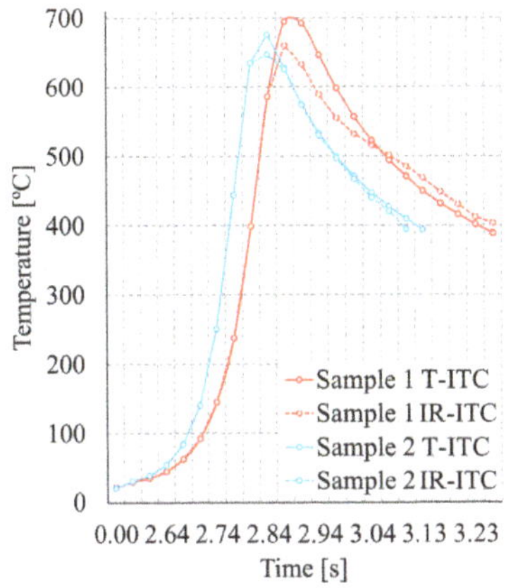

Figure 11. Comparison between the extracted thermal cycles from the MHS model (T-ITC) and the IR-guided thermal cycles (IR-ITC) for Sample 1 and Sample 2.

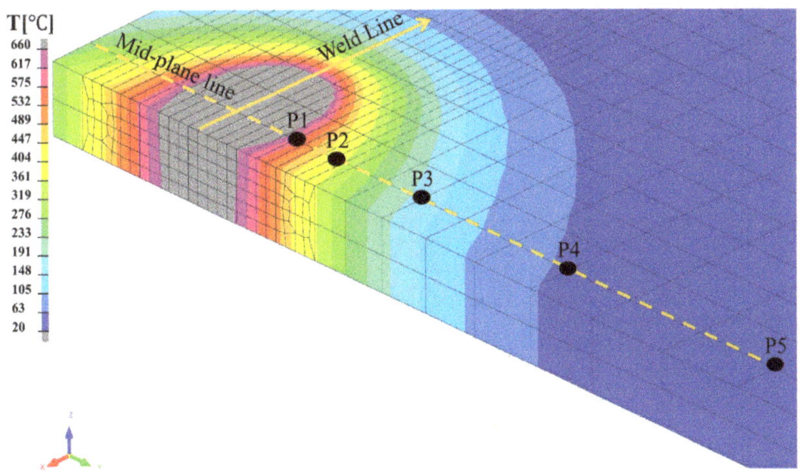

Figure 12. Selected points (P1–P5) along the mid-plane line used for quantitative thermal analysis.

Figure 13 confirms the monotonic temperature drop with increasing distance from the weld center, consistent with thermal-diffusion theory. The larger heat input in Sample 1 produces uniformly higher peak temperatures. To quantify thermo-metallurgical model fidelity, instantaneous temperatures at P1–P5 are compared with the experiment, and the average difference (Avg_Diff) is computed. Table 3 summarizes these results for all models, including IR-ITC with three LOAD options. It should be noted that the higher nodal temperatures observed in Sample 1 are due to points P1–P5 being located closer to the melt pool boundary, whereas in Sample 2, these points lie slightly farther from the fusion zone. Consequently, Sample 1 is expected to exhibit greater mechanical distortion than Sample 2. In both samples, the IR-ITC model with the narrowest LOAD (LOAD-1) gives the greatest thermal deviation; LOAD-2 shows a smaller but similar trend. In contrast, MHS, T-ITC, and IR-ITC with the widest LOAD (LOAD-3) produce comparable Avg_Diff values. Due to the minimal nodal temperature differences among these three models, significant microstructural variations are unlikely. Considering this, IR-ITC with LOAD-1 and LOAD-2 is likely to yield less accurate distortion predictions than the other models due to significant thermal deviations from experimental results and corresponding nodal temperature differences. Therefore, IR-ITC evaluated with LOAD-3 can be considered a reliable representative of both MHS and T-ITC models.

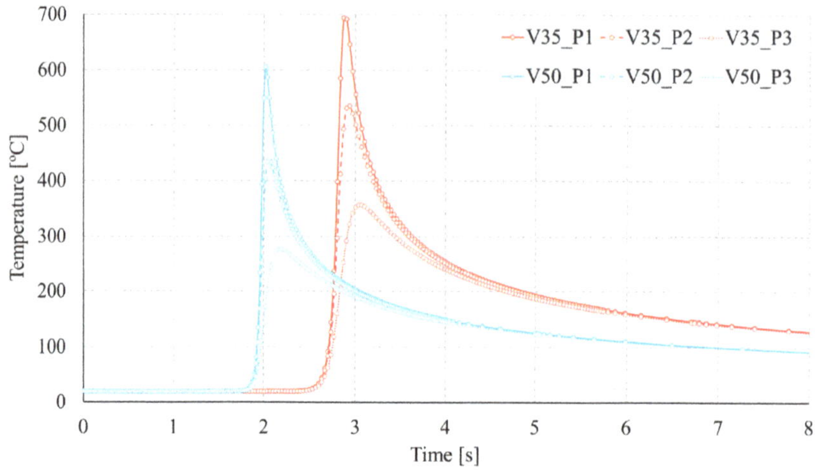

Figure 13. Temperature evolution at points P1–P3 along the central cross-section for Sample 1 and Sample 2.

Table 3. Comparison of simulated and experimental instantaneous temperatures at five selected locations (P1–P5) for different models and samples (units in °C).

Sample	Model	P1	P2	P3	P4	P5	Avg_Diff (°C)
1	Experiment	660	532	282	118	71	–
	MHS	695	494	253	97	39	31
	T-ITC	695	521	245	69	28	35
	IR-ITC (LOAD-1)	470	344	158	49	24	124
	IR-ITC (LOAD-2)	540	404	188	55	24	90
	IR-ITC (LOAD-3)	655	506	248	70	26	32
2	Experiment	628	442	201	86	64	–
	MHS	606	404	173	56	26	31
	T-ITC	620	431	159	40	24	29
	IR-ITC (LOAD-1)	423	277	102	31	21	113
	IR-ITC (LOAD-2)	507	344	127	34	21	77
	IR-ITC (LOAD-3)	600	435	168	39	21	31

4.5. Mechanical Analysis Results

During welding, the weldment temperature is raised above the fusion temperature to ensure proper melting of the joint area. The large temperature gradient experienced during the heating and cooling stages causes significant volumetric expansion and contraction, leading to metallurgical and mechanical transformations within the material. The thermal data computed from the thermo-metallurgical simulation serve as input for the mechanical analysis, enabling the prediction of welding-induced deformations, including potential buckling behavior and residual stresses, particularly for thin aluminium alloy components. To account for these effects accurately, all numerical models were computed with nonlinear geometry effects, as recommended in [47]. In this study, the mechanical performance of the numerical models is evaluated by analyzing the estimation errors between experimental and simulated distortion angles. Figure 14 illustrates the method used to assess the distortion angle from the simulations and to compare it with the actual distortions measured using a profilometer.

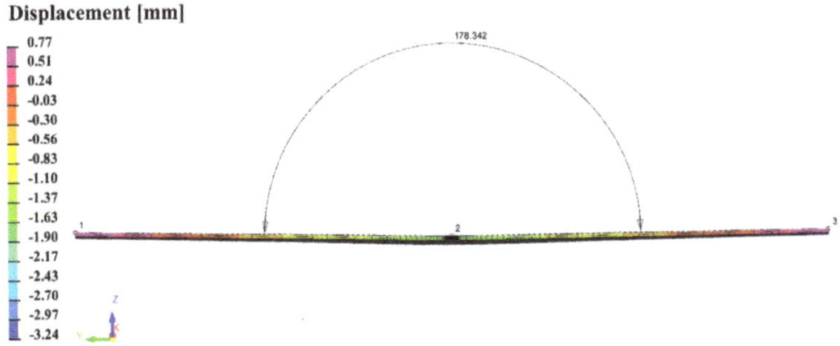

Figure 14. Distortion angle measurement from numerical models.

At first, LOAD sensitivity test was conducted for each LOAD configuration in the IR-ITC model under identical thermal cycles derived from IR camera measurements corresponding to each sample (refer to Figure 11). The highest distortion errors are observed with the narrowest LOAD configuration in both samples, while the lowest errors occur with the widest LOAD. Specifically, for Sample 1, angular distortion errors of 0.24% and 0.18% were achieved for LOAD-2 and LOAD-3, respectively; for Sample 2, the corresponding errors are 0.32% and 0.27%. Due to the poor performance of LOAD-1 configuration in predicting distortion, its results were excluded from the final analysis. This pattern indicates that as the fusion zone (FZ) dimensions predicted by the MHS model converge with those defined by the IR-ITC LOAD configurations, the mechanical distortion discrepancies are correspondingly minimized. Table 4 summarizes the comparison between the numerically predicted distortion angle and the experimental measurements.

Table 4. Comparison of distortion angle results between simulation models and experiments for Sample 1 and Sample 2.

Sample	Model	Angle (°)	Error (%)
1	MHS	177.98	0.15
	T-ITC	177.97	0.15
	IR-ITC	177.92	0.18
	Experiment	178.24	
2	MHS	178.43	0.22
	T-ITC	178.37	0.25
	IR-ITC	178.34	0.27
	Experiment	178.82	

For both samples, the numerical prediction of angular distortions matches the experimental measurements, with estimation errors consistently below 0.3%. A slightly more accurate distortion estimate is obtained for Sample 1, characterized by a lower welding speed. Notably, the distortion estimation errors are comparable between the traditional simulation methods and the proposed technique. This demonstrates the efficiency of the IR-ITC approach, as it provides reasonably similar results in about 90 min, significantly faster than the over 10 h required by traditional simulation methods.

The excellent agreement between simulation and experimental results across both thermal and mechanical parameters confirms the validity of our simulation approach, whether using conventional iterative calibration or the proposed method based on thermal camera data.

5. Conclusions

This study proposes an FEM simulation approach to predict the thermal and distortion behaviors during the laser welding of thin aluminium alloy sheets. A comparative analysis between experimental observations and simulation results demonstrates strong agreement, particularly in reproducing thermal cycles and distortion profiles.

A key innovation in this work is the use of IR thermal camera data to impose thermal cycles directly in the simulation, thereby eliminating the need for calibrating a moving heat source model. This significantly simplifies the computational process and speeds up the tuning procedure. Although the predictive accuracy of the proposed simulation method may not fully match that of the MHS approach, the advantageous trade-off between significantly reduced overall workflow time and satisfactory accuracy underscores the method's potential for practical engineering applications.

Experimental validation confirms that the proposed method achieves accuracy comparable to that of traditional simulation approaches, while reducing total time by more than 8.5 h in the present study. However, the magnitude of time and resource savings may vary depending on the model scale and other influencing parameters. This proves the efficiency of the proposed simulation approach.

It should be noted that the experimental validation in this study was limited to two welding speeds, specifically selected to induce noticeable distortion. This choice was deliberate, as our primary objective is to validate the proposed IR-ITC method under conditions known to produce significant deformation, which is particularly challenging to predict with FEM simulations accurately. However, the generalizability of our findings to a broader range of process parameters, such as different welding speeds, varying power inputs, and alternative joint configurations, still requires investigation. Future research efforts should focus on assessing the robustness and accuracy of the proposed IR-ITC methodology across a wider spectrum of welding conditions. Additionally, given that thermal cycle extraction via infrared thermography primarily relies on surface emissivity rather than being strictly material-dependent, the proposed IR-ITC method is expected to be broadly transferable to other materials and sheet thicknesses. Nevertheless, future studies should explicitly validate this methodology across different metals, alloys, and material thicknesses to fully establish its versatility and industrial applicability.

In summary, the integration of IR thermography with FEM simulations represents a meaningful advancement in the predictive modeling of laser welding processes. The proposed methodology is particularly relevant for industries such as automotive manufacturing, where it can streamline production workflows and enhance weld quality by reducing dependence on time-consuming and resource-intensive calibration procedures. Furthermore, the method is a practical engineering tool for optimizing manufacturing processes, especially in the development of lightweight and high-performance components.

Author Contributions: Conceptualization, M.D.M. and P.R.S.; methodology, M.D.M. and P.R.S.; software, H.M.; validation, H.M.; formal analysis, H.M., M.D.M., V.R. and L.S.; investigation, H.M., M.D.M. and D.B.; data curation, H.M., M.D.M., V.R. and L.S.; writing—original draft preparation, H.M., V.R. and L.S.; writing—review and editing, M.D.M., P.R.S. and D.B.; supervision, M.D.M. and P.R.S. All authors have read and agreed to the published version of the manuscript.

Funding: This research received no external funding.

Data Availability Statement: The original contributions presented in this study are included in the article. Further inquiries can be directed to the corresponding author.

Acknowledgments: The authors gratefully acknowledge the support of ESI Group and J-Tech@Polito, an advanced joining technologies research center at Politecnico di Torino (http://www.j-tech.polito.it/, accessed on 1 July 2025).

Conflicts of Interest: The authors declare that they have no known competing financial interests or personal relationships that could have appeared to influence the work reported in this paper.

References

1. Zhang, W.; Xu, J. Advanced lightweight materials for Automobiles: A review. *Mater. Des.* **2022**, *221*, 110994. [CrossRef]
2. Basile, D.; Sesana, R.; De Maddis, M.; Borella, L.; Russo Spena, P. Investigation of Strength and Formability of 6016 Aluminum Tailor Welded Blanks. *Metals* **2022**, *12*, 1593. [CrossRef]
3. Aminzadeh, A.; Silva Rivera, J.; Farhadipour, P.; Ghazi Jerniti, A.; Barka, N.; El Ouafi, A.; Mirakhorli, F.; Nadeau, F.; Gagné, M.O. Toward an intelligent aluminum laser welded blanks (ALWBs) factory based on industry 4.0; A critical review and novel smart model. *Opt. Laser Technol.* **2023**, *167*, 109661. [CrossRef]
4. Bunaziv, I.; Akselsen, O.M.; Ren, X.; Nyhus, B.; Eriksson, M. Laser Beam and Laser-Arc Hybrid Welding of Aluminium Alloys. *Metals* **2021**, *11*, 1150. [CrossRef]
5. Stavridis, J.; Papacharalampopoulos, A.; Stavropoulos, P. Quality assessment in laser welding: A critical review. *Int. J. Adv. Manuf. Technol.* **2018**, *94*, 1825–1847. [CrossRef]
6. Li, Y.; Wang, Y.; Yin, X.; Zhang, Z. Laser welding simulation of large-scale assembly module of stainless steel side-wall. *Heliyon* **2023**, *9*, e13835. [CrossRef] [PubMed]
7. Anca, A.; Cardona, A.; Risso, J.; Fachinotti, V.D. Finite element modeling of welding processes. *Appl. Math. Model.* **2011**, *35*, 688–707. [CrossRef]
8. Kik, T. Calibration of Heat Source Models in Numerical Simulations of Welding Processes. *Metals* **2024**, *14*, 1213. [CrossRef]
9. Mohan, A.; Franciosa, P.; Dai, D.; Ceglarek, D. A novel approach to control thermal induced buckling during laser welding of battery housing through a unilateral N-2-1 fixturing principle. *J. Adv. Join. Process.* **2024**, *10*, 100256. [CrossRef]
10. Chuang, T.C.; Lo, Y.L.; Tran, H.C.; Tsai, Y.A.; Chen, C.Y.; Chiu, C.P. Optimization of Butt-joint laser welding parameters for elimination of angular distortion using High-fidelity simulations and Machine learning. *Opt. Laser Technol.* **2023**, *167*, 109566. [CrossRef]
11. Yan, H.; Zeng, X.; Cui, Y.; Zou, D. Numerical and experimental study of residual stress in multi-pass laser welded 5A06 alloy ultra-thick plate. *J. Mater. Res. Technol.* **2024**, *28*, 4116–4130. [CrossRef]
12. Pyo, C.; Kim, J.; Kim, Y.; Kim, M. A study on a representative heat source model for simulating laser welding for liquid hydrogen storage containers. *Mar. Struct.* **2022**, *86*, 103260. [CrossRef]
13. Murua, O.; Arrizubieta, J.; Lamikiz, A.; Schneider, H. Numerical simulation of a laser beam welding process: From a thermomechanical model to the experimental inspection and validation. *Therm. Sci. Eng. Prog.* **2024**, *55*, 102901. [CrossRef]
14. Walker, T.; Bennett, C. An automated inverse method to calibrate thermal finite element models for numerical welding applications. *J. Manuf. Process.* **2019**, *47*, 263–283. [CrossRef]
15. Speka, M.; Matteï, S.; Pilloz, M.; Ilie, M. The infrared thermography control of the laser welding of amorphous polymers. *NDT E Int.* **2008**, *41*, 178–183. [CrossRef]
16. Bagavathiappan, S.; Lahiri, B.; Saravanan, T.; Philip, J.; Jayakumar, T. Infrared thermography for condition monitoring—A review. *Infrared Phys. Technol.* **2013**, *60*, 35–55. [CrossRef]
17. Santoro, L.; Sesana, R.; Molica Nardo, R.; Curá, F. Infrared in-line monitoring of flaws in steel welded joints: A preliminary approach with SMAW and GMAW processes. *Int. J. Adv. Manuf. Technol.* **2023**, *128*, 2655–2670. [CrossRef]
18. Santoro, L.; Sesana, R. A pilot study using flying spot laser thermography and signal reconstruction. *Opt. Lasers Eng.* **2025**, *188*, 108901. [CrossRef]
19. Sesana, R.; Santoro, L.; Curà, F.; Molica Nardo, R.; Pagano, P. Assessing thermal properties of multipass weld beads using active thermography: Microstructural variations and anisotropy analysis. *Int. J. Adv. Manuf. Technol.* **2023**, *128*, 2525–2536. [CrossRef]
20. Vasilev, M.; MacLeod, C.N.; Loukas, C.; Javadi, Y.; Vithanage, R.K.W.; Lines, D.; Mohseni, E.; Pierce, S.G.; Gachagan, A. Sensor-Enabled Multi-Robot System for Automated Welding and In-Process Ultrasonic NDE. *Sensors* **2021**, *21*, 5077. [CrossRef] [PubMed]
21. Nguyen, H.L.; Van Nguyen, A.; Duy, H.L.; Nguyen, T.H.; Tashiro, S.; Tanaka, M. Relationship among Welding Defects with Convection and Material Flow Dynamic Considering Principal Forces in Plasma Arc Welding. *Metals* **2021**, *11*, 1444. [CrossRef]
22. Zhu, C.; Cheon, J.; Tang, X.; Na, S.J.; Cui, H. Molten pool behaviors and their influences on welding defects in narrow gap GMAW of 5083 Al-alloy. *Int. J. Heat Mass Transf.* **2018**, *126*, 1206–1221. [CrossRef]
23. Filyakov, A.E.; Sholokhov, M.A.; Poloskov, S.I.; Melnikov, A.Y. The study of the influence of deviations of the arc energy parameters on the defects formation during automatic welding of pipelines. In Proceedings of the IOP Conference Series: Materials Science and Engineering, Nizniy Tagil, Russia, 24–29 February 2020; Volume 966. [CrossRef]
24. Aucott, L.; Huang, D.; Dong, H.B.; Wen, S.W.; Marsden, J.A.; Rack, A.; Cocks, A.C.F. Initiation and growth kinetics of solidification cracking during welding of steel. *Sci. Rep.* **2017**, *7*, 40255. [CrossRef] [PubMed]

25. Hong, Y.; Yang, M.; Chang, B.; Du, D. Filter-PCA-Based Process Monitoring and Defect Identification During Climbing Helium Arc Welding Process Using DE-SVM. *IEEE Trans. Ind. Electron.* **2023**, *70*, 7353–7362. [CrossRef]
26. D'Accardi, E.; Chiappini, F.; Giannasi, A.; Guerrini, M.; Maggiani, G.; Palumbo, D.; Galietti, U. Online monitoring of direct laser metal deposition process by means of infrared thermography. *Prog. Addit. Manuf.* **2024**, *9*, 983–1001. [CrossRef]
27. Razza, V.; Santoro, L.; De Maddis, M. Gradient-based image generation for thermographic material inspection. *Appl. Therm. Eng.* **2025**, *268*, 125900. [CrossRef]
28. Santoro, L.; Razza, V.; De Maddis, M. Frequency-based analysis of active laser thermography for spot weld quality assessment. *Int. J. Adv. Manuf. Technol.* **2024**, *130*, 3017–3029. [CrossRef]
29. Santoro, L.; Razza, V.; De Maddis, M. Nugget and corona bond size measurement through active thermography and transfer learning model. *Int. J. Adv. Manuf. Technol.* **2024**, *133*, 5883–5896. [CrossRef]
30. Zhang, C.; Li, X.; Gao, M. Effects of circular oscillating beam on heat transfer and melt flow of laser melting pool. *J. Mater. Res. Technol.* **2020**, *9*, 9271–9282. [CrossRef]
31. Zhou, Q.; Rong, Y.; Shao, X.; Jiang, P.; Gao, Z.; Cao, L. Optimization of laser brazing onto galvanized steel based on ensemble of metamodels. *J. Intell. Manuf.* **2018**, *29*, 1417–1431. [CrossRef]
32. Mirapeix, J.; García-Allende, P.; Cobo, A.; Conde, O.; López-Higuera, J. Real-time arc-welding defect detection and classification with principal component analysis and artificial neural networks. *NDT E Int.* **2007**, *40*, 315–323. [CrossRef]
33. Zhang, H.; Chen, Z.; Zhang, C.; Xi, J.; Le, X. Weld Defect Detection Based on Deep Learning Method. In Proceedings of the 2019 IEEE 15th International Conference on Automation Science and Engineering (CASE), Vancouver, BC, Canada, 22–26 August 2019; pp. 1574–1579. [CrossRef]
34. Sarkar, S.S.; Das, A.; Paul, S.; Mali, K.; Ghosh, A.; Sarkar, R.; Kumar, A. Machine learning method to predict and analyse transient temperature in submerged arc welding. *Measurement* **2021**, *170*, 108713. [CrossRef]
35. Bergman, T.L.; Lavine, A.S.; Incropera, F.P.; DeWitt, D.P. *Fundamentals of Heat and Mass Transfer*, 8th ed.; John Wiley & Sons, Inc.: Hoboken, NJ, USA, 2018.
36. Kik, T. Computational Techniques in Numerical Simulations of Arc and Laser Welding Processes. *Materials* **2020**, *13*, 608. [CrossRef] [PubMed]
37. Unni, A.K.; Vasudevan, M. Computational fluid dynamics simulation of hybrid laser-MIG welding of 316 LN stainless steel using hybrid heat source. *Int. J. Therm. Sci.* **2023**, *185*, 108042. [CrossRef]
38. Rong, Y.; Huang, Y.; Wang, L. Evolution Mechanism of Transient Strain and Residual Stress Distribution in Al 6061 Laser Welding. *Crystals* **2021**, *11*, 205. [CrossRef]
39. Tian, X.; Liao, J.; Cheng, P.; Ling, Y. Element Simulation of Welding Residual Stresses and Distortion in 5083 Incorporating Metallurgical Phase Transformation. In Proceedings of the 2017 International Conference on Applied Mathematics, Modeling and Simulation (AMMS 2017), Shanghai, China, 26–27 November 2017; Atlantis Press: Dordrecht, The Netherlands,2017; pp. 164–168. [CrossRef]
40. Schieck, B.; Stumpf, H. Deformation analysis for finite elastic-plastic strains in a lagrangean-type description. *Int. J. Solids Struct.* **1993**, *30*, 2639–2660. [CrossRef]
41. Lima, T.R.; Tavares, S.M.; de Castro, P.M. Residual stress field and distortions resulting from welding processes: Numerical modelling using Sysweld. *Ciência Tecnol. Dos Mater.* **2017**, *29*, e56–e61. [CrossRef]
42. Kik, T. Heat Source Models in Numerical Simulations of Laser Welding. *Materials* **2020**, *13*, 2653. [CrossRef] [PubMed]
43. Battista, F.R.; Ambrogio, G.; Giorgini, L.; Guerrini, M.; Costantino, S.; Ricciardi, F.; Filice, L. Prediction of the keyhole TIG welding-induced distortions on Inconel 718 industrial gas turbine component by numerical-experimental approach. *Int. J. Adv. Manuf. Technol.* **2024**, *134*, 4593–4608. [CrossRef]
44. Vargas, J.A.; Torres, J.E.; Pacheco, J.A.; Hernandez, R.J. Analysis of heat input effect on the mechanical properties of Al-6061-T6 alloy weld joints. *Mater. Des. (1980–2015)* **2013**, *52*, 556–564. [CrossRef]
45. Marques, E.S.V.; Silva, F.J.G.; Pereira, A.B. Comparison of Finite Element Methods in Fusion Welding Processes—A Review. *Metals* **2020**, *10*, 75. [CrossRef]
46. Long, H.; Gery, D.; Carlier, A.; Maropoulos, P. Prediction of welding distortion in butt joint of thin plates. *Mater. Des.* **2009**, *30*, 4126–4135. [CrossRef]
47. Vinoth, A.; Sivasankari, R. Numerical Simulation Studies in Tungsten Inert Gas Welding of Inconel 718 Alloy Sheet. *J. Mater. Eng. Perform.* **2024** . [CrossRef]
48. Kožar, I.; Rukavina, T.; Ibrahimbegović, A. Method of Incompatible Modes—Overview and Application. *Građevinar* **2018**, *70*, 19–29. [CrossRef]

metals

MDPI

Article

Analysis of Temperature and Stress Fields in the Process of Hot-Rolled Strip Coiling

Meng Dai [1,2], Yuting Hu [1,2], Yanchao Hao [1,2], Ping Qiu [2,*] and Hong Xiao [1,2,*]

1 National Engineering Research Center for Equipment and Technology of Cold Strip Rolling, Yanshan University, Qinhuangdao 066004, China; dai199509@stumail.ysu.edu.cn (M.D.); hyt88571@163.com (Y.H.); haoyanchao@stumail.ysu.edu.cn (Y.H.)

2 School of Mechanical Engineering, Yanshan University, Qinhuangdao 066004, China

* Correspondence: qiuping@ysu.edu.cn (P.Q.); xhh@ysu.edu.cn (H.X.)

Abstract: During the coiling process of a hot-rolled strip, with the increasing layers the temperature and stress distribution inside the coil constantly change and interact with each other. Due to the contact with the sleeve and the transition of the heat exchange state, it is inaccurate to consider the temperature of the whole coil as the coiling temperature set by the process requirement. Meanwhile, due to the periodic interlayer contact in the radial direction, the relation between stress and deformation is nonlinear. For the coiling process, it is difficult to consider the above factors using conventional methods. Therefore, an incremental model has been established to couple the temperature and stress of the coil. In order to obtain the mechanical properties of the strip and radial elastic modulus of the coil, tensile tests and laminated compression experiments are conducted at different temperatures. The effects of changes in strip thickness, coiling tension, and initial temperature of the sleeve on the stress and the temperature inside the coil are studied. Finally, by comparing the model results with measurements and analytical solutions, the effectiveness of the incremental coupled model is verified and the errors caused by the analytical method are analyzed.

Keywords: hot-rolled strip coiling; stress field analysis; temperature field analysis; incremental method

1. Introduction

As the final process of the hot-rolling production line, coiling directly affects the quality of products [1–3]. The coiling temperature mainly depends on the control of the microstructure and properties of the coil [4–7]. However, there is a significant temperature difference between the cooled strip and the sleeve, resulting in severe heat transfer at the contact position. The temperature drop near the inner ring of the coil can cause changes in the microstructure. During repeated heating and cooling processes, surface cracks and other forms of failure may appear on the sleeve. The defects in the coil are determined by the internal stress state. Excessive tension can lead to defects such as interlayer bonding, deformation of the sleeve, and interlayer slip damage, while insufficient tension can lead to collapse after unwinding [8–10]. Therefore, understanding the temperature and stress distribution inside the coil can provide a theoretical basis to optimize process parameters and reduce defects. Therefore, it is important to establish a model to calculate temperature and stress during the coiling process.

The coil is composed of repeated periodic contact interface layers, and the heat transfer state is significantly different from that of the strip itself [11]. In previous studies, for the

 https://doi.org/10.3390/met15020111

anisotropy caused by interlayer contact, the effective thermal conductivity in the radial direction was commonly used [12–14]. Considering strip thickness, surface characteristics and contact pressure, Park, S.J. [15] established a unit thermal resistance model. Based on this model, Park, S.J. conducted a finite element simulation to analyze the effects of different sizes and cooling methods on the cooling process of the coil. Park, S.J.'s model has been widely applied in research. The thermal resistance model was adopted by Yang, Y.J. [16] to perform differential calculations on the temperature inside the coil, and the thermal deformation during the cooling process was analyzed. Saboonchi, A. [17,18] established a temperature field model for coils stored in the warehouse, and the effects of different stacking methods were discussed. In Cheng, J.F.'s works [19], material density and thickness were considered as the factors affecting thermal conductivity, and the effects of different element compositions of the strip on the temperature field were studied. Witek, S. [20] established a temperature–stress coupled model during the cooling process and analyzed the residual stress distribution after cooling. The above models all focus on the cooling process, and the coiling temperature is considered as the starting point of the cooling process. In fact, due to the heat loss at the axial end faces and the heat exchange between the coil and sleeve, the overall temperature of the coil after coiling is no longer uniform and there is a certain gap with the coiling temperature. Therefore, when using the above thermal resistance model, the real-time changes in the radial thermal conductivity caused by the stress as well as the heat transfer between the coil and the sleeve should be considered. For solving the above problems, an incremental method is more suitable.

The analytical methods are widely used to calculate the stress distribution during the coiling process. The axisymmetric assumption is adopted, and the coiling process is equivalent to the thin-walled cylinders nested layer by layer on a thick-walled cylinder. Using the elastic deformation of the sleeve and coiling tension as boundary conditions, the internal stress of the coil can be calculated by superposition. This method requires the condition that the coil and sleeve are isotropic and have uniform properties at all positions. However, similar to the temperature conduction process, the layered structure of coils can also cause changes in radial mechanical properties, so the radial elastic modulus is normally modified in research [21–23]. Altmann, H.C. [24] considered the anisotropy of coils and established a planar model to solve the stress. In subsequent studies, it was found that the radial elastic modulus was related to interlayer pressure. Hakiel, Z. [25] established a function of radial elastic modulus using radial pressure as a variable to reflect nonlinearity. Based on the model established by Hakiel, Z., Benson, R.C. [26] described the relationship between radial elastic modulus and interlayer pressure using an exponential function. Li, S.P. [27] assumed that the radial elastic modulus was a quadratic function of interlayer pressure and determined the coefficients by laminated compression experiments. For hot-rolled coils, the temperature distribution is not uniform in the radial and axial directions. The above models are limited to the plane assumption and cannot consider the influence of temperature, which may cause certain errors. In order to consider the influence of the temperature field on physical properties, Park, W.W. [28] fitted an axial temperature curve of the strip based on measurements and assumed that each layer of the coil maintained this temperature distribution during the coiling process. In fact, the temperature at different positions of the coil continues to change during the coiling process, which causes changes in physical properties. Therefore, the influence of a dynamic temperature field on stress should be considered.

In summary, the temperature field and stress field during the coiling process interact with each other, and calculating temperature or stress separately may cause certain errors. Meanwhile, due to the real-time changes in temperature and stress state, the incremental model is more in line with the dynamic process. Therefore, a new incremental model for temperature–stress coupled calculation is established to simulate the coiling process in this article. In order to consider the anisotropy of coils, laminated compression experiments are conducted at different temperatures to obtain the nonlinear relationship between pressure and deformation. Finally, the temperature field results of the model are compared with the measurements to verify the accuracy of the model. The stress field results are compared with the analytical solutions. The differences between the two solutions and the errors caused by analytical methods are analyzed.

2. Materials and Methods

2.1. Establishment of Temperature Field Model

The process of coiling is simplified into thin-walled cylinders nested layer by layer on a thick-walled cylinder. The surface of the coil is covered by a layer of high-temperature strip and forms a new coil. After a period of heat exchange, the new coil is covered by a new layer of high-temperature strip to exchange heat. This process is repeated until the end of coiling, and the heat transfer state of each layer is different and changes in real time during the process. According to the radial heat transfer state, the strips can be divided into two categories: (a) The inner side of the strip is in the contact heat conduction state, while the outer side is in the heat dissipation state exposed to the air; (b) Both the inner and outer sides of the strip are in the contact heat conduction state. The temperature field model is shown in Figure 1. When coiling the $(n + 1)$ layer, the n-th layer is in the heat exchange state (a) and the other layers are in state (b). The duration is the time required to coil the $(n + 1)$ layer. After the coil is covered by the $(n + 1)$ layer, the $(n + 1)$ layer is in state (a) and other layers are in state (b). The duration is the time required to coil the $(n + 2)$ layer. By repeating this process, the temperature distribution at any time during the coiling process can be calculated based on the heat exchange state and time of each layer.

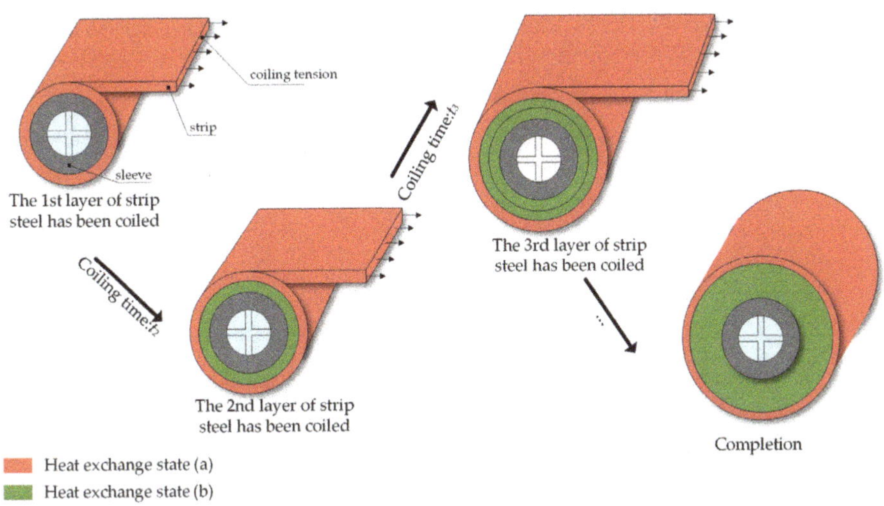

Figure 1. Schematic diagram of heat exchange state of each layer of the coil.

Both the coil and the sleeve can be regarded as axisymmetric thick-walled cylinders, and the thermal conductivity differential equations can be presented as Equation (1).

$$C_p\rho\frac{\partial T}{\partial t} = \lambda_r\left(\frac{\partial^2 T}{\partial r^2} + \frac{1}{r}\frac{\partial T}{\partial r}\right) + \lambda_s\frac{\partial^2 T}{\partial z^2}, \quad (1)$$

where C_p is the specific heat capacity, ρ is the density, T is the temperature, λ_r is the radial thermal conductivity and λ_s is the thermal conductivity of the material itself (for the sleeve, $\lambda_r = \lambda_s$).

To simplify the calculation process, half of the circumferential section of the coil and the sleeve is taken for analysis. Figure 2 shows the circumferential section and the mesh division.

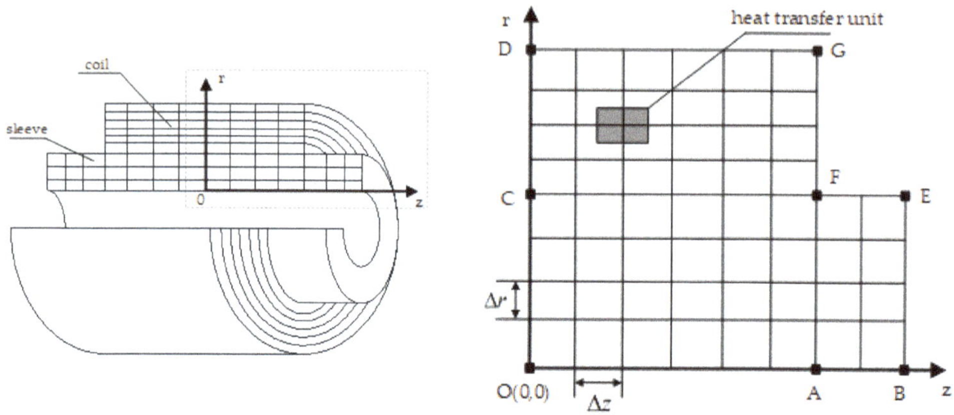

(**a**) (**b**)

Figure 2. (**a**) Circumferential section of the coil and the sleeve; (**b**) Division of units on the section.

The radial layered structure can be regarded as a continuous superposition of thermal conduction units, as indicated by the shaded areas in Figure 2. The thermal conduction unit is composed of the strip itself, the oxide layer, and the air layer at the contact interface. The thermal resistance of the thermal conduction unit can be presented as Equation (2). The composition of the thermal conduction unit is shown in Figure 3a.

$$R_t = R_s + R_i + R_o, \tag{2}$$

where R_t is the thermal resistance of the thermal conduction unit, and R_s, R_i and R_o are the thermal resistance of the strip, air layer, and oxide layer, respectively.

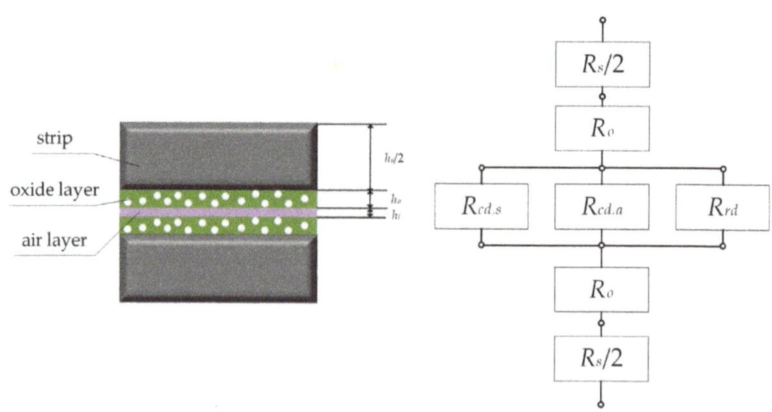

(**a**) (**b**)

Figure 3. (**a**) Composition of heat transfer unit; (**b**) Thermal resistance of heat transfer unit.

The air layer is the gap left due to insufficient contact between the surfaces of the strips, and there are three ways of heat transfer in the air layer, namely, direct-contact heat transfer on rough surfaces, radiation heat transfer on non-contact parts, and heat transfer of the air in the gap. The total thermal resistance can be calculated as follows:

$$R_i = \frac{1}{\frac{1}{R_{cd,s}} + \frac{1}{R_{rd}} + \frac{1}{R_{cd,a}}}, \tag{3}$$

where $R_{cd,s}$, R_{rd} and $R_{cd,a}$ are contact thermal resistance, radiation thermal resistance, and air layer thermal resistance, respectively, all of which are related to interlayer contact pressure and can be calculated by Equation (4) [29], Equation (5) [15] and Equation (6) [14]:

$$R_{cd,s} = \frac{\sigma_p}{1.13\lambda_s K}\left(\frac{H+P}{P}\right)^{0.94}, \tag{4}$$

$$R_{rd} = \frac{1}{4\left(1 - \frac{P}{H+P}\right)\varepsilon \times 5.67 \times 10^{-8} T^3}, \tag{5}$$

$$R_{cd,a} = \frac{42.7 \times 10^{-6} \exp\left(-5 \times 10^{-2} P\right)}{\left(1 - \frac{P}{H+P}\right)\lambda_a}, \tag{6}$$

where σ_p is the standard deviation of contour height, K is the absolute value of the average slope of the contour, H is the microhardness, P is the interlayer pressure, λ_a is the thermal conductivity of the air layer and ε is the radiance.

Figure 3b shows the thermal resistance of the thermal conduction unit. By substituting Equations (3)–(6) into Equation (2), the effective thermal conductivity in the radial direction of the coil can be presented as follows:

$$\lambda_r = \frac{\Delta r}{\frac{h_s}{\lambda_s} + 2\frac{h_o}{\lambda_o} + \left[\frac{1}{R_{cd,s}} + \frac{1}{R_{rd}} + \frac{1}{R_{cd,a}}\right]^{-1}}, \tag{7}$$

By substituting the effective thermal conductivity into the thermal conductivity differential equation, the temperature distribution of the coil and sleeve can be obtained. It is worth noting that the effective thermal conductivity is closely related to contact pressure. Therefore, the variation of radial stress distribution inside the coil must be considered during the calculation process.

2.2. Establishment of Stress Field Model

When coiling a layer of strip, it can be regarded as adding a layer of elements on the outer diameter side of the entire grid. The circumferential stress of the new layer of elements is equivalent to the coiling tension. When coiling the $(n + 1)$ layer of strip, material properties and radial elastic modulus can be calculated according to the temperature and interlayer pressure of the first n layers of the coil. After applying the boundary conditions of the outermost tension, the temperature and stress increment can be calculated. By superimposing the increment onto the initial condition, the initial condition for coiling the $(n + 2)$ layer of strip can be obtained. By repeating this process until the end of coiling, the stress distribution of the coil and sleeve can be calculated. Due to the short coiling time, the heat loss of the coil is not significant. The area with a large temperature drop in the coil is relatively small. Compared with the mechanical stress, the thermal stress generated by temperature changes is also relatively small. Therefore, this article mainly focuses on the mechanical stress inside the coil, and the thermal stress is not considered. In addition, the strip hardly slides in the axial and circumferential directions during the coiling process.

Compared to the circumferential stress, the interlayer friction force is smaller and it is not considered in the model. Due to the continuous application of significant coiling tension on the strip in the circumferential direction within a short coiling time, the time–stress relaxation is also not considered. Figure 4 shows the stress distribution inside the coil in the model.

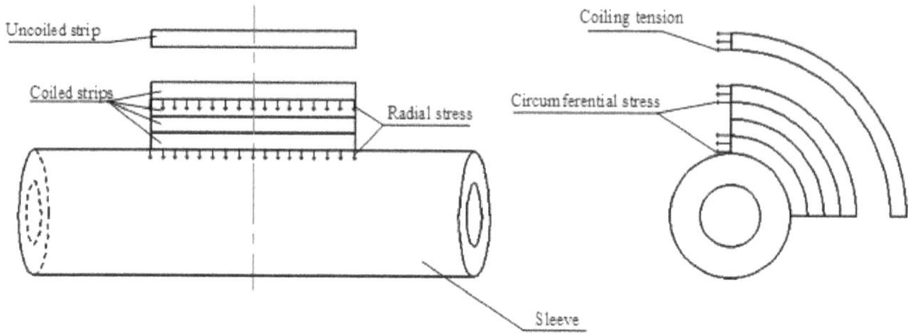

Figure 4. The stress distribution inside the coil in the model.

The completion time of coiling the nth and $(n + 1)$ layer is t_n and t_{n+1}, respectively. In the axisymmetric model, the incremental equilibrium equations for the coil and sleeve can be presented as Equations (8) and (9).

$$\frac{\partial \Delta\sigma_r}{\partial r} + \frac{\partial \Delta\tau_{rz}}{\partial z} + \frac{\Delta\sigma_r - \Delta\sigma_\theta}{r} = 0, \tag{8}$$

$$\frac{\partial \Delta\sigma_z}{\partial z} + \frac{\partial \Delta\tau_{rz}}{\partial r} + \frac{\Delta\tau_{rz}}{r} = 0, \tag{9}$$

where $\Delta\sigma_r$, $\Delta\tau_{rz}$, $\Delta\sigma_\theta$ and $\Delta\sigma_z$ are the increments of radial stress, shear stress, circumferential stress, and axial stress from t_n to t_{n+1}, respectively.

The constitutive equation can be presented as follows:

$$\begin{bmatrix} \Delta\varepsilon_r \\ \Delta\varepsilon_\theta \\ \Delta\varepsilon_z \\ \Delta\gamma_{rz} \end{bmatrix} = \begin{bmatrix} \frac{1}{E^*} & -\frac{\mu}{E} & -\frac{\mu}{E} & 0 \\ -\frac{\mu}{E} & \frac{1}{E} & -\frac{\mu}{E} & 0 \\ -\frac{\mu}{E} & -\frac{\mu}{E} & \frac{1}{E} & 0 \\ 0 & 0 & 0 & \frac{2(1+\mu)}{E} \end{bmatrix} \begin{bmatrix} \Delta\sigma_r \\ \Delta\sigma_\theta \\ \Delta\sigma_z \\ \Delta\tau_{rz} \end{bmatrix} \tag{10}$$

where $\triangle\varepsilon_r$, $\triangle\varepsilon_\theta$, $\triangle\varepsilon_z$, and $\triangle\gamma_{rz}$ are the increments of radial strain, circumferential strain, axial strain, and shear strain of the sleeve unit from t_n to t_{n+1}, respectively, E is the elastic modulus, and E^* is the radial elastic modulus. When the structure is isotropic, $E = E^*$.

By taking strain increment as a known quantity and solving the stress increment in Equation (10), constitutive equations in the sleeve and coil can be obtained.

The sleeve is isotropic, and the incremental constitutive equation can be presented as follows:

$$\begin{bmatrix} \Delta\sigma_r \\ \Delta\sigma_\theta \\ \Delta\sigma_z \\ \Delta\tau_{rz} \end{bmatrix} = \begin{bmatrix} A_1^n & A_2^n & A_2^n & 0 \\ A_2^n & A_1^n & A_2^n & 0 \\ A_2^n & A_2^n & A_1^n & 0 \\ 0 & 0 & 0 & G^n \end{bmatrix} \begin{bmatrix} \Delta\varepsilon_r \\ \Delta\varepsilon_\theta \\ \Delta\varepsilon_z \\ \Delta\gamma_{rz} \end{bmatrix}, \tag{11}$$

where E_t^n is the elastic modulus at t_n. A_1, A_2 and G are the parameters related to elastic modulus and Poisson's ratio, $A_1^n = \frac{(1-\mu_t)E_t^n}{(1-2\mu_t)(1+\mu_t)}$, $A_2^n = \frac{\mu_t E_t^n}{(1-2\mu_t)(1+\mu_t)}$ and $G^n = \frac{E_t^n}{2(1+\mu_t)}$.

The coil is anisotropic and the incremental constitutive equation can be presented as follows:

$$\begin{bmatrix} \Delta\sigma_r \\ \Delta\sigma_\theta \\ \Delta\sigma_z \\ \Delta\tau_{rz} \end{bmatrix} = \begin{bmatrix} B_1^n & B_2^n & B_3^n & 0 \\ B_2^n & B_4^n & B_5^n & 0 \\ B_2^n & B_5^n & B_4^n & 0 \\ 0 & 0 & 0 & G^n \end{bmatrix} \begin{bmatrix} \Delta\varepsilon_r \\ \Delta\varepsilon_\theta \\ \Delta\varepsilon_z \\ \Delta\gamma_{rz} \end{bmatrix}, \tag{12}$$

where E_s^n and E_r^n are the elastic modulus of the strip and the radial elastic modulus of the coil at t_n, respectively. B_1, B_2, B_3, B_4 and B_5 are the parameters related to elastic modulus and Poisson's ratio, $B_1^n = \frac{E_s^n \cdot E_r^n (1-\mu_s)}{E_s^n(1-\mu_s)-2\mu_s^2 E_r^n}$, $B_2^n = \frac{E_s^n \cdot E_r^n \mu_s}{E_s^n(1-\mu_s)-2\mu_s^2 E_r^n}$, $B_3^n = \frac{E_s^n \cdot E_r^n (1+\mu_s)}{E_s^n(1-\mu_s)-2\mu_s^2 E_r^n}$, $B_4 = \frac{(E_s^n)^2 - E_s^n \cdot E_r^n \mu_s^2}{E_s^n(1-\mu_s^2)-2\mu_s^2(1+\mu_s)E_r^n}$, and $B_5 = \frac{(E_s^n)^2 + E_s^n \cdot E_r^n \mu_s^2}{E_s^n(1-\mu_s^2)-2\mu_s^2(1+\mu_s)E_r^n}$.

When the elastic modulus of the coil is equal in the radial, axial, and circumferential directions, parameter A_1^n is equal to parameter B_1^n and B_4^n, and parameter A_2^n is equal to parameters B_2^n, B_3^n and B_5^n. By substituting the constitutive equation into the equilibrium equation and combining the boundary conditions with the incremental model mentioned above, the internal stress of the coil can be calculated.

Using the temperature and stress at time t_n as initial conditions, the physical properties and thermal parameters can be calculated. The partial differential in the model can be converted into differential form and discretized according to the grid. Then, the equation system can be solved based on the boundary conditions to obtain the temperature increment and stress increment, which can be added to the initial state to obtain the distribution of temperature and stress at time t_{n+1}. The above process is repeated until the end of the coiling process to obtain the distribution of temperature and stress at any time during the coiling process. The numerical calculations are implemented by writing calculation programs in the C programming language. The software used for compilation is Microsoft Visual Studio 2010.

3. Results and Discussion

3.1. Calculation Parameters

According to the actual parameters on the production line, the material of the sleeve is 45 # cast steel and the structure is the four-sided pyramid. The steel grade is Q235B, and the elemental composition is shown in Table 1. Tables 2–4 show the relevant thermal property parameters of the strip, the parameters related to radial thermal conductivity and the parameters of coiling equipment and process in the incremental coupled model, respectively.

Table 1. The elemental composition of materials.

Element	C	Si	Mn	P	S
Content (%)	0.15	0.29	1.2	0.041	0.037

Table 2. Thermal property parameters of the strip.

T (°C)	λ_s (W·m^{-1}·K^{-1})	ρ (kg·m^{-3})	C_p (J·g^{-1}·K^{-1})
30	49.61	7810	0.45
150	47.86	7774	0.50
250	45.45	7742	0.54
350	42.56	7709	0.59
450	39.43	7674	0.65
550	36.41	7638	0.73
650	33.69	7600	0.87

Table 3. The parameters related to radial thermal conductivity, respectively.

h_o (μm)	λ_o (W·m^{-1}·K^{-1})	λ_a (W·m^{-1}·K^{-1})	H (MPa)	σ_p (μm)	K	ε
7	3	0.0048	1133.86	3.22	0.086	0.9

Table 4. The parameters of coiling equipment and process.

Parameters	Value
Thickness h	2.0 mm
Width W	685.0 mm
Coiling speed v	12,700.0 mm/s
Number of layers N	255
Coiling tension σ_T	27.0 MPa
Outer diameter of the sleeve R_m	752.0 mm
Coiling temperature T_s	620.0 °C
Initial temperature of the sleeve T_{t0}	200.0 °C

In order to study the effect of temperature on the elastic modulus of a strip, samples are collected from the site and tensile tests are conducted at the temperature nodes of room temperature, 150 °C, 250 °C, 350 °C, 450 °C, 550 °C, and 650 °C. Figure 5a shows the dimensions of the specimen, and the edges of the specimen are polished. In Figure 5b, the testing machine model is Zwick 150 kN. Figure 5c shows the specimen clamping inside the heating furnace. The temperature inside the heating furnace is raised to the temperature node and held for 60 s. Then, the tensile test is repeated three times at each temperature node and the deformation of the specimen is recorded with an extensometer.

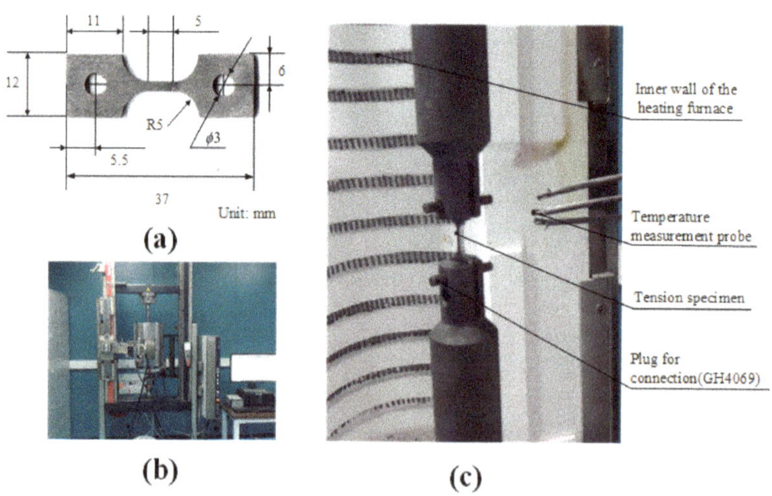

Figure 5. (**a**) Dimension of the tensile specimen; (**b**) Tensile testing machine; (**c**) Clamping of the tensile specimen.

To obtain the radial elastic modulus of the coil, laminated compression experiments are conducted. The strip is cut into circular pieces with a diameter of 20 mm. Each set of the laminates consists of 40 specimens, and compression experiments are conducted at the same temperature node as the tensile tests. The testing machine model is Inspekt Table 100 kN. In order to eliminate the influence of various possible factors such as flatness defects, impurities, and improper operation, each set of the tests is repeated three times. According to the experimental results, the relationship between displacement and pressure is converted into the relationship between the elastic modulus of the laminated material and the compressive stress.

The trend of elastic modulus variation of a strip is shown in Figure 6a. Figure 6b shows the relationship between the elastic modulus of the laminates and stress at different temperatures. During calculation, the elastic modulus between temperature nodes can be obtained through interpolation.

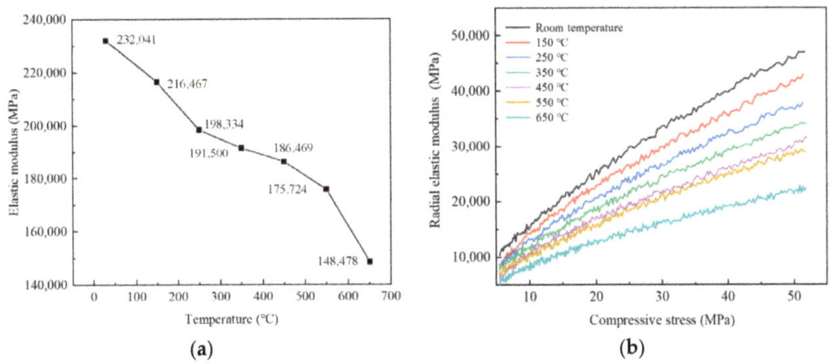

Figure 6. (**a**) Elastic modulus of strip at different temperatures; (**b**) Radial elastic modulus of coil at different temperatures.

3.2. Analysis of Temperature Field Calculation Results

Figure 7 shows the temperature distribution of the circumferential section of the coil and sleeve. From Figure 7a, it can be seen that in the radial direction, because of the heat loss caused by the contact between the coil and the sleeve, there is a significant decrease in temperature near the inner ring of the coil. The number of layers with a temperature drop is 30, with the maximum temperature drop at around 270 °C in the innermost ring. The temperature at the rest positions is close to the coiling temperature. The temperature drop is most significant at the inner ring edges in the axial direction, with a difference of 50 °C from the middle. This is because the heat at the inner ring edges can be transferred along the axial direction of the sleeve. With the increasing of distance from the contact surface, the temperature difference between the edges and the middle decreases by about 18 °C. This is due to the low efficiency and limited time of convective and radiative heat transfer between the axial end faces and the air. The temperature of the sleeve in Figure 7b corresponds to the temperature drop of the coil. There is a severe temperature rise in the contact area, with a maximum value of 180 °C. The thickness of the area where the temperature rises accounts for about 30% of the thickness of the sleeve.

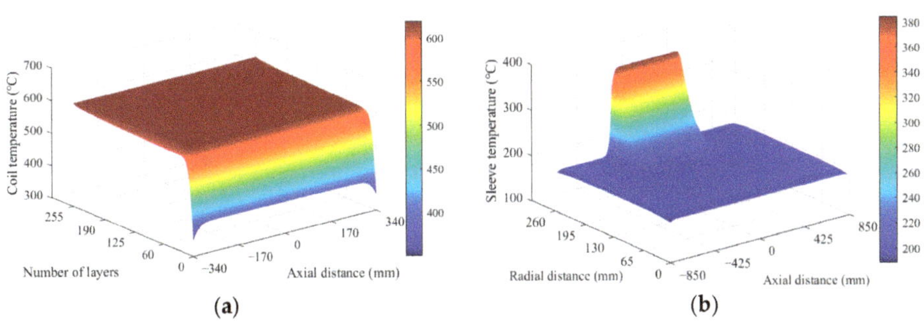

Figure 7. (**a**) Temperature distribution of circumferential section of the coil when the initial temperature of the sleeve is 200 °C; (**b**) Temperature distribution of circumferential section of the sleeve when the initial temperature of the sleeve is 200 °C.

3.3. Analysis of Stress Field Calculation Results

During the coiling process, the radial stress inside the coil is related to the anisotropy, the tightness of the coil and the stress on the sleeve. The circumferential stress and radial stress are interrelated. The coiling tension can be controlled based on the circumferential stress to avoid situations such as loose coiling. Excessive local circumferential stress may cause plastic deformation, resulting in defects in the shape. Therefore, the distribution of radial and circumferential stresses in the coil are mainly discussed. Figure 8 shows the stress distribution of the coil and sleeve after coiling. From Figure 8a, it can be seen that the radial stress is highest at the inner ring and gradually decreases with the increase in radius. The circumferential stress in Figure 8b is relatively high at the inner ring and gradually decreases as it moves away from the contact surface, even transforming into compressive stress near the lowest point. Then, with the increase in radius, the circumferential stress gradually increases to the coiling tension. This distribution pattern is due to the significant difference in radial stiffness between the coil and the sleeve, which limits the radial deformation of the strip. The limit effect is most significant in the innermost ring, resulting in the smallest increment in circumferential compressive stress. When combined with the initial coiling tension, larger circumferential tensile stress is presented. With the increasing of distance from the contact surface, the limit effect gradually weakens and the circumferential compressive stress gradually increases, causing a gradual decrease in circumferential tensile stress. When the compressive stress is larger than the tensile stress provided by the initial coiling tension, the circumferential stress is converted into compressive stress. As the radius increases, the circumferential compressive stress gradually weakens due to the decrease in the number of stacked layers, and the circumferential tensile stress gradually increases to the coiling tension. The difference in the stress of the coil is not significant in the axial direction. This is because the axial temperature difference of the coil is small, resulting in a small difference in mechanical properties.

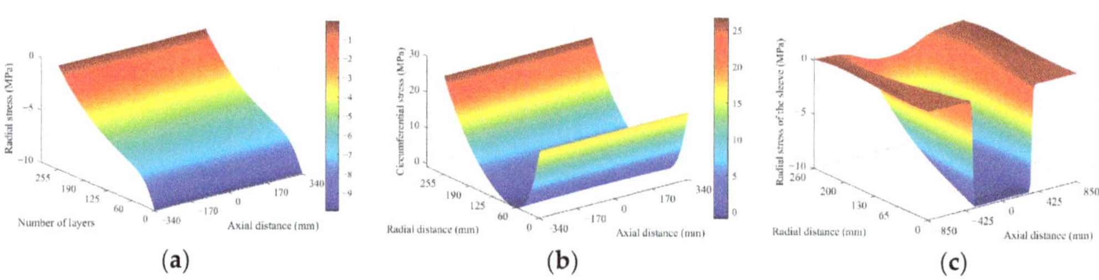

Figure 8. (**a**) Radial stress of the coil; (**b**) Circumferential stress of the coil; (**c**) Radial stress of the sleeve.

3.4. Comparative Analysis of Coupled and Uncoupled Results

During the dynamic coiling process, because of the mutual influence of temperature and stress, the time increment step is reduced and the temperature field and stress field are coupled in the calculation. For the convenience of observing temperature differences, Figure 9 shows the temperature distribution on the axial symmetry plane (OD section in Figure 2). For the coil, the layer range is from 0 to 60 layers, while for the sleeve, the radius range is from R = 260 mm to R = 376 mm.

As shown in Figure 9a, the temperature of the coupled model is relatively high near the contact surface. As the radius increases, both results gradually approach the coiling temperature, but the temperature of the coupled model is smaller. The difference is largest around the 15th layer, about 60 °C. The reason is that the radial stress in the coupled model is relatively high, which increases the radial thermal conductivity. Within the same coiling time, more heat flows from the coil to the sleeve, resulting in a decrease in temperature of the coil. In Figure 9b, compared with the uncoupled model, the surface temperature rise of the sleeve in the coupled model is larger and the temperature difference is about 50 °C.

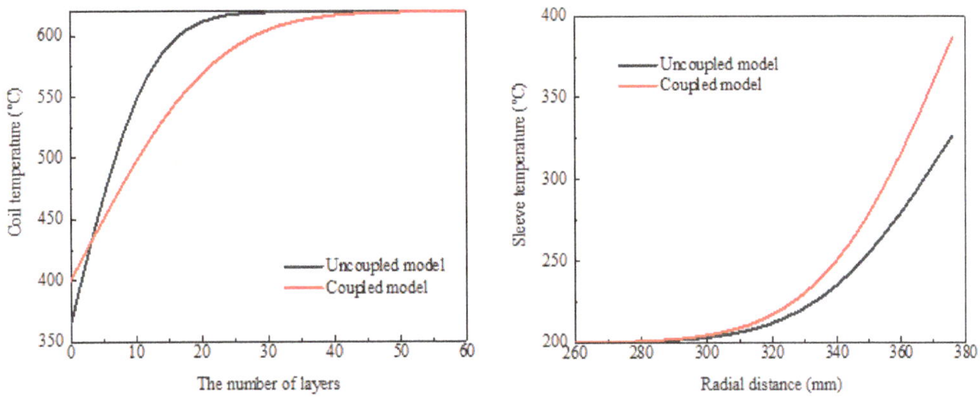

(a) (b)

Figure 9. (**a**) Coil temperature; (**b**) Sleeve temperature.

Figure 10 shows the stress comparison between the uncoupled model and the coupled model. The radial stress in the coupled model is relatively large overall. The difference between the two models is largest near the inner ring, about 25%. This is because the influence of temperature on material properties is considered in the coupled model, resulting in a decrease in the elastic modulus. It is more difficult for the sleeve to deform. Corresponding to Figure 10a, circumferential stress of the coupled model in Figure 10b is larger, but the difference between the two models is not significant.

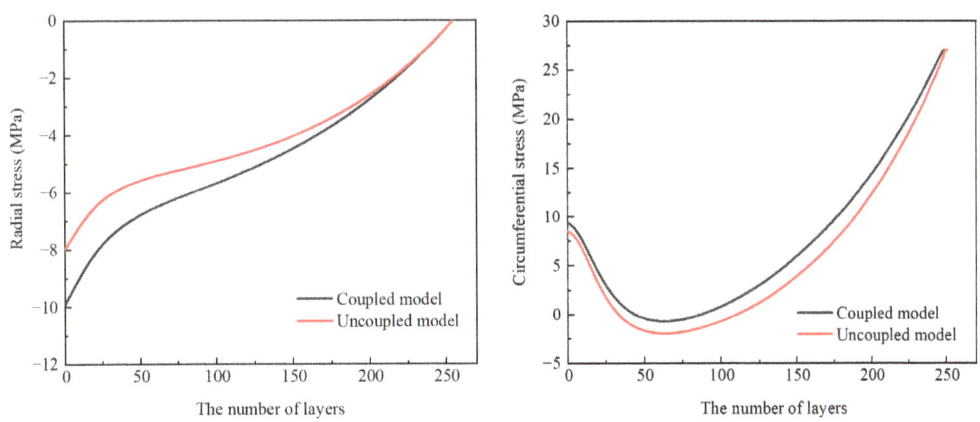

(a) (b)

Figure 10. (**a**) Radial stress of the coil in coupled model and uncoupled model; (**b**) Circumferential stress of the coil in coupled model and uncoupled model.

3.5. The Influence of Coiling Process Parameters on Temperature and Stress Distribution

The temperature and stress fields of the coil are affected by the coiling process parameters. Therefore, in this article, the thickness of the strip, the coiling tension, and the initial temperature of the sleeve are taken as variables to study the temperature and stress fields.

Figure 11 shows the calculation results of stress. It can be seen that as the thickness of the strip and the coiling tension increase, and as the initial temperature of the sleeve decreases, the radial stress inside the steel increases. This is because the increase in thickness of the strip and coiling tension both lead to an increase in the total tensile force of the circumferential section, resulting in an increase in the radial stress increment. By reducing the initial temperature of the sleeve, the difference in elastic modulus between the coil and the sleeve is increased. It is more difficult to deform for the sleeve, and the reduction in radial displacement of the coil can lead to an increase in radial stress. In the case of the same radius, the circumferential stress corresponds to the slope of the radial stress curve. This rule is reflected in Figure 11d,f. However, since the horizontal axis represents the number of layers rather than the radius, the rule is not applicable in Figure 11b.

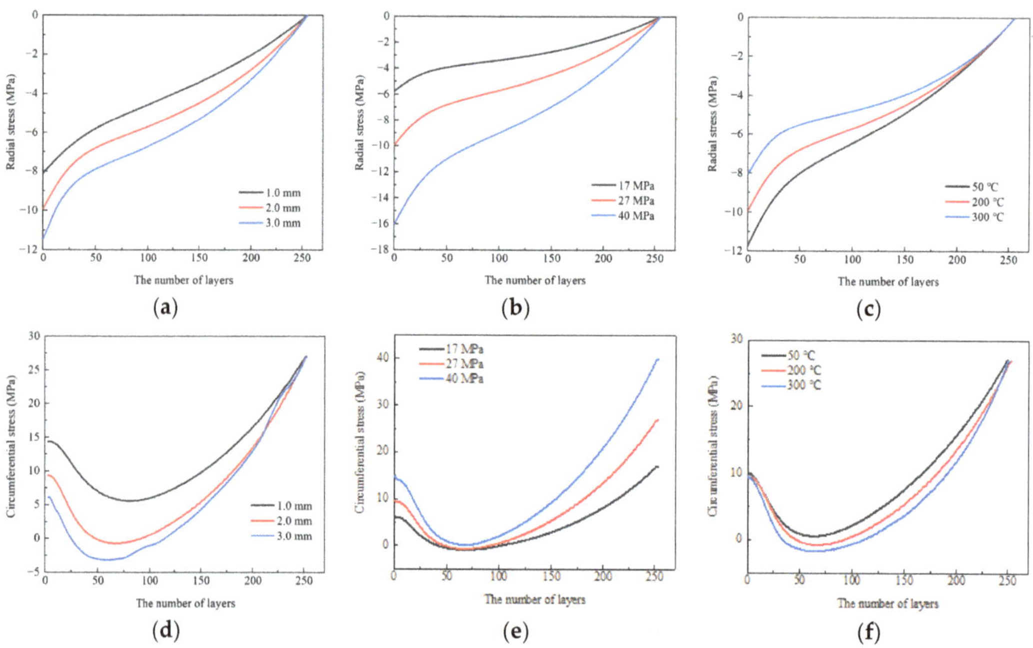

Figure 11. (**a**) The radial stress distribution inside the coil when the thickness of the strip changes; (**b**) The radial stress distribution inside the coil when the coiling tension changes; (**c**) The radial stress distribution inside the coil when the initial temperature of the sleeve changes; (**d**) The circumferential stress distribution inside the coil when the thickness of the strip changes; (**e**) The circumferential stress distribution inside the coil when the coiling tension changes; (**f**) The circumferential stress distribution inside the coil when the initial temperature of the sleeve changes.

Figure 12 shows the temperature calculation results, which indicate that as the thickness of the strip and the coiling tension increase, the temperature rise of the sleeve increases. The reason is that the increase in radial stress leads to an increase in the effective thermal conductivity in the radial direction, causing more heat to flow from the coil to the sleeve, resulting in a corresponding decrease in the temperature of the coil. However, the temperature of the strip of the thickness of 3 mm in Figure 12a is still higher because a thicker strip has a higher total heat content. In Figure 12e, as the initial temperature of the sleeve decreases, the temperature drop of the coil increases. This is because the heat transfer efficiency is improved by the increase in temperature difference on both sides of the contact surface, which is also the reason for the increase in the temperature rise of the sleeve in Figure 12f.

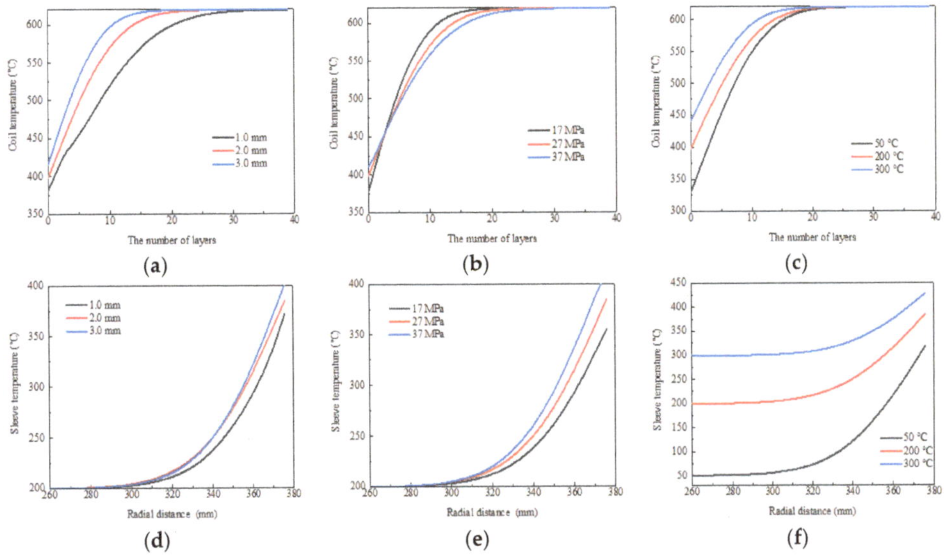

Figure 12. (**a**) Temperature distribution of the coil when the thickness of the strip changes; (**b**) Temperature distribution of the coil when the coiling tension changes; (**c**) Temperature distribution of the coil when the initial temperature of the sleeve changes; (**d**) Temperature distribution of the sleeve when the thickness of the strip changes; (**e**) Temperature distribution of the sleeve when the coiling tension changes; (**f**) Temperature distribution of the sleeve when the initial temperature of the sleeve changes.

3.6. Verification and Analysis of Simulation Results

3.6.1. Comparison Between the Calculated Results and the Measured Temperature

In actual production, due to the direct exposure to air, the temperature of the strip before coiling will be lower than the temperature set by the process and there will be a certain temperature drop at the edges of the strip. Therefore, in this article, the temperature before coiling on the hot-rolling line is measured and the measurement results are fitted. The temperature field is calculated based on the fitting results and compared with measurements on site to verify the correctness of the model. The initial temperature of the sleeve is set to 50 °C, and the other conditions are the same as the parameters in Table 4. Figure 13 shows the temperature distribution of the strip before coiling. The surface temperature of the strip is directly captured by a thermal imaging system, and the model is FLIR-E95.

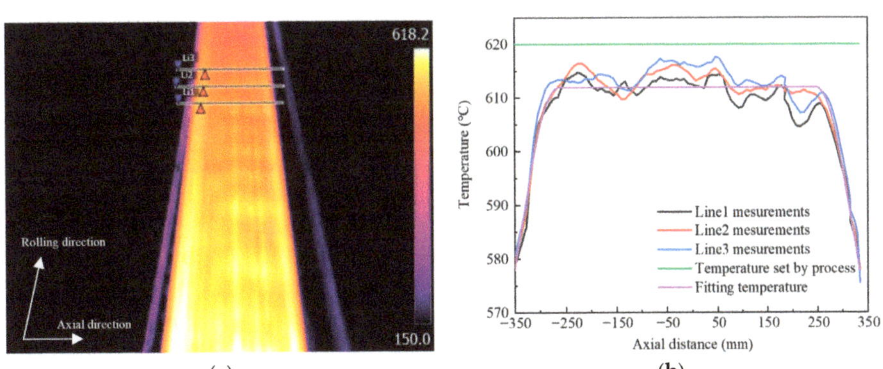

Figure 13. (**a**) Temperature distribution before coiling; (**b**) Measurement results and the fitting curve.

Figure 14 shows the temperature distribution of the circumferential section of the coil and sleeve. Corresponding to the temperature of the strip before coiling, the overall temperature of the coil in Figure 14a is lower than 620 °C and there are obvious low-temperature areas at the edges. The maximum axial temperature difference at the contact surface is about 90 °C. As the position moves away from the contact surface, the temperature difference gradually decreases to the initial temperature difference of 35 °C.

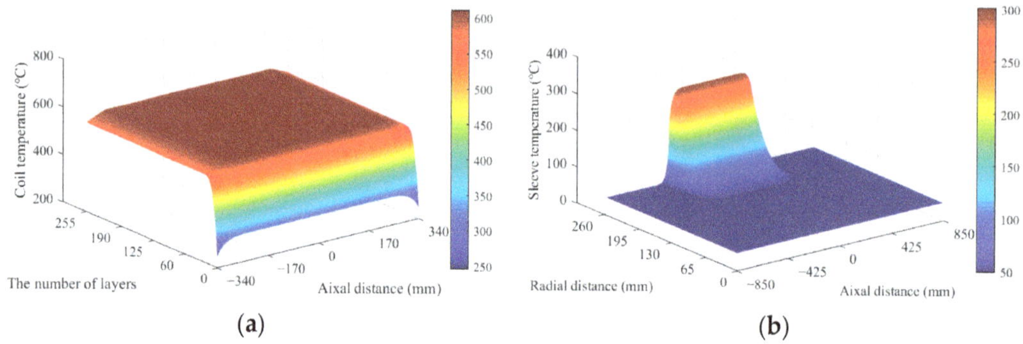

Figure 14. (**a**) Temperature distribution of circumferential section of the coil; (**b**) Temperature distribution of circumferential section of the sleeve.

Due to the difficulty in measuring the internal temperature of the coil, the temperature measurements at the axial end face of the coil and the surface of the sleeve are compared with the calculated results. Figures 15 and 16 are the comparison charts, which show that the calculated results tend to be consistent with the measurements. The error of the surface temperature of the sleeve is less than 30 °C, and the maximum error occurs at the boundary between the contact area and the non-contact area. The reason is that the surface temperature of the sleeve can only be measured after unloading. Due to the heat transfer in the sleeve during the unloading process, the temperature difference between the contact area and the non-contact area is reduced. The maximum temperature error of the coil is less than 20 °C. The above results indicate that the temperature field calculation model established is feasible.

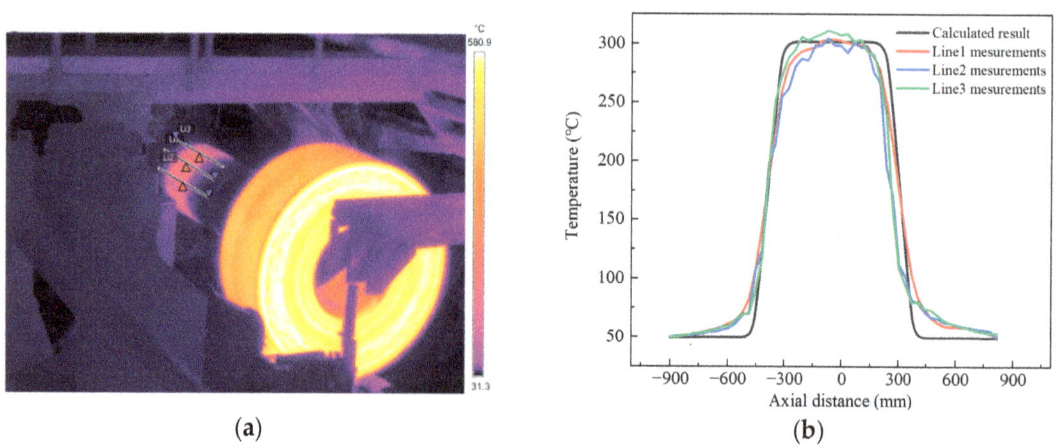

Figure 15. (**a**) The surface of the sleeve after coiling; (**b**) Comparison between calculated result and the measurements.

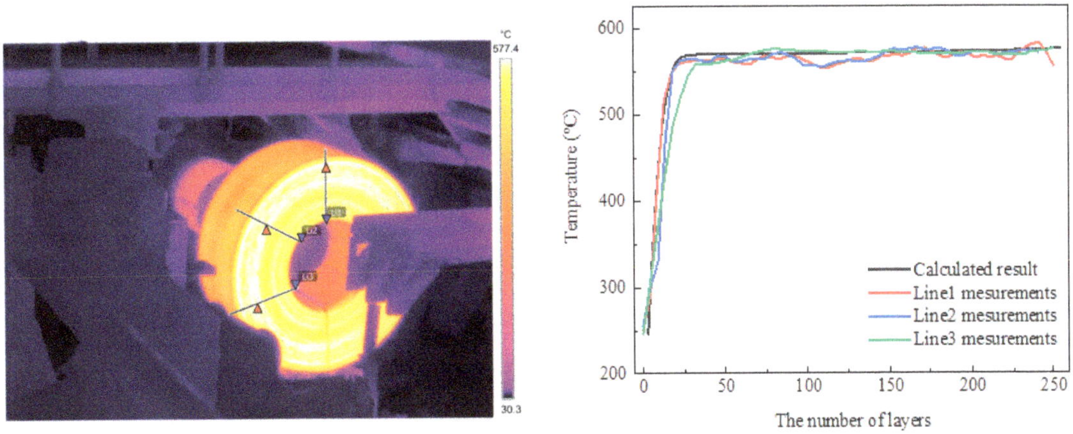

(**a**) (**b**)

Figure 16. (**a**) The axial end face of the coil after coiling; (**b**) Comparison between calculated result and the measurements.

3.6.2. Comparison of Incremental Coupled Solution and Analytical Solution in Stress

Figure 17 shows the radial and circumferential stress of the coil after coiling. It can be seen that the radial stress near the inner ring is larger in the middle and smaller at the edges in the axial direction. This is because the temperature drop at the edges is relatively large, which leads to the higher elastic modulus and the larger radial displacement at the edges. Corresponding to the radial stress distribution, the circumferential stress in the middle of the coil is larger than that at the edges. In the axial direction, the difference in stress is not significant and gradually disappears with the increase in radius. It is predicted that if the temperature drop at the edges is larger, the stress difference in the axial direction will be more significant.

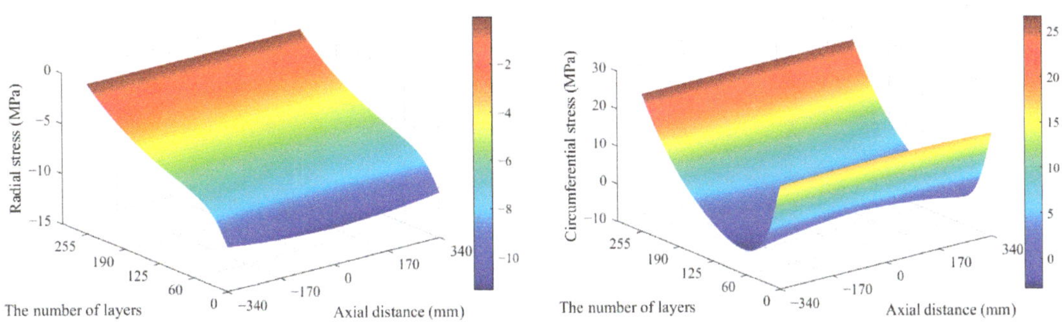

(**a**) (**b**)

Figure 17. (**a**) Radial stress of the coil; (**b**) Circumferential stress of the coil.

Figure 18 shows the stress difference between the middle and edge of the coil more clearly. During the coiling process, it is very difficult to measure the stress distribution inside the coil. Therefore, it is not possible to prove the validity of the calculated results through experiments. However, analytical methods have been accepted by scholars in academic research and practical production. Therefore, in Figure 18, the analytical solution is compared with the calculated results to verify the rationality, and the differences between the two methods are analyzed. The analytical solution can be represented by Equations (13) and (14) [30].

$$\sigma_{r,i} = \frac{-\sigma_T \cdot h}{R_i} + \sum_{t=i+1}^{N} \left[\frac{-\sigma_T \cdot h}{R_t} - \frac{\left(\frac{R_t}{R_i}\right)^2 - 1}{\left(\frac{R_t}{R_m}\right)^2 - 1} \cdot \left(p_{1,t} - \frac{\sigma_T \cdot h}{R_t} \right) \right] \tag{13}$$

$$\sigma_{\theta,i} = \sigma_T + \sum_{t=i+1}^{N} \left[\frac{-\sigma_T \cdot h}{R_t} - \frac{\left(\frac{R_t}{R_i}\right)^2 + 1}{\left(\frac{R_t}{R_m}\right)^2 - 1} \cdot \left(p_{1,t} - \frac{\sigma_T \cdot h}{R_t} \right) \right] \tag{14}$$

where $\sigma_{r,i}$ and $\sigma_{\theta,i}$ are radial stress and circumferential stress of the i-th layer and coiling tension. t is the number of coiled layers; R_i and R_t are the radius of the i-th layer and the radius of the coiled layers; $p_{1,t}$ is the contact pressure of the innermost layer when the t-th layer strip has been rolled, which can be obtained from the elastic deformation of the sleeve.

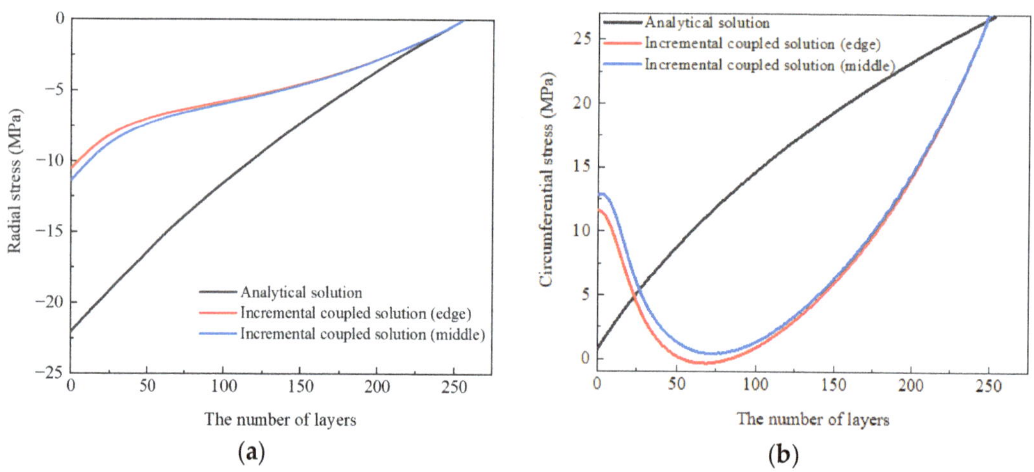

Figure 18. (**a**) Comparison of incremental coupled solution and analytical solution in radial stress; (**b**) Comparison of incremental coupled solution and analytical solution in circumferential stress.

The radial stress trends of the two solutions are the same. But numerically, the incremental coupled solution is relatively small. This is because in the analytical model, the coil and sleeve are considered isotropic. The real-time changes in the radial stiffness caused by interlayer contact and the influence of temperature on the physical properties cannot be considered in the analytical model. According to the comparison, it is safe to design the sleeve and other components based on analytical solutions. However, the temperature results indicate that there is a certain area of temperature rise in the sleeve. If there are defects on the sleeve, such as surface cracks or fatigue damage, the influence of temperature on its physical properties should be considered in the design.

In terms of circumferential stress, the incremental coupled solution is smaller than the analytical solution. This is because the decrease in radial stiffness of the coil in the incremental coupled solution increases the radial deformation of the strip. The circumferential compressive stress increases, and after being added to the coiling tension, the circumferential stress decreases. In the analytical model, if there are enough layers, the circumferential stress near the inner ring may be converted into compressive stress. In fact, because the radial deformation near the inner ring is limited by the sleeve, the circumferential stress remains tensile stress and maintains a certain value. It is worth noting that the high temperature of the coil leads to a decrease in the yield limit. If the internal stress state is analyzed based on analytical solutions, the plastic deformation generated near the inner ring may be ignored.

4. Conclusions

(1) An incremental calculation model to couple the temperature field and the stress field during the coiling process of hot-rolled strip is proposed. The mutual influence of temperature and stress, the anisotropy of the coil, the uneven axial temperature of the strip and the contact between the coil and sleeve can be considered. The calculation of temperature field and stress field can be closer to actual working conditions.

(2) In production, if the thickness of the strip and the coiling tension are increased and the initial temperature of the sleeve is reduced, the radial stress inside the coil and the surface pressure of the sleeve will be increased to a certain extent. If there are defects such as plastic deformation of the sleeve or the inner layer of the coil, the coiling process can be adjusted according to the above rules. For example, when the strip thickens or the coiling tension increases, the cooling time of the sleeve can be appropriately reduced to increase the initial temperature, thereby reducing the sleeve pressure and internal stress of the coil.

(3) A comparison is made between the incremental coupled solution and the analytical solution. The results indicate that the actual pressure on the sleeve should be much smaller than the analytical solution. Therefore, when the process adjustment increases the pressure, the sleeve designed by the analytical method is safe and can withstand a certain range of increase in pressure. At this point, there is no need to replace the sleeve with a stronger one.

Author Contributions: All authors contributed to the study conception and design. Methodology, H.X. and M.D.; software, Y.H. (Yuting Hu); validation, M.D. and Y.H. (Yanchao Hao); formal analysis, M.D., Y.H. (Yuting Hu) and P.Q.; investigation, M.D. and Y.H. (Yanchao Hao); resources, P.Q.; data curation, Y.H. (Yuting Hu); writing—original draft preparation, M.D.; writing—review and editing, P.Q., H.X. and M.D.; visualization, Y.H. (Yanchao Hao); supervision, H.X.; project administration, H.X.; funding acquisition, H.X. All authors have read and agreed to the published version of the manuscript.

Funding: This work was supported by the National Natural Science Foundation of China (Funder: Hong Xiao and the funding number: 51474190).

Data Availability Statement: The data presented in this study are available on request from the corresponding author due to privacy.

Acknowledgments: The authors would like to thank the National Engineering Research Center for Equipment and Technology of Cold Rolled Strip in Yanshan University for assistance in the tests.

Conflicts of Interest: The authors declare no conflicts of interest.

References

1. Zhou, C.; Le, Q.C.; Wang, T.; Liao, Q.Y.; Zhu, Y.T.; Zhao, D.Z.; Bao, L.; Jia, W.X. Effect of asymmetry on microstructure and mechanical behavior of as-rolled AZ31 magnesium alloy medium plates during coiling at warm temperatures. *Mater. Sci. Eng. A.* **2024**, *894*, 146474. [CrossRef]
2. Wang, C.; Wu, H.B.; Zhang, Y.Y. Structural transformation behavior of oxide scale during coiling of 0.9 wt% Cr-containing high-strength steel. *J. Mater. Res. Technol.* **2024**, *30*, 840–853. [CrossRef]
3. Dai, M.; Liang, S.J.; Qiu, P.; Xiao, H. Simulation of cold rolling strip coiling process considering additional contact deformation. *Iron Steel* **2023**, *58*, 80–86.
4. Gu, C.; Scott, C.; Fazeli, F.; Gaudet, M.J.; Su, J.; Wang, X.; Bassim, N.; Zurob, H. Evolution of the microstructure and mechanical properties of a V-containing microalloyed steel during coiling. *Mater. Sci. Eng. A* **2023**, *880*, 145332. [CrossRef]
5. Gu, C.; Scott, C.; Fazeli, F.; Wang, X.; Bassim, N.; Zurob, H. Site-specific analysis of precipitates during the coiling of an HSLA steel containing V and Nb. *J. Mater. Res. Technol.* **2023**, *27*, 6308–6318. [CrossRef]

6. Chen, D.M.; An, Z.Z.; Wang, G.D.; Liu, T.H. Effects of hot rolling and coiling temperatures on microstructure, texture and magnetic properties of 1.6 wt% Si non-oriented silicon steel. *Mater. Today Commun.* **2022**, *31*, 103807. [CrossRef]
7. Xiao, B.; Yu, Y.S.; Hu, B.; Wang, H.R.; Wang, W.; Liu, S.L.; Misra, R.D.; Liu, W.Q. On balanced strength and ductility synergy in low alloy steels through multiphase heterostructure involving cumulative process of hot rolling, coiling and tempering. *Mater. Sci. Eng. A* **2024**, *891*, 145987. [CrossRef]
8. Wu, H.; Sun, J.; Peng, W.; Jin, L.; Zhang, D.H. A symplectic analytical approach for thermal-metallurgical coupling problems: A case study of hot-rolled coil cooling. *Int. J. Heat Mass Transfer.* **2024**, *223*, 125218. [CrossRef]
9. Wang, X.D.; Wang, Z.Y.; Tang, W.; Li, J.; Liu, B.; Bai, H.K. Study on heart shaped buckling mechanism and prevention method of thin silicon steel. *J. Mech. Eng.* **2023**, *59*, 123–131.
10. Dai, M.; Liang, S.J.; Qiu, P.; Xiao, H. Efficient finite element simulation of cold rolled strip coiling process considering additional contact deformation between layers. *ISIJ Int.* **2024**, *64*, 1037–1046. [CrossRef]
11. Yang, L.Z.; Du, S.H.; He, F.M.; Zhang, Z.J. Three-dimensional thermoelastic analysis of orthotropic laminated cylindrical shells subjected to uniformly localized thermal boundary conditions. *J. Therm. Stresses* **2023**, *46*, 1227–1247. [CrossRef]
12. Nam, A.; Prüfert, U.; Pietrzyk, M.; Kawalla, R. Coil model for magnesium alloy strips and its heat transfer analysis. *Procedia Manuf.* **2018**, *15*, 185–192. [CrossRef]
13. Mats, K. Modelling of the temperature distribution of coiled hot strip products. *ISIJ Int.* **2011**, *51*, 416–422.
14. Baik, S.C.; Kwon, O.; Park, S.J.; Hong, B.H.; Oh, K.H. Analysis of heat transfer in hot rolled coils for optimum condition of forced cooling. *Met. Mater.* **1999**, *5*, 369–375. [CrossRef]
15. Park, S.J.; Hong, B.H.; Baik, S.C.; Oh, K.H. Finite element analysis of hot rolled coil cooling. *ISIJ Int.* **1998**, *38*, 1262–1269. [CrossRef]
16. Jung, Y.J.; Lee, G.T.; Kang, C.G. Coupled thermal deformation analysis considering strip tension and with/without strip crown in coiling process of cold rolled strip. *J. Mater. Process. Technol.* **2002**, *130–131*, 195–201. [CrossRef]
17. Saboonchi, A.; Hassanpour, S. Heat transfer analysis of hot-rolled coils in multi-stack storing. *J. Mater. Process. Technol.* **2007**, *182*, 101–106. [CrossRef]
18. Saboonchi, A.; Hassanpour, S. Simulation-based prediction of hot-rolled coil forced cooling. *Appl. Therm. Eng.* **2008**, *28*, 1630–1637. [CrossRef]
19. Cheng, J.F.; Liu, Z.D.; Dong, H.; Gan, Y. Analysis of the factors affecting thermal evolution of hot rolled steel during coil cooling. *J. Univ. Sci. Technol. Beijing* **2006**, *13*, 139–143. [CrossRef]
20. Witek, S.; Milenin, A. Numerical analysis of temperature and residual stresses in hot-rolled steel strip during cooling in coils. *Arch. Civ. Mech. Eng.* **2018**, *48*, 659–668. [CrossRef]
21. Liu, M.L. A nonlinear model of center-wound rolls incorporating refined boundary conditions. *Comput. Struct.* **2009**, *87*, 552–563. [CrossRef]
22. Wang, Y.Q.; Li, L.; Yan, X.C.; Luo, Y.X.; Wu, L. Modeling of stress distribution during strip coiling process. *J. Iron Steel Res. Int.* **2012**, *19*, 6–11. [CrossRef]
23. Xiao, H.; Xu, Z.A.; Zhang, C.J. Incremental solution of internal stress incoiling process of cold rolled strip. *Iron Steel* **2020**, *55*, 82–87.
24. Altmann, H.C. Formulas for computing the stresses in center-wound rolls. *Tappi J.* **1968**, *51*, 176–179.
25. Hakiel, Z. Nonlinear model for wound roll stresses. *Tappi J.* **1987**, *70*, 113–117.
26. Benson, R.C. A nonlinear wound roll model allowing for large deformation. *J. Appl. Mech.* **1995**, *62*, 853–859. [CrossRef]
27. Li, S.P.; Cao, J. A hybrid approach for quantifying the winding process and material effects on sheet coil deformation. *Eng. Mater. Technol.* **2004**, *126*, 303–313. [CrossRef]
28. Park, W.W.; Kim, D.K.; Im, Y.T.; Kwon, H.C.; Chun, M.S. Effects of processing parameters on elastic deformation of the coil during the thin-strip coiling process. *Met. Mater. Int.* **2014**, *4*, 719–726. [CrossRef]
29. Mikić, B.B. Thermal contact conductance: Theoretical considerations. *Int. J. Heat Mass Transfer.* **1974**, *17*, 205–214. [CrossRef]
30. Doghri, I. *Mechanics of Deformable Solids: Linear and Nonlinear Analytical and Computational Aspects*, 1st ed.; Springer: Berlin/Heidelberg, Germany, 2000; pp. 193–232.

metals MDPI

Article

Computational Thermochemistry for Modelling Oxidation During the Conveyance Tube Manufacturing Process

Megan Kendall [1,*], Mark Coleman [2], Hollie Cockings [3], Elizabeth Sackett [1], Chris Owen [4] and Michael Auinger [5]

1 Department of Materials Science and Engineering, Swansea University, Swansea SA1 8EN, UK; e.sackett@swansea.ac.uk
2 Oxford Instruments NanoAnalysis, Halifax Road, High Wycombe, Bucks HP12 3SE, UK; mark.coleman@oxinst.com
3 Frazer-Nash Consultancy, Hill Park South, Springfield Drive, Leatherhead, Surrey KT22 7LH, UK; hollie.cockings@fnc.co.uk
4 Tata Steel (Tubes), Corby NN17 5UA, UK
5 WMG, University of Warwick, Coventry CV4 7AL, UK; m.auinger@warwick.ac.uk
* Correspondence: m.kendall.2134577@swansea.ac.uk

Abstract: Conveyance tube manufacturing via a hot-finished, welded route is an energy-intensive process which promotes rapid surface oxidation. During normalisation at approximately 950 °C to homogenise the post-weld microstructure, an oxide mill scale layer grows on tube outer surfaces. Following further thermomechanical processing, there is significant yield loss of up to 3% of total feedstock due to scale products, and surface degradation due to inconsistent scale delamination. Delaminated scale is also liable to contaminate and damage plant tooling. The computational thermochemistry software, Thermo-Calc 2023b, with its diffusion module, DICTRA, was explored for its potential to investigate oxidation kinetics on curved geometries representative of those in conveyance tube applications. A suitable model was developed using the Stefan problem, bespoke thermochemical databases, and a numerical solution to the diffusion equation. Oxide thickness predictions for representative curved surfaces revealed the significance of the radial term in the diffusion equation for tubes of less than a 200 mm inner radius. This critical value places the conveyance tubes' dimensions well within the range where the effects of a cylindrical coordinate system on oxidation, owing to continuous surface area changes and superimposed diffusion pathways, cannot be neglected if oxidation on curved surfaces is to be fully understood.

Keywords: oxidation; modelling; steel; diffusion; heat treatment

1. Introduction

Low carbon steel thick-walled conveyance tubes are employed in wide-ranging building and industrial service applications, e.g., Heating, Ventilation and Air Conditioning (i.e., HVAC systems), and petrochemical transport, where high pressures, e.g., refrigeration, and temperatures, e.g., a steam system, are possible. The manufacturing process must be designed to achieve the material and mechanical properties necessary for these service conditions. In the UK alone, there are approximately 425,000 km of mains water infrastructure supplying domestic and commercial settings [1] and, according to the 2021 Census, approximately 1.1 million homes in Wales depend on a mains gas or oil supply for central heating [2]. Component damage and failure due to surface defects associated with phenomena such as oxidation could therefore have far-reaching social and economic consequences [3]. However, it is a challenge for plants to optimise their processes to maximise product quality, capacity, process flexibility, and cost effectiveness whilst high temperature oxidation on curved surfaces is still relatively poorly understood.

1.1. Conveyance Tube Manufacturing Process

Mild steel is delivered as a coil formed from a continuous cast billet. The steel sheet then passes through an accumulator series and is cold worked by forming rolls into the required hollow, cylindrical geometry. At this stage, the sheet still has a fine, homogenous microstructure, ensuring technical standard-defined mechanical properties. High frequency induction (HFI) welding is used to join the formed ends of the steel sheet creating a heat affected zone (HAZ) and some residual stresses. Therefore, the tube is subsequently heated to a temperature of over 950 °C for 2–4 min in a walking beam furnace, which restores the bulk to the desirable austenitic phase, before cooling in air. Tubes produced in this manner have several advantages over their cold-forged equivalents, including improved toughness, ductility, and pressure integrity, giving them an overall higher factor of safety. However, rapid growth of a multi-phase layer of dark-coloured, brittle iron oxide scale is also observed. Further thermomechanical processing (TMP) is used to produce a range of tube wall thicknesses and tube diameters, achieved by a multi-pass stretch reduction of 3–5% reduction per stand, with the greatest reduction occurring in the first few stands). The processing temperature is defined by achieving a balance between strain hardening and softening to minimise deformation-induced point and line defects (which are also defined by chemical composition, initial microstructure and deformation condition, deformation rate), and maintaining the material in the austenitic region. Typically, hydraulic descaling is applied in a post-normalisation, pre-reduction 'descaling box' for its combined benefits of reduced scale compaction and partial cooling to the ideal stretch reduction temperature of 850 °C. A summary of the manufacturing process is shown in Figure 1.

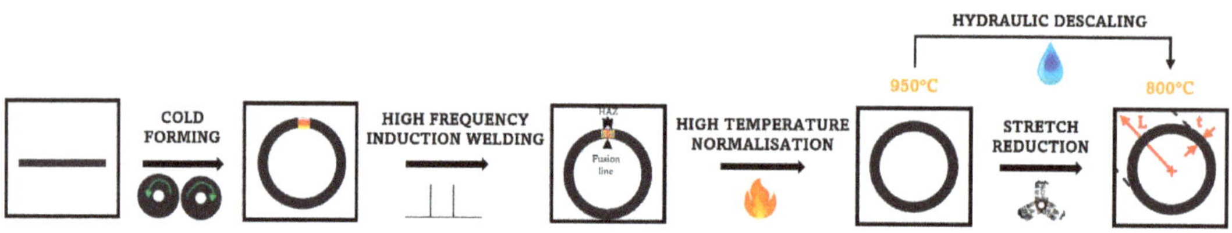

Figure 1. Schematic sequence of the conveyance tube manufacturing process.

Stretch reduction, and its associated high mechanical strain, causes the remaining scale to spall, leading to substrate exposure and surface wear. This has environmental and economic consequences as losses of up to 3% by weight occur due to scaling alone (equivalent to approximately 500 tonnes of feedstock loss per warehouse batch) [4]. There are also aesthetic implications due to either compaction of scale on the substrate surface ('rolled-in effect') and/or inadequate adherence of value-added coatings, e.g., anti-corrosion paint, and in-service surface performance problems. The resulting inconsistent scale adhesion that not only appears to give a poor surface finish but may also lead to complications for those customers who wish to apply additional paint or coat the tubes. Furthermore, scale removed either by descaling or spalling tends to disperse into the manufacturing environment, causing damage and contamination to product and plant alike.

1.2. Scale Management

Growth stress, arising from volumetric increase associated with oxygen uptake, depends on whether scale is internal or external and isotropic or radial, and is directly related to scale failure [5]. Matsuno et al. [6] specifically explored the impact of growth (oxidation) stress on descalability during a high-temperature hydraulic descaling operation. They also highlighted the Pilling and Bedworth ratio (PBR), which represents the ratio of oxide to metal volume as a measure of stress arising from volume dilatation during conversion of a metal to its oxide [7–9], which can be measured via deformation analysis. Ultimately, they proposed that oxidation stress is the dominant cause of adhesive failure within the combined effect of stress and adhesion vs. cohesion strength dominance, since temperature will accelerate oxidation and cause great compressive volume dilatation. For austenitic steels, such as are seen during conveyance tube normalisation, the dominance of Cr^{3+} migration favours the development of external oxide, i.e., oxygen ions are diffusing as well as iron and chromium (III) for ferritic martensitic steels so they experience both external and internal oxidation [10]. Furthermore, assuming radial (anisotropic) oxide growth and absence of axial strain, growth stresses are a purely geometric, mono-axial effect only observed on curved surfaces (where planar surfaces have an infinite radius of curvature which drives quotient-based strain equations to 0) [11]. This supports Sabau and Wright's conclusions on the significant impact of geometry on spallation [12], thus showing that geometry has an impact on scale management practice. Scale can be managed via a reactive approach which aims to eliminate the problem after detection by identifying its root causes and amplifiers after repeated incidence. This can take the form of anti-oxidation coatings [13–15], the details of which are beyond the scope of this paper. However, a more proactive approach using modelling would allow the industry to account for scale kinetics phenomena during thermomechanical process design. Pillai et al. agreed in their review of models for oxidation-based high-temperature degradation that the mechanism and extent of high-temperature oxidation of low carbon steel is characterised as a function of alloy composition, temperature, time, growth kinetics, geometry, and service environment [16]. Computational process modelling, when optimally designed and validated, can identify routes to minimise costs, improve product quality, and increase output, based on an accurate and fast assessment of scale and its associated trends. There are several manufacturing benefits, including faster and cheaper product and process innovations, reduced downtime and maintenance, reduced material usage, and, overall, more robust processes. Furthermore, experimental validation of curved surface oxidation is challenging, particularly due to the limitations of equipment used in in situ oxidation investigations. Thermogravimetric furnaces, which provide continuous mass gain data, are often limited in their sample size capacity so only very small tubes can be used, or tubes have to be sectioned which causes machining-induced residual stress [17,18]. However, there are issues with TGA measurements, as they give the total oxidation on a sample and neglect the effect of the gas flow on the extent of oxidation in each part of the sample geometry, as shown in the work by Mori et al. [5]. Therefore, any experimental value derived from TGA curves is an average over the entire geometry which does not address the differences between the side facing towards the flow and the one facing away. This is a strength of modelling; its ability to distinguish between the two and the reason for focusing on the modelling studies and then subsequently correlate these with the total oxidation thickness measured after heat exposure (the latter being subject to errors due to spallation of poorly adherent oxides). Furnaces with a larger sample size capacity do not usually have TGA equipment so rely on pre- and post-heat treatment high resolution mass balance measurements of the sample. High temperature oxides are known to be brittle, as seen in observation by the authors and as reported by the literature [5,19–21]. This brittle nature makes scale spalling inevitable so that material is lost between transferring the sample from the furnace to a mass balance, in addition to losses during sample cooling before it is removed from the furnace due to thermal mismatch between the metal and oxide. Typical oxide thickness predictions can be derived from input of temperature-time profiles based on analytical or computational predictions of thermofluid phenomena. However, this method still requires considerable involvement via definition of any equations and models, and translation of their results into functions for parameters used to calculate rate constants such as in Sun et al.'s work [22]. CALPHAD-based ('Computer Coupling of Phase Diagrams and Thermochemistry') software can investigate the thermodynamics and material properties of an oxidising system via phase equilibria outputs. The CALPHAD methodology can obtain a feasibility description for alloy system reactions and their most likely phase product via the Second Law of Thermodynamics and a Gibbs Free Energy minimisation [23]. Such a system relies on a comprehensive thermodynamic database for all the elements concerned which undergoes assessment, model selection, optimisation, storage, and validation. As a minimum, it must be possible to describe the enthalpy, entropy, and Gibbs' Free Energy of a solution (generally described using the sub-lattice modelling-based Compound Energy Formalism (CEF) for solid oxide systems) as a function of temperature and composition in order to perform accurate and versatile thermodynamic calculation [24]. Consequently, as with any computational model, descriptive accuracy must be balanced with mathematical feasibility based on the Central Processing Unit (CPU) resources available so that computation times are not excessive. However, ever-increasing computational power availability is accompanied by options for increasingly complex feasible models. Poerschke argues that understanding and prediction of spall-based surface degradation relies on an understanding of the 'interplay' between substrate and scale (as well as any applied coatings) and the thermodynamic information provided by computational materials engineering modelling when systems become complex, e.g., non-planar geometries [25]. The Thermo-Calc DICTRA commercial package has existed since 1994, but there have been few attempts to extend its application of the moving phase boundary approach beyond phase transformation applications [26]. Furthermore, where surface oxide layer growth has been explored using this software, the geometry has been limited to planar cases [3]. The use of DICTRA for surface oxide has also been inhibited by a lack of attempts to expand on existing commercial thermodynamic and mobility data sets to incorporate oxide phases.

1.3. Geometry Effects

Despite many experimental and computational investigations into oxidation kinetics and oxide adhesion during thermomechanical processes involving slab and/or strip steel [27–29], there is far less literature concerning oxidation phenomena on non-planar surfaces [11,12], such as those found in conveyance tube manufacturing. Works using a computational approach are rarer still and seldom focus on oxidation as the main phenomenon [30–32]. Nevertheless, investigations of wall temperature, reaction environment, and thermal stress are still valuable as precursory information for heavily temperature-dependent oxidation thermochemistry. Landfahrer et al. numerically modelled a full-scale reheating furnace for investigation of the effect of scale on heat transfer in low carbon steel tubes via extension of previous study [30]. The key advantage of their model was the coupling of two independent simulations (stationary combustion and single-tube transient heating) in place of a more computationally demanding fully transient, chemical kinetics analysis. However, even in these rare cases, existing works are dominated by studies of steam superheater systems where parallels can only be drawn to conveyance tube manufacturing in the case of 'fireside' corrosion of tubes within these systems. Equation (2) gives the diffusion equation in cylindrical coordinates, as is appropriate for conveyance tube geometry, where c_k is the species concentration as a function of time, t, and position, (r, ϕ, z). Dependency on the azimuthal, ϕ, and vertical, z, coordinates is eliminated by assuming rotational symmetry during oxidation, and changes in oxidation behaviour are only observed along the radial coordinate, r (see (2)).

$$\frac{\partial c_k}{\partial t} = \frac{1}{r}\frac{\partial}{\partial r}\left(r\frac{\partial c_k}{\partial r}\right) + \frac{1}{r^2}\frac{\partial^2 c_k}{\partial \theta^2} + \frac{\partial^2 c_k}{\partial z^2} \tag{1}$$

$$\frac{\partial c_k}{\partial t} = D\left(\frac{1}{r}\frac{\partial c_k}{\partial r} + \frac{\partial^2 c_k}{\partial r^2}\right) \tag{2}$$

Changes in radius are accompanied by a change in available reacting surface area and surface area to volume ratio (see Figure 2), and highlight how an increase in radius is accompanied by an increase in surface area, A, (see (3)) and a decrease in surface area to volume ratio, $A : V$ (see (4)). Therefore, the surface area increases during inner surface oxidation (consumption of the substrate widens the cylinder bore, i.e., the inner radius) lead to an overall slower rate of reaction and therefore thinner oxide thickness after a given time and, as such, the converse is true of the outer surface.

$$A = 2\pi r_{3_{t=t}} z \tag{3}$$

$$A : V = \frac{2r_{3_{t=t}}}{{r_{3_{t=t}}}^2 - {r_{2_{t=t}}}^2} \tag{4}$$

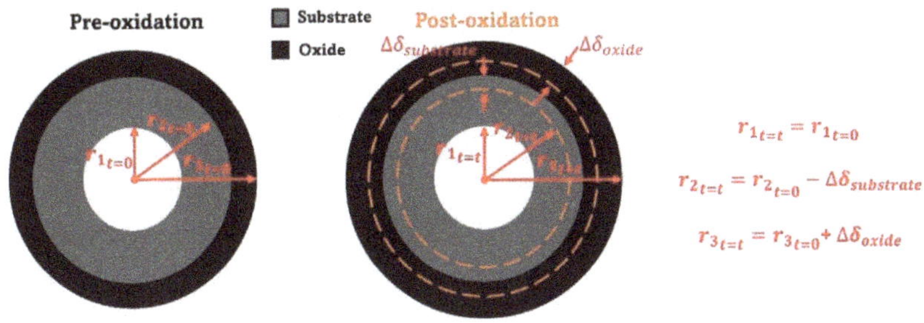

Figure 2. Diagram showing changes in cylinder outer radius during surface oxidation both pre- and post-oxidation (adapted from source [33]).

1.4. Oxidation Theory for a Computational Approach

Low carbon steel oxidation kinetics are defined by either a linear or parabolic rate law (dependent on instantaneous oxidation time and temperature considered) [34]. As most steel reheat treatments are long, high-temperature processes, including tube normalisation, a logarithmic rate law description is not appropriate. Initially, external oxidation of steel is governed by phase boundary processes at the substrate–scale (SSI) and scale–gas (SGI) interfaces. For the SGI, the rate of mass transport of oxygen-containing gas to the surface for adsorption, splitting to free oxygen, chemisorption, lattice incorporation, nucleation, and growth, is the most critical interface phenomenon and governs the linear rate law, i.e., it is assumed that, initially, kinetics are dominated by gas transport from the environment to the interface [35]. However, during the later stages of oxidation, kinetics are dominated by ion diffusion phenomena in the oxide. To maintain equilibrium and high metal activity as oxidation proceeds, there must be electron mobility, and ion flux levels must be maintained as the flux path lengthens due to scale thickening. Surface reaction rate, i.e., SGI activity, must fall to satisfy this constraint via a reduction in iron ion diffusion, making it the rate-controlling step. This kinetic change triggers a transition from a linear to parabolic rate law profile. Hence, kinetics are controlled by the dynamic balance of oxygen adsorption and iron ion diffusion during scale thickening and established oxidation is characterised by diffusion. Iordanova et al. argue that the strong influence of ion diffusion phenomena on oxidation further supports the assumption of a parabolic rate law for high-temperature investigations [34]. The relative thicknesses of the three phases (wustite, magnetite, and haematite in a 95:4:1 ratio) typically present on heat-treated low carbon steel also reflect this assumption of bulk diffusion-controlled (i.e., parabolic) oxidation and local thermodynamic equilibrium at phase interfaces, as wustite has a much greater iron diffusion coefficient than magnetite and haematite [36]. Highly defective crystal lattices, generated by divalent and trivalent interstitial iron cations, are the source of both outward cation and inward anion diffusion. Vacancies allow ions to 'jump' and thus move through the crystal lattice unimpeded, provided enough energy is supplied. Modelling of this process demands an understanding of the concept of sublattices, which are either tetrahedral or octahedral in the Fe-O system. Scale formed on grades similar to those used in tube manufacturing is dominated by wustite, a non-stoichiometric halitic phase $Fe_{1-x}O$ [37]. Hidayat et al. argued that there is insufficient reliable quantitative data on sublattice site occupancy, defect composition, and defect clustering to develop a comprehensive thermodynamic model of the halitic wustite structure (ideal, stoichiometric, or otherwise) [38]. Nevertheless, such a model is required to generate comprehensive analytical and computational models, and they therefore accept the assumption made by Sundman et al. that iron cations and vacancies can only mix on octahedral sublattice sites, i.e., iron cations present on tetrahedral sites are ignored. The resulting simplified sublattice formalism for wustite is given below [39].

$$\left(Fe^{+2}, Fe^{+3},\ Va\right)_1\left(O^{-2}\right)_1$$

Hallstrom et al. asserted that diffusion of Fe can be modelled without consideration of its valency as crystal lattice electrons are significantly more mobile in wustite than Fe cations at high temperatures (T > 600 °C) [40]. The formula unit can therefore be simplified as below,

$$(Fe,\ Va)_1(O)_1$$

which provides the necessary information to analytically define the system, specifically the flux, J, and bulk chemical diffusion coefficient, D, of both the iron and oxygen species. Without a driving force, the energy barrier to diffusion (specifically that required to allow an atom to jump to an empty lattice site) is defined by the free energy of migration, G_m. (Then 5 shows how diffusivity is defined in part by atomic mobility, M, due to continuous random movement in the absence of a driving force in this case (where R and T are the molar gas constant (8.31 $J{\cdot}K^{-1}{\cdot}mol^{-1}$) and temperature (K), respectively).

$$D = RTM \tag{5}$$

Hallstrom et al. [3] used Thermo-Calc's Diffusion Module ('DICTRA') to numerically solve the 1D diffusion equation, i.e., for a planar geometry, and computationally predict oxidation kinetics. Xia et al. argue that the use of a 1D system ignores the realistic 2D composition distribution of the material and restricts diffusion coefficient calculations to fixed compositions or those that only vary compositionally in a single direction [41]. When diffusion across a non-planar boundary, such as is found on a curved surface, is under investigation, the computational domain must be complexified to accurately account for two dimensions. Since both species flux and temporal concentration gradient, $\frac{\partial c_k}{\partial t}$, are unknown variables, the equation for mass conservation in the absence of chemical reactions must be coupled to Fick's First Law, which relates flux and spatial concentration gradient, $\frac{\partial c_k}{\partial z}$, via a chemical diffusion coefficient, to derive Fick's Second Law (see (6)).

$$\frac{\partial c_k}{\partial t} = D\frac{\partial^2 c_k}{\partial z^2} \tag{6}$$

Computational solutions for the independent flux of metal cations and oxygen anions for the two phases can be supplied to Fick's Second Law, alongside volume and molar fractions which are defined experimentally (via advanced imaging techniques) and computationally (via phase diagram simulations), respectively. A moving phase boundary ('Stefan problem') formulation for diffusion is used and the output of (6) is a description of the migration rate of the boundary between two phases, i.e., metal and oxide, when the flux of species k out of one phase, α, and into another, β, respectively, is not balanced, i.e., $J_k^{\alpha} - J_k^{\gamma} \neq 0$ [42]. The presence of the moving boundary renders this problem non-linear, combining PDEs and complicated interphase geometrical movement, and it is difficult to obtain analytical solutions without oversimplification. The moving phase boundary approach is most suitable for isobarothermal binary systems where the interfacial compositions are fixed so the numerical efficiency of the approach can be maintained [43]. Thus, oxide thickness, $\Delta\delta_{t_{step}}$, after a given timestep, t_{step}, can be predicted using (7), and the cumulation of values from each timestep shows the time-dependent kinetic trends (for a given temperature).

$$\Delta\delta_{t_{step}} = V_m^{\alpha} \frac{J_k^{\alpha} - J_k^{\gamma}}{x_k^{\alpha} - x_k^{\gamma}} t_{step} \quad (7)$$

As previously mentioned, diffusion in interstitial alloys with small migrating interstitials, e.g., oxygen in iron, in dynamic equilibrium with each other is characterised by a continual vacancy migration mechanism. In (5), the authors show that the variables for a successful computation of diffusion depends on accurate expressions for the bulk mobility of all involved species. The species molar fraction to substitutional element (oxygen) contribution, u_k, and chemical potential derivative, $\frac{\partial \mu_k}{\partial u_k}$, are derived from the literature. Application of (7) relies on an assumption of a planar diffusion front (long diffusion period and small grains) to ensure diffusion field overlap around neighbouring grain boundaries. This can be reasonably assumed in compounds such as wustite where the grains are small compared to the oxide scale thickness and therefore the metal–oxide interface is straight at the relevant length scales [3]. The final key assumption is that of an Arrhenius-type expression for bulk mobility, including a pre-exponential coefficient, M_0, and activation energy, Q (see (8)).

$$M_k = M_0 e^{-\frac{Q}{RT}} \quad (8)$$

It is noted that, alongside temperature, mobility is also influenced by oxygen partial pressure when considering oxide phases (unlike diffusion due to the strength of atomic binding) and the dopant effect of certain alloy elements. However, accounting for these effects is beyond the scope of this work. Overall, oxidation as a phenomenon is complex and dependent on multiple parameters. However, oxidation in the context of the tube production process is further complicated by manufacturing parameters and tube geometry. The increased computational resources available in the 21st century could be used to improve the depth and accuracy of knowledge surrounding scale kinetics, morphology, and adhesion on curved surfaces across a range of conditions not limited to the design envelope. The work within this paper aims to explore the potential for using a computational thermodynamics approach to understand scale kinetics phenomena during tube normalisation. In particular, the Stefan problem moving phase boundary methodology is applied to a new context, i.e., tube normalisation, so hollow cylindrical geometry and high temperature low carbon steel and scale phases and appropriate tube manufacturing-friendly thermodynamic and kinetic phase databases are developed. A high accuracy Simultaneous Thermal Analysis (STA) approach, comprising Thermogravimetric Analysis (TGA) and Differential Scanning Calorimetry (DSC) techniques, is used for validation to accelerate the analysis of tube oxidation under normalisation conditions.

2. Materials and Methods

2.1. Materials

The materials under investigation during in situ oxidation tests for validation purposes were the two grades of low carbon, low alloy steel P235GH and P265GH (compliant with the material specification BS EN10217-2 [44]). The specification given in BS EN10217-2 is outlined in Table 1. Nose crop samples of both grades were supplied by Tata Steel Tubes UK Ltd in Corby, UK.

Table 1. Alloy element composition limits specified by BS EN 10217-2 (welded steel tubes for pressure purposes—technical delivery conditions [part 2: electric welded non-alloy and alloy steel tubes with specified elevated temperature properties]) for steel grades P235GH and P265GH.

% by Mass of Element	C	Si (max.)	Mn	P (max.	S (max.)	Cr (max.)	Mo	Ni (max.)	Al (total)	Cu (max.)	Nb (max.)	Ti (max.)	V (max.)	Cr + Cu + Mo + Ni (max.)
P235GH	≤0.16	0.35	≤1.20	0.025	0.020	0.30	≤0.08	0.30	0.020	0.30	0.010	0.03	0.02	0.70
P265GH	≤0.20	0.40	≤1.40	0.025	0.020	0.30	≤0.08	0.30	0.020	0.30	0.010	0.03	0.02	0.70

2.2. Computational Methodology

Although sublattice models exist for all three oxide phases found on low carbon steel, for simplicity of the initial model pre-processing, the entire scale layer was assumed to be wustite. Field Emission Gun Scanning Electron Microscopy (FEGSEM) imaging, coupled with Energy Dispersive Spectroscopy (EDS) of the oxide on a normalised tube sample revealed a comparatively compact, homogeneous scale (see Figures 3–5).

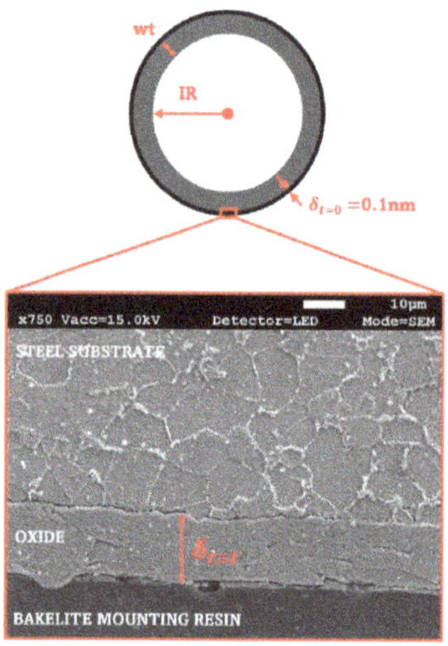

Figure 3. FEGSEM image of oxide found at the outer surface of the tube wall for sample with a 20.5 mm internal diameter.

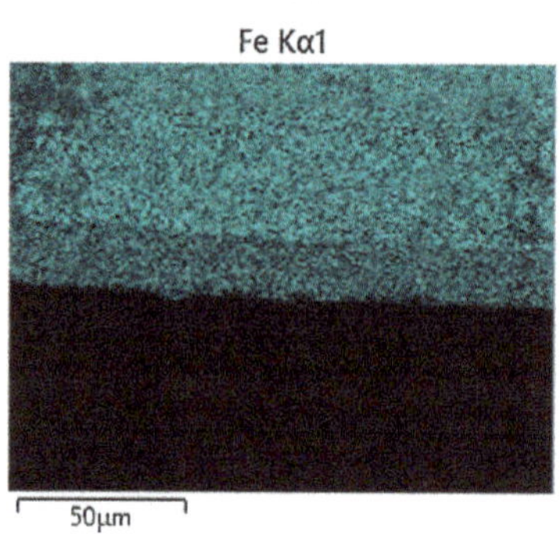

Figure 4. EDS map of iron (Fe) contribution for section of oxide formed on outer surface of a 15 nb tube sample post-normalisation in industry.

The Fe-O binary system phase diagram was used to determine the phases present and their relevant mole fractions at the normalisation temperature of 1223 K. Initial compositions of oxygen and iron were supplied as left and right interface boundary conditions for each region (austenite and wustite), respectively. Composition can only vary along one spatial coordinate, and only substitutional elements contribute to volume. For the overall system of finite volume cells, composition gradients are only considered between cells rather than inside. The metal and scale regions are joined by a finite interfacial region comprising two cells (one from each region). Local equilibrium is assumed to hold between the two regions, with constant volume fractions of each phase.

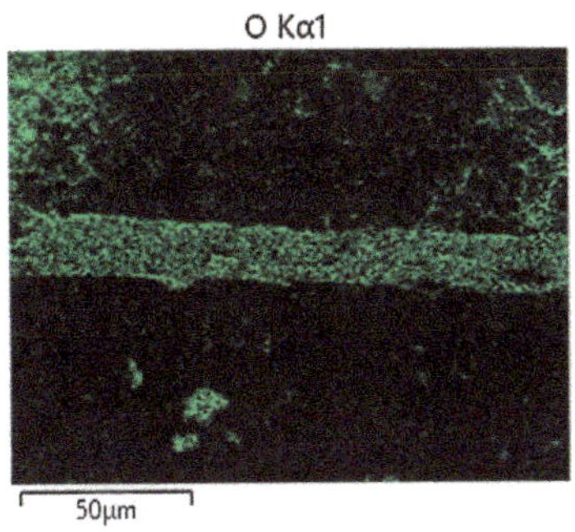

Figure 5. EDS map of oxygen (O) contribution for section of oxide formed on outer surface of a 15 nb tube sample post-normalisation in industry.

A cylindrical cross-sectional geometry of infinite length was used (see Figure 3). Inner radius and wall thickness were set at values associated with typical conveyance tube dimensions. Control of these two parameters makes the model well-suited to hot-finished conveyance tubes, which are manufactured with a controlled inner radius, IR, and a variable wall thickness, wt. The initial wustite thickness, $\delta_{t=0}$, was unchanged at 0.1 nm.

The inability to define the oxide region as halite (the most accurate microstructural representation of the wustite scale phase [39]) prevented the use of the commercial TCFE12 and MOBFE7 databases available in Thermo-Calc (although data for magnetite and haematite were used as part of the subsequent bespoke database development). Inspection of the phase and phase composition inventory within the system definer revealed that mobility data, necessary to define diffusion, was not available in the Thermo-Calc database for common oxide scale microstructures (halite, spinel, corundum, etc.).

Initial database set-up relies on the definition of constituent elements (Fe and O in this case, along with vacancy and electron additions) in terms of their Stable Element Reference (SER) state (298.15 K, 1 bar) and enthalpy and entropy of formation. Stoichiometric formulae are also needed for each species to include atomic and ionic states as well as compounds. Gibbs Free Energy functions must also be defined ready for minimisation, which is an essential part of computational thermodynamics. These functions (see (9)) were taken from Sundman [39] and Kowalski and Spencer's [45] works describing the Fe-O system.

$$G_{FeO} = -279318 + 252.848T - 46.12826Tln(T) - 0.0057402984T^2 \quad (9)$$

Overall, wustite can have complex physical properties due to its non-stoichiometry, variable site distribution of ferric iron, clustering of defects, long- and short-range ordering, and exsolution. The level of non-stoichiometry in iron oxides is controlled by the concentration of cation (metal) vacancies evolved during the diffusion-controlled oxidation process, and heavily influenced by temperature and partial pressure. However, Hazen and Jeanloz remarked that wustite thermal expansion is not significantly affected by stoichiometry [46]. Furthermore, deviation of stoichiometry in wustite only becomes significant at temperatures lower than 800 °C, i.e., below typical normalisation temperatures.

2.2.1. Mobility Database Definition

Elements, species, constituents, and phases in the kinetic database were defined identically to those in the thermodynamic database. Volume data took a standard value for all relevant volume-contributing substitutional phases (including oxygen gas) of 1×10^{-5} m^3mol^{-1}, as defined in the Thermo-Calc database management documentation [47]. Interstitial species (O, Fe^{2+}, Fe^{3+}) were assigned as zero-volume species. Data for the FCC_A1 (austenite) phase (iron and oxygen diffusivity as described for ferrite) came from [48,49] in a diffusivity form, as did that of halite (wustite) from the widely accepted work of Kofstad [50].

An implicit Euler method was used for time discretisation where the unknown, concentrations were defined by coupled sets of equations. The simulation length, 180 s, was defined by the mean of the heat treatment period range in the normalisation furnace. The initial and minimum acceptable timesteps were both set at 0.1 ms, respectively. A convergence study was performed to assess the maximum timestep which provided the optimal balance of accuracy and computational efficiency. Although the default maximum timestep value for Thermo-Calc is 10 s, the use of a multi-resolution mesh (geometric for the austenite region and double geometric for the wustite region), to reflect the steep gradients of ion concentrations due to fluxing at both the steel–scale and scale–gas interfaces, increases the computational demand. For example, using a linear and geometric mesh for the austenite and wustite regions, respectively, led to a CPU demand of 77 s to solve for 204 datapoints. However, using the more appropriate geometric-double geometric approach (see Figure 6), making both the steel–scale and scale–gas interfaces regions of interest with the most elements, demanded 259 s to solve 324 datapoints. For timesteps greater than 10 s, there was a negligible change in CPU demand and a consistent < 1% decrease in maximum predicted oxide thickness. A maximum allowable timestep of 10 s (~3% of the total simulation test time) was therefore selected as the best balance of accuracy and CPU demands.

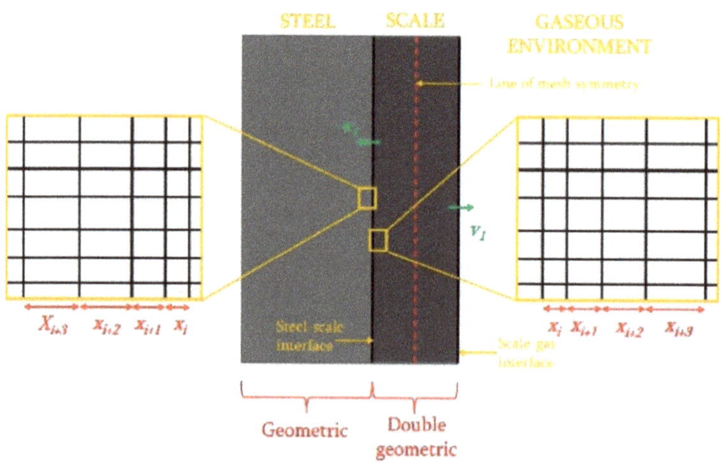

Figure 6. Geometric mesh structure for model domain.

2.2.2. Initial and Boundary Conditions

To solve the system boundary value PDE, i.e., (8), a fully implicit system was selected. Fully implicit systems carry the advantage of being unconditionally stable across all values of the mesh ratio parameter. Numerical integration was performed using the Crank–Nicolson method (see (10)).

$$-rc_{i+1}^{n+1} + (1+2r)c_i^{n+1} - rc_{i-1}^{n+1} = rc_{i+1}^{n} + (1-2r)c_i^{n} + rc_{i-1}^{n}, \text{ where } r = \frac{D\Delta t}{2(\Delta x)^2}. \quad (10)$$

The value for initial oxide thickness must be minimal but non-zero (<0.1 nm). The system is closed by assuming a 'closed' system on the metal side of the metal–gas interface, i.e., no diffusion of iron (Fe) in or out of that side of the system, and applying a fixed oxygen flux to the oxide–gas interface, corresponding to the experimental atmosphere. The combination of these conditions allows the oxide layer to grow outwards, i.e., external oxidation is assumed. To further reflect the assumption of wustite as the oxide phase, an oxygen flux of 4.5×10^{-4} $m^2 \cdot s^{-1}$ was applied by selecting the maximum value for oxygen permeability before accumulation to prevent phase transitions to magnetite and haematite, characterised by higher oxygen content. Standard reference conditions of 298 K and 1 bar were applied for the gaseous oxygen. Both regions were discretised as a finite number of nodes (within automatically assigned meshes, as described above)

2.3. In Situ Oxidation Experiments

Complex systems where multiple reactions are anticipated, e.g., different oxide phases evolved and alternative reaction schemes are possible, are significantly influenced by sample size, form, and environment. Simultaneous Thermal Analysis (STA), a combination of Thermogravimetric Analysis (TGA) and Differential Scanning Calorimetry (DSC) uses these complementary quantitative and qualitative techniques, respectively, suitable for all solid–gas reaction studies. Wire Electrical Discharge Machining (EDM) was used to generate as geometrically consistent $3 \times 3 \times 3$ mm cubic coupons of E24 and E41 tube grade steel as possible to ensure comparable STA data. The heating action of the EDM approach introduced further oxidation to the machined coupon surfaces, which were therefore cleaned using 4000 grit polishing paper and isopropyl alcohol. Each prepared coupon was placed in the centre of an alumina crucible (which exhibits high thermal conductivity and inert behaviour towards oxygen at high temperatures). Replica heat treatment in a 10% O_2-90% N_2 atmosphere was applied sequentially to each crucible. A ramp of 5 $K \cdot s^{-1}$ was used for both the heating and cooling periods. Before running the thermal cycle, a 'blank' run was performed using an empty crucible to establish a buoyancy effect correction factor. Hence, uncertainty is accounted for pre-measurement. Given that the mass gain data was supplied by the STA experiment but thickness data was the output of the model, one data set required conversion. Converting the STA mass gain data demands the assumption of a fully dense oxide layer and the method itself is complexified by recognition of the fact the sample surface area is constantly changing as oxidation proceeds, i.e., sample dimensions are not fixed due to the moving steel–scale phase boundary. The alternative approach converts the model thickness data to mass gain data (see (11)).

$$\Delta m = \rho A x_1 \tag{11}$$

The value of oxide (wustite) density, ρ_{oxide}, was calculated from its molar volume, V_m, defined in the thermodynamic database, and molar mass, M_r (see (12)). The resulting value of 5914 $kg \cdot m^{-3}$ agreed with values found in the literature [36]. This was arguably a more accurate approach as, unlike the STA samples, within the model definition there was the existing parameter which justified the assumption of equal oxide growth across the surface, and only a single surface area, A, to consider, unlike the cubic STA samples. Ref. [11] was used to convert the model oxide thickness predictions to a total mass gain value, Δm, for the isothermal test simulated. Extension of the 1D model to 3D was achieved by extending backwards to 3 mm to match the STA samples' depth, x_1.

$$\rho_{oxide} = \frac{M_r}{V_m} \tag{12}$$

When applying the same conversion methodology to cylindrical samples, the moving phase boundary for both the SSI and SGI must be accounted for, i.e., not only is the scale layer growing radially outwards on the external surface, but the steel substrate is retreating as it is consumed during the ongoing oxidation reaction (see (13)).

$$\Delta m = \rho \pi L \left(OR^2 - (IR - \delta_{ox})^2 \right) \tag{13}$$

3. Results and Discussion

3.1. Planar Geometry

A plot of the initial validation data in the 1000 ± 1 °C for both E24 and E41 (see Figure 7) was made so that a direct comparison could be made against the identical conditions in the Thermo-Calc isothermal simulation (planar geometry heated at 1000 °C for 5 min). Figure 7 shows the comparison between the mass gain in the isothermal region of STA testing, for both grades, compared to iron oxidised in the Thermo-Calc model (converted from oxide thickness to mass gain using wustite density). Each plot is annotated with its respective parabolic rate constant.

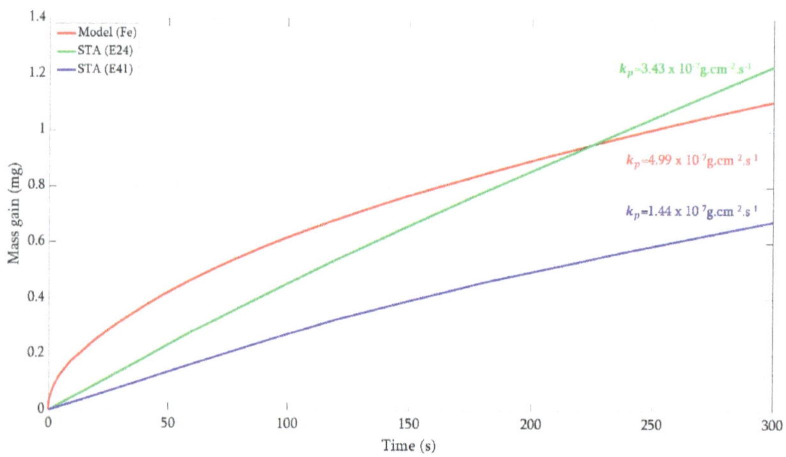

Figure 7. Comparative plot of mass gain following heating to 1000 °C for 5 min for Thermo-Calc model (red) vs. STA testing for E24 (green) and E41 (blue) tube grades, annotated with parabolic rate constant, k_p.

Although the model's predicted total mass gain lay within the range observed for the STA samples, suggesting a sufficiently accurate pre-processing methodology, its oxidation rate was higher overall. Both experimental data sets also had minimal correlation to a parabolic profile (inherent in the model due to its theoretical foundations discussed in the introduction of this work). The discrepancy between the computational and experimental results was assumed to be multi-factorial. Firstly, whilst the experimental data can be obtained directly from the equipment in a mass gain format, the model data must be converted from the thickness gain form, owing to the methodology used to make computational predictions. Whilst the experimental samples are 3 × 3 × 3 mm cubes, i.e., with six surfaces sharing the same material volume, only a single surface, i.e., 1D geometry, can be modelled, hence that single surface of equivalent area has access to six times as much material as each individual surface in the experimental samples. Hence, the mass gain associated with the model was divided by six to allow for an equivalent comparison. Furthermore, a key assumption of the model and its post-processing calculations, particularly those concerning density, was that of a perfectly compact scale layer, i.e., no porosity. However, porosity is a key property affecting diffusion mechanisms at the microstructural level. This is discussed in greater detail during geometric validation in Section 3.2, but, ultimately, it is concluded that only a modified diffusion coefficient can account for porosity and its associated effects, rather than in the pre-processing of the model [51]. In this manufacturing case, porosity was most likely to arise due to thermal deformation mismatch between the steel and scale, mechanical strength differences between phase layers (in the real case where multiple oxide phases form), the degree of vacancy annihilation (Kirkendall effect), and growth stresses due to oxidation-induced volumetric changes. Furthermore, it should be noted that Wagner's descriptive and predictive model for parabolic oxidation, which uses ionic and electronic charge transfer and balance during simultaneous metal oxidation and free oxygen reduction to characterise alloy element distribution [10], assumes a scale with a perfect microstructure and adhesion, i.e., they ignore the possibility of defects, e.g., cracks, porosity, etc., whose propagation could be accelerated by the high temperatures and mechanical work of tube stretch reduction. This is particularly important since high temperatures, such as those used in tube normalisation, are associated with greater development of defects during cooling [18]. Such propagation has been shown to interrupt oxidation diffusion mechanisms [16], and its effect on kinetics and morphology must therefore be considered in a non-idealised approach. However, this was beyond the scope of this investigation. The STA data alone reveal a 13% difference in total mass gain between the two grades tested. This was attributed to the interactions of the dominant non-carbon alloy elements, manganese [52] and silicon [53], with iron and oxygen to form intermetallic compounds which affect oxidation kinetics and adhesion.

3.2. Cylindrical Geometry

As with the planar case, the CPU demand is low with 120 datapoints and approximately 115 s run time for each test (using a maximum timestep of 1000 s; interestingly, decreasing the maximum timestep to 100 s has no effect on the predicted oxide thickness but increases CPU demand to 899 datapoints and 684 s run time). It has been discussed that the 1D approach to diffusion modelling does not reflect the multi-axial nature of oxidation. Beyond the differences arising due to the link between geometry stress states and porosity evolution, there is also the consideration of how geometry affects diffusion pathways. The temporal concentration gradient becomes a function of $\frac{1}{r}$. Accordingly, assuming the concentration only changes due to diffusion in the radial direction, i.e., the 1D form, the radius is the discretised dimension in the numerical solution. It should be noted that, in this case, the concentration in each segment, i, is an average. When setting initial concentration conditions, $c_{i,t=0}$, and calculating average i^{th} segment concentration, every point of the circumference at radius, r, has a different weight dependent on which segment it occupies. The area of the ring with an average concentration corresponding to $c_{i,t=0}$ is defined and integration of the concentration at individual points around the ring results in an instantaneous initial average concentration. An increase in wall thickness, whilst controlling outer diameter in the same way as is carried out during tube stretch reduction and product specification, revealed a decrease in oxide thickness and the corresponding parabolic oxidation rate constants. This is due to the extension of the diffusion pathway so fewer metal cations diffuse to the SSI to react per unit time. It has also been suggested that oxidation kinetics depend on the oxide scale stress state due to the negative correlation between stress magnitude and the mole fraction of intrinsic defects within the oxide [54]. Higher stress is associated with a lower vacancy concentration, which can inhibit iron diffusion via iron vacancies. Additionally, Asensio-Jimenez et al. cited the degree of oxide growth stress relaxation by plastic deformation as a source of oxidation kinetics variation across their samples (more influential than minor alloy element depletion, which was also discussed) [55]. For example, if a sample is exposed to conditions which promote creep, the compressive and tensile stress evolved in the scale and metal, respectively, will eventually reach a level that causes substrate creep and therefore growth stress relaxation in the oxide. A thinner sample will reach this critical value sooner, i.e., have a shorter 'incubation period', and therefore initiate a period of enhanced oxidation associated with stress state changes. The difference in stress state for samples with only a small thickness difference can be significant due to the power law relationship between creep rate and the specimen thickness (although this is for flat surfaces). Although the incubation period for observing any difference in oxidation rate due to sample thickness can be far longer than is observed in the tube manufacturing process compared to a planar surface, curved surfaces will experience additional stresses associated with the forming process [56]. Asesnio-Jimenez et al. quoted as much as a 60-fold increase in substrate creep rate by decreasing specimen thickness from 0.5 to 0.3 mm. Clearly, stress cannot be directly accounted for in the thermochemical model developed; however, it should nevertheless be considered, particularly when analysing validatory data. Previous authors have investigated oxide formation on a range of tubes with varying dimensions but do not isolate a single dimension to investigate its isolated effect on oxide thickness [20,31]. This reinforces the advantages of a computational approach, as representative samples in the context of manufacturing components are often limited to their commercial range and can be restricted at a given time to only the specifications required to meet current demand. Furthermore, the 2D diffusion pathway associated with the converging and diverging radial path of oxygen and iron ions, and the fact that their concentrations are defined normal to the surface, leads to an overlap of ion pathways, unlike in the planar case. However, the relative effect of these phenomena is dependent on the radial scale, i.e., very large radii may evolve a solution which approaches a flat plate solution as the $\frac{1}{r}$ term tends to 0. The scale at which radial effects become relevant for diffusion must be investigated. Conversely, an increase in OD, whilst controlling wall thickness, has no effect on predicted oxide thickness and parabolic rate constants. It was noted that the material was consumed approximately twice as quickly in the cylindrical case than the planar case (30 h vs. 60 h).

3.3. Radial Dependency Study

Since volume and mass gain were revealed in this work to be dependent on inner radius during oxidation, it was hypothesised that there would be a critical radius value at which the effect of cylindrical coordinates becomes influential. The effect of cylindrical geometry on the analytical theory of diffusion was outlined earlier in this work, and has been explored by other authors in an analytical context [57–60]. An initial investigation using the model explored the extremes of tube inner radial dimensions, 0.5 mm $\leq IR \leq$ 1000 mm. Volume gain, a calculation modified to reflect the cylindrical moving phase boundary and the presence of an inner, ID, and outer diameter, OD (see (13)), and its dependence on radius, was explored first and showed a strong linear relationship between volume gain and tube radius. This demonstrated that, even with the geometric conversion steps where inner radius is not involved, the physical diffusion phenomena predicted by the model are not lost, i.e., the reciprocal relation between concentration and tube radius. Further investigations were performed by exploring the relationship between mass-based parabolic rate constant (normalised against sample surface area) and tube radius (see Figure 8).

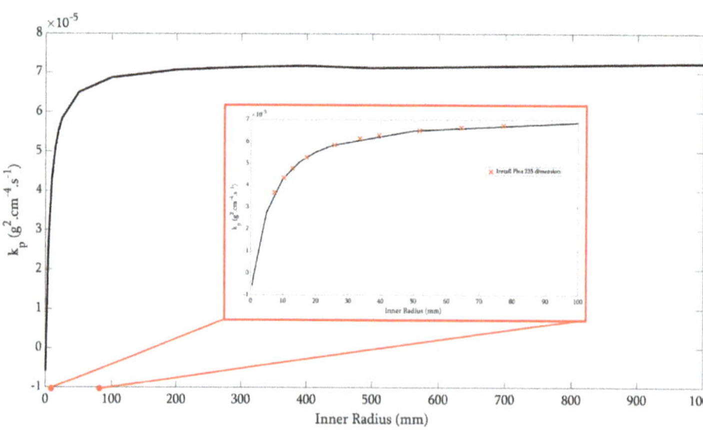

Figure 8. Plot of mass-based, surface area normalised parabolic rate constant against tube inner radius. The industrial relevant inner radius range is magnified (red box) and highlighted (red dashed line).

Crucially, the parabolic profile reveals a critical radius where radius no longer has a significant effect, associated with changes in surface area and surface area to volume ratio, as demonstrated by (2) and Figure 2, on oxidation kinetics. This change in kinetic dependency occurs at approximately 200 mm. This is an industrially relevant conclusion for typical conveyance tube applications. Furthermore, 98% of the change in thickness gain occurred in the radius range below 100 mm, again demonstrating the influence of cylindrical coordinates on conveyance tube product oxidation behaviour. This radius region shows the sharpest change in oxidation rate. This conclusion is also supported by equivalent plots of thickness gain (see Figure 9), again suggesting a critical inner radial value of 200 mm.

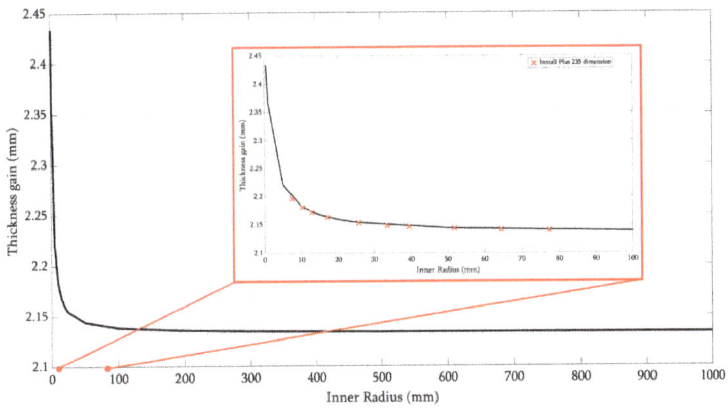

Figure 9. Plot of oxide thickness against tube inner radius. The industrial relevant inner radius range is magnified (red box) and highlighted (red dashed line).

Changes in radius are accompanied by a change in available reacting surface area and surface area to volume ratio (see Figure 2) and highlight how an increase in radius is accompanied by an increase in surface area, A, (see (3)) and a decrease in surface area to volume ratio, $A : V$ (see (4)). Therefore, the surface area increases during inner surface oxidation (consumption of the substrate widens the cylinder bore, i.e., the inner radius), which leads to an overall slower rate of reaction and therefore thinner oxide thickness after a given time and, as such, the converse is true of the outer surface.

This computational approach has enabled the extremes of potential tube dimensions to be investigated beyond that which time and sample access allows within an empirical approach. The critical radius result is industrially relevant to the range of conveyance tubes manufactured in this context, and indicates that oxidation kinetics on tubes cannot be inferred from planar components subjected to equivalent thermomechanical processing conditions. However, there are many high-temperature-application cylindrical and curved components within this dimensional range, both hollow and solid, which could benefit from the ability to understand and predict oxidation kinetics to streamline their respective manufacturing processes.

3.4. Further Work

There is little empirical data for the incorporation of porosity into the model via a modified diffusion coefficient. Porosity can influence diffusion, and subsequently oxidation, via its effect on ion migration through vacancies. The mechanically intensive processes which accompany tube manufacturing could encourage pores to propagate, amongst other defects, and capturing the effect of porosity within the model is therefore paramount to accurate oxide thickness predictions.

Attempts were made to modify the Thermo-Calc model to reflect the unavoidable heat ramp of the STA validation experiment and allow a more direct comparison. However, this came at a significant computational cost. The model complexity was also exacerbated by the introduction of a non-isothermal period due to the temperature dependence of so many of the parameters in both the thermodynamic and kinetic databases, thus introducing many more calculations to achieve a solution. The computational thermochemistry approach does not yet lend itself to efficiency over its empirical counterpart when considering non-isothermal scenarios. Hence, this work is limited to consideration of isothermal oxidation.

Although validation work was performed for the planar geometry iteration of the model using STA, the specific information surrounding curved surfaces demands its own validation using a specialist furnace to apply a representative heat treatment to tube samples.

Further techniques, such as indentation, could also be used to validate stress state analyses for both the pre- and post-normalised states.

4. Conclusions

It has been shown in this work that the undesirable, diffusion-controlled oxidation reaction, occurring on external surfaces during hot-finished conveyance tube manufacturing, and specifically the normalisation stage, can be computationally modelled. Thermo-Calc's DICTRA commercial software package can be used to solve the Stefan problem-based moving phase boundary equation for a curved surface and predict surface oxide thickness using numerical techniques, provided that bespoke thermodynamic and kinetic databases can be developed from empirical data. The following conclusions were drawn during model development:

- External surface oxide thickness on cylindrical geometries, representative of those used in conveyance applications, can be predicted quickly with ~120 datapoints.
- Dependence on the radial coordinate within the cylindrical coordinate form of the diffusion equation leads to a more complex and faster oxidation process due to continuous changes in surface area and overlapping normal diffusion pathways.
- There is a critical radius, estimated at ~200 mm, where the curvature is low enough to no longer affect oxidation, however this is still larger than the conveyance product range.

Overall, oxidation of curved surfaces appears to be more complex and unstable than its planar equivalent, a topic which has seen much greater attention in the literature. Optimal understanding and control of oxidation during conveyance tube manufacturing, and that of equivalent products, demands a separate scientific consideration.

Author Contributions: Conceptualization, M.K., H.C. and C.O.; methodology, M.K.; software, M.K.; validation, M.K.; formal analysis, M.K.; investigation, M.K.; resources, M.C., H.C., E.S., C.O. and M.A.; writing—original draft preparation, M.K.; writing—review and editing, M.C., E.S., C.O. and M.A.; visualization, E.S.; supervision, M.C., H.C., E.S., C.O. and M.A. All authors have read and agreed to the published version of the manuscript.

Funding: This research was funded by COATED M2A from the European Social Fund via the Welsh Government (c80816), the Engineering and Physical Sciences Research Council (Grant Ref: EP/S02252X/1), and Tata Steel.

Data Availability Statement: The data sets presented in this article are not readily available because this work is part of an ongoing study and is currently commercially sensitive. Requests to access the data sets should be directed to M.K.

Conflicts of Interest: M.C. was employed by the company Oxford Instruments NanoAnalysis, H.C. was employed by Frazer-Nash Consultancy, and C.O. was employed by Tata Steel. The remaining authors declare that the research was conducted in the absence of any commercial or financial relationships that could be construed as a potential conflict of interest.

References

1. Water UK. Companies Are Using New Techniques to Find and Fix More Leaks. Available online: https://www.water.org.uk/water-supply/leakage (accessed on 8 July 2024).

2. Office for National Statistics. Central Heating. Available online: https://www.ons.gov.uk/datasets/TS046/editions/2021/versions/4 (accessed on 8 July 2024).
3. Hallstrom, S.; Halvarsson, M.; Hoglund, L.; Jonsson, T.; Agren, J. High temperature oxidation of chromium: Kinetic modeling and microstructural investigation. *Solid State Ion.* **2013**, *240*, 41–50. [CrossRef]
4. Grant, J. The Development of Novel Coating Solutions for the Improvement of Pre/Post Heat Treatment Performance Properties of Carbon Steel Conveyance Tubes. Ph.D. Thesis, Department of Materials Science and Engineering, Swansea University, Swansea, Wales, UK, 2024. [CrossRef]
5. Schutze, M. The Role of Stresses in High-Temperature Corrosion: The Potential of Quantitative Approaches. *High Temp. Corros. Mater.* **2023**, *100*, 365–397. [CrossRef]
6. Matsuno, F. Blistering and Hydraulic Removal of Scale Films of Rimmed Steel at High Temperature. *Iron Steel Inst. Jpn. Int.* **1980**, *20*, 413–421. [CrossRef]
7. Birks, N.; Meier, G.H. Oxidation of Alloys. In *Introduction to High Temperature Oxidation of Metals*; Edward Arnold Ltd.: London, UK, 1983; pp. 91–130.
8. Zeng, C.; Neils, A.; Lesko, J.; Post, N. Machine learning accelerated discovery of corrosion-resistant high-entropy alloys. *Comput. Mater. Sci.* **2024**, *237*, 413–421. [CrossRef]
9. Wang, F.; Zou, D.; Wang, Q.; Zhang, X.; Tong, L.; Liang, X.; Li, Y.; Jiang, Y. The negative effect of V content on high temperature oxidation resistance of austenitic stainless steels at 850C. *Mater. Charact.* **2024**, *212*, 113967. [CrossRef]
10. Espevik, S.; Rapp, R.A.; Daniel, P.L.; Hirth, J.P. Oxidation of Ni-Cr-W Ternary Alloys. *Oxid. Met.* **1980**, *14*, 85–108. Available online: https://link.springer.com/article/10.1007/BF00603987 (accessed on 5 December 2024). [CrossRef]
11. Manning, M.I. Geometrical effects on oxide scale integrity. *Corros. Sci.* **1981**, *21*, 301–316. [CrossRef]
12. Sabau, A.S.; Wright, I.G. On the estimation of thermal strains developed during oxide growth. *J. Appl. Phys.* **2009**, *106*, 023503. [CrossRef]
13. Yu, B.; Liu, Y.; Wei, L.; Zhang, X.; Du, Y.; Wang, Y.; Ye, S. A Mechanism of Anti-Oxidation Coating Design Based on Inhibition Effect of Interface Layer on Ions Diffusion within Oxide Scale. *Coatings* **2021**, *11*, 454. [CrossRef]
14. Fu, G.Y.; Wei, L.Q.; Zhang, X.M.; Cui, Y.B.; Lv, C.C.; Ding, J.; Yu, B.; Ye, S.F. A high-silicon anti-oxidation coating for carbon steel at high temperature. *Surf. Coat. Technol.* **2017**, *310*, 166–172. [CrossRef]
15. Wu, H.; Chen, H.; Lu, J.; Feng, Y.; Zhao, S.; Guo, E. Superior anti-oxidation layer induced by PVA-boehmite sol-gel film on AISI304 steel. *Ceram. Int.* **2024**, *50*, 35273–35286. [CrossRef]
16. Pillai, R.; Chyrkin, A.; Quadakkers, W.J. Modeling in High Temperature Corrosion: A Review and Outlook. *Oxid. Met.* **2021**, *96*, 385–436. Available online: https://link.springer.com/article/10.1007/s11085-021-10033-y (accessed on 5 December 2024). [CrossRef]
17. Shajan, N.; Kumar, R.; Manik, R.; Asati, B.; Dhagde, S.; Dhangi, D.; Kumar, S.; Mahapartra, M.M.; Arora, K.S. Role of residual stress in the failure of HF-ERW welded tubes. *Eng. Fail. Anal.* **2024**, *161*, 108342. [CrossRef]
18. Soori, M.; Arezoo, B. A Review in Machining-Induced Residual Stress. *J. New Technol. Mater.* **2022**, *12*, 64–83. Available online: https://hal.science/hal-03679993v1 (accessed on 5 December 2024).
19. Wang, C.; Wu, H.; Zhang, P.; Li, Z.; Cao, R.; Shang, C. Analysis of the Local Plastic Deformation of the Surface Oxide Layer formed on Low-Alloy High-strength Steel. *Trans. Indian Inst. Met.* **2022**, *75*, 1441–1450. [CrossRef]
20. Wassilkowska, A.; Wojciech, D. Silicon as a component of ferric oxide scale covering ductile iron pipes after annealing. *Eng. Fail. Anal.* **2021**, *125*, 105381. [CrossRef]
21. Picque, B.; Bouchard, P.-O.; Montmitonnet, P.; Picard, M. Mechanical behaviour of iron oxide scale: Experimental and numerical study. *Wear* **2006**, *260*, 231–242. [CrossRef]
22. Sun, L.; Yan, W. Estimation of Oxidation Kinetics and Oxide Scale Void Position of Ferritic-Martensitic Steels in Supercritical Water. *Adv. Mater. Sci. Eng.* **2017**, *2017*, 9154934. [CrossRef]
23. Spencer, P.J. A brief history of CALPHAD. *Comput. Coupling Phase Diagr. Thermochem.* **2008**, *32*, 1–8. [CrossRef]
24. Hillert, M. The compound energy formalism. *J. Alloys Compd.* **2001**, *320*, 161–176. [CrossRef]
25. Poerschke, D.L. Developments in Thermodynamic Models of Deposit-Induced Corrosion of High-Temperature Coatings. *J. Miner. Met. Mater. Soc.* **2022**, *24*, 260–273. [CrossRef]
26. Thermo-Calc Software AB. Diffusion Module (DICTRA) Documentation Set. ed, 2023, pp. 23–34. Available online: https://thermocalc.com/support/documentation/ (accessed on 7 August 2023).
27. Wilkstrom, P.; Blasiak, W.; Du, S.C. A study on oxide scale formation of low carbon steel using thermo gravimetric technique. *Ironmak. Steelmak.* **2008**, *35*, 621–632. [CrossRef]
28. Lee, J.H.; Noh, W.; Kim, D.; Lee, M. Spallation analysis of oxide scale on low carbon steel. *Mater. Sci. Eng. A* **2016**, *676*, 385–394. [CrossRef]
29. Basabe, V.; Szpunar, J. Phase Composition of Oxide Scales during Reheating in Hot Rolling of Low Carbon Steel. *Steel Res. Int.* **2006**, *27*, 818–824. Available online: https://onlinelibrary.wiley.com/doi/epdf/10.1002/srin.200606467 (accessed on 5 December 2024). [CrossRef]
30. Landfahrer, M.; Schluckner, C.; Prieler, R.; Gerhardter, H.; Zmek, T.; Klarner, J.; Hochenauer, C. Numerical and experimental investigation of scale formation on steel tubes in a real-size reheating furnace. *Int. J. Heat Mass Transf.* **2019**, *129*, 460–467. [CrossRef]

31. Purbolaksono, J.; Khinani, A.; Rashid, A.Z.; Ali, A.A.; Nordin, N.F. Prediction of oxide scale growth in superheater and reheater tubes. *Corros. Sci.* **2009**, *51*, 1022–1029. [CrossRef]
32. Sun, L.; Yan, W. Prediction of wall temperature and oxide scale thickness of ferritic–martensitic steel superheater tubes. *Appl. Therm. Eng.* **2018**, *134*, 171–181. [CrossRef]
33. Hsueh, C.H.; Evans, A.G. Oxidation induced stresses and some effects on the behavior of oxide films. *J. Appl. Phys.* **1983**, *54*, 6672–6686. [CrossRef]
34. Iordanova, I.; Surtchev, M.; Forcey, K.S.; Krastev, V. High-temperature surface oxidation oflow-carbon rimming steel. *Surf. Interface Anal.* **2000**, *30*, 158–160. [CrossRef]
35. Jang, J.H.; Lee, D.E.; Kim, M.Y.; Kim, H.G. Investigation of the slab heating characteristics in a reheating furnace with the formation and growth of scale on the slab surface. *Int. J. Heat Mass Transf.* **2010**, *53*, 4326–4332. [CrossRef]
36. Chen, R.Y.; Yeun, W.Y.D. Review of the High-Temperature Oxidation of Iron and Carbon Steels in Air or Oxygen. *Oxid. Met.* **2003**, *59*, 433–468. Available online: https://link.springer.com/article/10.1023/A:1023685905159 (accessed on 5 December 2024). [CrossRef]
37. Birks, N.; Meier, G.H. Mechanisms of Oxidation. In *Introduction to High Temperature Oxidation of Metals*; Edward Arnold Ltd.: London, UK, 1983; pp. 31–65.
38. Hidayat, T.; Shishin, D.; Jak, E.; Decterov, S.A. Thermodynamic reevaluation of the Fe–O system. *Calphad* **2015**, *48*, 131–144. [CrossRef]
39. Sundman, B. An Assessment of the Fe-O system. *J. Phase Equilibria* **1991**, *12*, 127–140. Available online: https://link.springer.com/article/10.1007/BF02645709 (accessed on 5 December 2024). [CrossRef]
40. Hallstrom, S.; Hoglund, L.; Agren, J. Modeling of iron diffusion in the iron oxides magnetite and hematite with variable stoichiometry. *Acta Mater.* **2010**, *59*, 53–60. [CrossRef]
41. Xia, C.; Xia, S.; Li, Y.; Lu, X. Evaluating atomic mobility and interdiffusivity based on two-dimensional diffusion simulations and diffusion triple experiments. *Scr. Mater.* **2020**, *188*, 124–129. [CrossRef]
42. Koga, S.; Krstic, M. Phase Change Model: Stefan Problem. In *Materials Phase Change PDE Control & Estimation*; Birkhäuser: Cham, Switzerland, 2020; pp. 1–13. [CrossRef]
43. Larsson, H. A model for 1D multiphase moving phase boundary simulations under local equilibrium conditions. *Calphad* **2014**, *47*, 1–8. [CrossRef]
44. Welded Steel Tubes for Pressure Purposes—Technical Delivery Conditions (Part 2: Electric Welded Non-Alloy and Alloy Steel Tubes with Specified Elevated Temperature Properties), BS EN10217-2, 2019. [Online]. Available online: https://bsol.bsigroup.com/Bibliographic/BibliographicInfoData/000000000030438140 (accessed on 5 December 2024).
45. Kowalski, M.; Spencer, P.J. Thermodynamic reevaluation of the C-O, Fe-O and Ni-O systems: Remodelling of the liquid, BCC and FCC phases. *Calphad* **1995**, *19*, 229–243. [CrossRef]
46. Hazen, R.M.; Jeanloz, R. Wüstite (Fe1-x O): A review of its defect structure and physical properties. *Rev. Geophys.* **1984**, *22*, 37–46. [CrossRef]
47. Thermo-Calc Software AB. Database Manager User Guide. In *Thermo-Calc Documentation Set*; Thermo-Calc Software AB: Solna, Sweden, 2023; pp. 1743–1840. Available online: https://thermocalc.com/support/documentation/ (accessed on 5 December 2024).
48. Gamsjager, E.; Svoboda, J.; Fischer, F.D. Austenite-to-ferrite phase transformation in low-alloyed steels. *Comput. Mater. Sci.* **2005**, *32*, 360–369. [CrossRef]
49. Bornstein, L. *Numerical Data and Functional Relationships in Science and Technology: Group III*; Springer: Berlin/Heidelberg, Germany, 1990; Volume 26.
50. Kofstad, P. Oxides of Group VII and VIII Elements. In *Nonstoichiometry, Diffusion, and Electrical Conductivity in Binary Oxides*; John Wiley & Sons: London, UK, 1972; pp. 213–264.
51. Imbrie, P.K.; Lagoudas, D.C. Morphological Evolution of TiO2 Scale Formed on Various 1D and 2D Geometries of Titanium. *Oxid. Met.* **2001**, *55*, 359–399. Available online: https://link.springer.com/article/10.1023/A:1010368412822 (accessed on 5 December 2024). [CrossRef]
52. Aghaeian, S.; Sloof, W.G.; Mol, J.M.C.; Bottger, A.J. Initial High-Temperature Oxidation Behavior of Fe–Mn Binaries in Air: The Kinetics and Mechanism of Oxidation. *Oxid. Met.* **2022**, *98*, 217–237. [CrossRef]
53. Birosca, S.; Dingley, D.; Higginson, R.L. Microstructural and microtextural characterization of oxide scale on steel using electron backscatter diffraction. *J. Microsc.* **2004**, *213*, 235–240. [CrossRef] [PubMed]
54. Simon, D.; Gorr, B.; Christ, H.J. Effect of Atmosphere and Sample Thickness on Kinetics, Microstructure, and Compressive Stresses of Chromia Scale Grown on Ni–25Cr. *Oxid. Met.* **2017**, *87*, 417–429. [CrossRef]
55. Asensio-Jimenez, C.; Niewolak, L.; Hattendorf, H.; Kuhn, B.; Huczkowski, P.; Singheiser, L.; Quadakkers, W.J. Effect of Specimen Thickness on the Oxidation Rate of High Chromium Ferritic Steels: The Significance of Intrinsic Alloy Creep Strength. *Oxid. Met.* **2012**, *79*, 15–28. [CrossRef]
56. Nygren, K.E.; Yu, Z.; Rouillard, F.; Couet, A. Effect of sample thickness on the oxidation and carburization kinetics of 9Cr-1Mo steel in high and atmospheric pressure CO_2 at 550 °C. *Corros. Sci.* **2020**, *163*, 108292. [CrossRef]
57. Auinger, M.; Buchler, M.; Schoneich, H.; Gierl-Mayer, C.; Danninger, H. Hydrogen accumulation and diffusion in cylindrical-shaped pipeline steels with coating defects. *Int. J. Hydrogen Energy* **2023**, *48*, 34454–34462. [CrossRef]

58. Degasperi, A.; Calder, M. Relating PDEs in Cylindrical Coordinates and CTMCs with Levels of Concentration. *Electron. Notes Theor. Comput. Sci.* **2010**, *268*, 49–59. [CrossRef]
59. Entchev, P.B.; Lagoudas, D.C.; Slattery, J.C. Effects of non-planar geometries and volumetric expansion in the modeling of oxidation in titanium. *Int. J. Eng. Sci.* **2001**, *39*, 695–714. [CrossRef]
60. Wang, P.; Du, K.; Yin, H.; Wang, D. Enhancing oxide scale growth and adhesion via electrochemically regulating ion diffusion. *J. Mater. Sci. Technol.* **2023**, *158*, 133–144. [CrossRef]

metals

MDPI

Article

Effects of a Welding Wire Containing Er or Sc on the Microstructure, Mechanical Properties, and Corrosion Resistance of the 5xxx Aluminum Alloy MIG Joint

Cunwei Zou [1], Ruizhi Wu [2,*], Xinhe Yang [3,*], Zhikun Ma [2] and Legan Hou [2]

1 School of New Energy Vehicle, Nanning Vocational and Technical University, Nanning 530008, China; 13581560129@163.com
2 Key Laboratory of Superlight Materials & Surface Technology, Ministry of Education, Harbin Engineering University, Harbin 150001, China; zkma@hrbeu.edu.cn (Z.M.); houlegan@hrbeu.edu.cn (L.H.)
3 School of New Energy and Mining, Xinjiang University of Technology, Hetian 848000, China
* Correspondence: rzwu@hrbeu.edu.cn (R.W.); yangxinhe@hrbeu.edu.cn (X.Y.)

Abstract: The development of MIG (metal inert gas) welding for five-series aluminum alloys primarily involves the improvement and optimization of welding processes. Building upon research findings regarding the enhancement of aluminum alloy properties through the use of scandium (Sc) and erbium (Er), our study incorporates Sc and Er into the welding wire to examine their impact on welding quality. The results show that the introduction of Er and Sc results in grain refinement from 47 μm to 29 μm and 31 μm, respectively. Grain refinement is mainly attributed to the heterogeneous nucleation of submicron-sized, coherent Al_3Er and Al_3Sc phases with L12 structure. The ultimate tensile strength (UTS), fracture elongation EI [%], and microhardness of joints welded with Er-containing and Sc-containing filler wires exhibit significant enhancements due to the refinement strengthening and dispersion strengthening. Joints welded with the filler wires containing Er and Sc display reduced corrosion current density and higher corrosion potential. The enhanced corrosion resistance comes from the formation of a denser oxide film and the equilibrium in the potential difference between the precipitated phases (Al_3Er and Al_3Sc) and the matrix. Filler wires containing Er and Sc have almost similar effects on improvements of the MIG welding joints.

Keywords: MIG welding; Er; Sc; mechanical properties; corrosion resistance

1. Introduction

Scandium (Sc) and zirconium (Zr) are commonly added to aluminum–magnesium alloys to enhance their mechanical properties through the formation of core–shell $Al_3(Sc,Zr)$ phases, which act as effective strengthening agents [1–3]. However, the high cost of scandium has prompted the search for more economical alternatives. Neodymium (Er) has emerged as a promising substitute, forming similar core–shell $Al_3(Er,Zr)$ phases that offer comparable benefits, including grain refinement, recrystallization inhibition, and precipitation strengthening [4–8]. The application of aluminum alloys in shipbuilding heavily relies on effective joining techniques. Fusion welding methods, such as metal inert gas (MIG) and tungsten inert gas (TIG) welding, are widely used for joining 5000 series aluminum alloys [9,10]. In particular, MIG welding provides deep penetration, high deposition rates, and enhanced production efficiency, making it suitable for welding thicker plates (thickness $\geq$ 1.6 mm) [11,12]. The corrosion resistance of 5000 series aluminum alloy welds is a critical factor in their application, particularly in marine environments. The primary failure mechanisms of these alloys are often associated with localized corrosion, such as pitting and intergranular corrosion (IGC), which can be exacerbated by microstructural inhomogeneities introduced during the welding process [13,14]. The presence of intermetallic compounds also strongly influences the electrochemical behavior of the alloys [15,16]. Recent studies have emphasized the importance of controlling the microstructure to enhance the corrosion resistance of 5000 series aluminum alloys. For instance, the addition of scandium has been shown to significantly reduce IGC susceptibility by altering the distribution and morphology of intermetallic phases, hindering recrystallization, and reducing continuous precipitation [17,18]. Furthermore, the use of optimized welding parameters can modify the distribution of grain boundary phases, affecting the intergranular corrosion resistance of different regions and thereby improving the overall corrosion resistance of the weld [16].

The selection of appropriate welding filler materials is crucial for achieving high-quality welds. However, research on welding filler materials for 5000 series aluminum alloys, particularly those containing scandium or neodymium, remains limited. In this study, three types of welding filler wires—Al-Mg-Mn-Zr, Al-Mg-Mn-0.2%Sc-Zr, and Al-Mg-Mn-0.2%Er-Zr—were designed for the MIG welding of cold-rolled Al-Mg-Mn-Er-Zr plates with a thickness of 4 mm. By examining the interactions between microstructural characteristics and corrosion, the effects of adding scandium and neodymium to the welding filler materials on the microstructure, mechanical properties, and corrosion resistance of Al-Mg-Mn-Er-Zr welds were systematically investigated.

2. Materials and Methods

The base metal and three filler wire materials (Wire-1: Al-Mg-Mn-Zr filler wire, Wire-2: Al-Mg-Mn-Er-Zr filler wire, Wire-3: Al-Mg-Mn-Sc-Zr filler wire) were sourced from Northeast Light Alloy Co., Ltd. (Harbin, China). Their compositions were determined via inductively coupled plasma–atomic emission spectroscopy(ICP-AES), as listed in Table 1. Prior to conducting MIG welding, we cut the sample using wire cutting, polished it with metallographic sandpaper of varying particle sizes, and then polished it with water on a PG-1A polishing machine. Subsequently, the surface of the sample was corroded with Keller's reagent, the surface was cleaned sequentially with purified water and anhydrous ethanol, and finally, it dried quickly with a hair dryer. During the MIG welding process, pure argon gas was used as shielding gas. MIG welding was executed on the butt of the cold-rolled Al-6Mg-0.8Mn-0.2Er-0.15Zr plates with a thickness of 4mm, with the welding direction perpendicular to the rolling direction, As shown in Figure 1, through welding wire overlay experiments [19], the welding current and welding speed were explored, and the final welding parameters obtained are shown in Table 2.

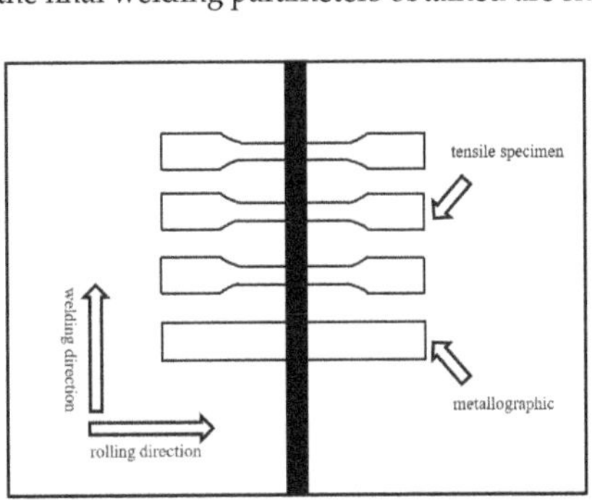

Figure 1. Welding wire performance test diagram.

Table 1. Chemical compositions (wt%) of the BM and filler wires.

	Mg	Mn	Sc	Er	Zr	Al
Wire-1	5.7	0.83	-	-	0.15	Bal.
Wire-2	5.9	0.78	-	0.19	0.16	Bal.
Wire-3	6.2	0.76	0.23	-	0.19	Bal.
BM	6.1	0.81		0.2	0.15	Bal.

Table 2. Welding parameters.

Parameter	Value
Welding current (A)	110
Welding voltage (V)	17.3
Welding speed (mm/s)	0.5
Weld width (mm)	1.2
Wire diameter (mm)	1.2
Shielding gas	Ar
Gas flow (L/min)	12

All the specimens were gained from the central region of the joint and perpendicular to the weld seam. For metallographic observations, the samples were etched for 15 s using Keller's reagent solution (2.5HNO_3:1.5HCl:1HF:95H_2O) and were observed using an optical microscope (OM, LEICA DM IRM, Leica Microsystems GmbH, Wwtzlar, Germany). The microstructures and fracture surfaces of the joints were observed using a scanning electron microscope (SEM, Quanta-200, Thermo Fisher Scientific, Hillsboro, OR, USA) and a transmission electron microscope (TEM, JEM-2100 EX, JEOL Ltd., Tokyo, Japan), both equipped with energy-dispersive X-ray spectroscopy (EDS, EDAX Genesis, EDAX Inc., Princeton, NJ, USA).

The mechanical properties of MIG-welded joints were tested using a universal testing machine (WDW 3050, Cangzhou Zhongke Beigong Test Instrument Co., Ltd., Cangzhou, China). Tensile specimens were perpendicular to the weld seam, with the weld axis positioned at the midpoint of the parallel section of the specimen. Tensile testing was conducted at a rate of 0.5 mm/min, and an average value was obtained from five parallel tests. Microhardness measurements were performed using an microhardness tester (HXS-1000Z, Guiyang Milite Instrument Co., Ltd., Guiyang, China), with a load of 0.98 N and a dwell time of 15 s. Measurements were taken at 1 mm intervals from the center of the weld seam towards both sides.

The corrosion resistance of the welded joints was tested using an electrochemical workstation (Im6/Im6ex). For the electrochemical measurements, a saturated calomel electrode, platinum sheet, and the specimens served as the reference electrode, auxiliary electrode, and working electrode, respectively, while a 3.5% NaCl (by mass) solution was utilized as the electrolyte. During the polarization curve test, the open circuit potential was initially determined. Subsequently, the upper and lower limit potentials were set at $\pm$0.3 V relative to the open circuit potential. A scan rate of 0.1 mV/s was employed to generate Tafel curves, from which corrosion current density and corrosion potential were derived through curve fitting. Each sample required 3 separate samples to be tested, and a total of 9 samples were tested.

3. Results and Discussion

3.1. Macroscopic Observation of Weld Seam Formation

Figure 2 illustrates the surface morphology of the weld seams on both the front and back sides of the butt MIG-welded joints. Subject to identical welding process parameters, the quality of the weld seams remains consistent. On the front side, uniform and finely detailed fish-scale patterns are evident, while on the back side, there is ample weld penetration, resulting in the creation of teardrop-shaped weld pools. This morphology shows that it allows single-side welding with double-sided formation, producing welds characterized by density, esthetic appeal, minimal spatter, and narrow weld bead widths. The back side of the weld seam displays complete fusion, devoid of any discernible defects such as weld cracks and incomplete fusion or biting edges. These observations attest to the stability of the welding process and the high-quality formation of the weld seams.

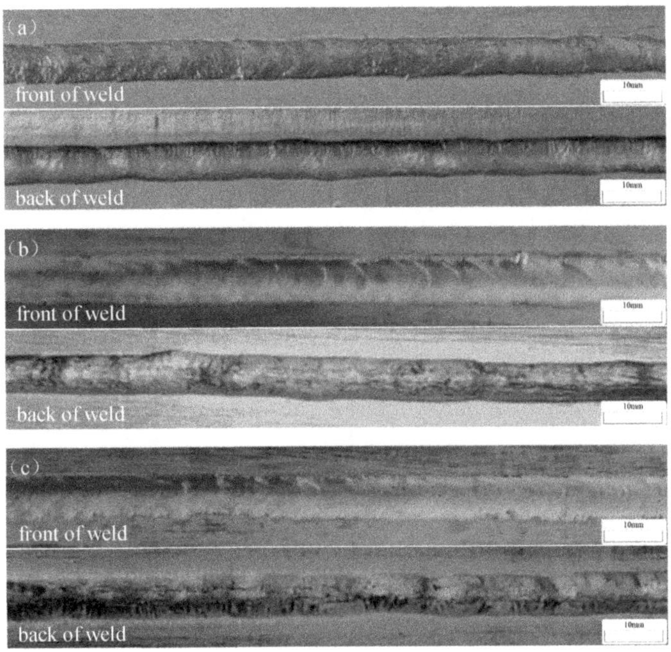

Figure 2. Surface morphology of welded joints: (**a**) Wire-1; (**b**) Wire-2; and (**c**) Wire-3.

3.2. Microstructure Observation

Figure 3a illustrates the low-magnification microstructure of the Wire-2 joint, which can be primarily divided into four typical regions: the base metal region (BM), the heat-affected zone (HAZ), the fusion zone (FZ), and the welding zone (WZ). In the BM, a fibrous structure resulting from rolling remained. Aluminum alloys containing Er and Zr can form numerous fine $Al_3(Er,Zr)$ particles during the welding process. These nanoscale particles effectively suppress recrystallization [20], resulting in the retention of more deformation structures during welding. The HAZ exhibits a microstructure with partial recrystallization due to the influence of the welding heat input, resulting in relatively coarser fibrous structures [21]. During welding, the temperature gradient typically leads to higher temperatures at the center of the weld. A transitional region, known as the fusion zone, is located between the welding zone (WZ) and the heat-affected zone (HAZ). It is characterized by grains with a distinct crystallographic orientation that extends from the welding zone towards the heat-affected zone. These grains form elongated structures with non-uniform sizes. The morphologies of the welding joints for the three types of welding wires are similar; therefore, only the low-magnification microstructure of the Wire-2 joint is presented here.

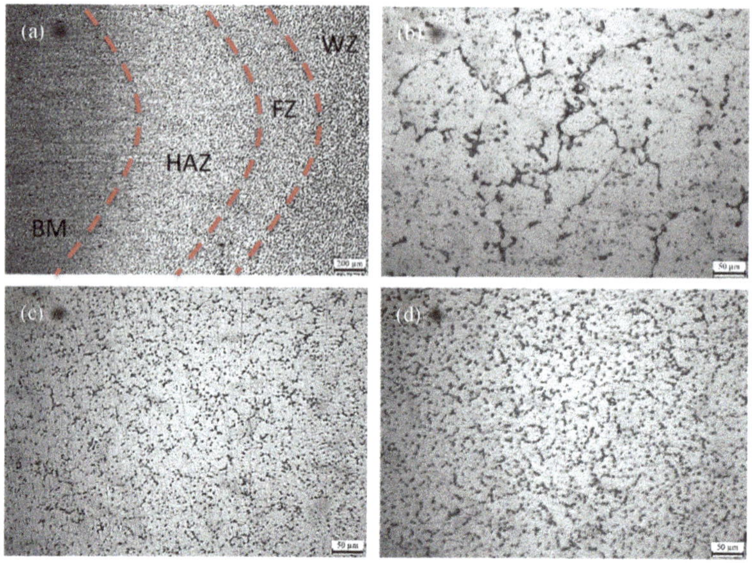

Figure 3. Optical microstructure of cross-section of welded joint: (**a**) appearance of welded joint of Wire-2; (**b**) WZ of Wire-1; (**c**) WZ of Wire-2; and (**d**) WZ of Wire-3.

Figure 3b–d depict the metallographic microstructures of the welding zones in the joints of Wire-1, Wire-2, and Wire-3, respectively. They all exhibit typical cast microstructures characterized by equiaxed grains. However, there are significant variations in grain size. All the samples were treated by electrolytic polishing and then observed by SEM, as shown in Figure 4. According to the calculations using IPP software (Image-Pro Plus 8.0), the Wire-1 joint exhibits the largest grain size, approximately 47 μm. In contrast, the joints with Wire-2 and Wire-3 have comparable grain sizes, approximately 29 μm and 31 μm, respectively. This discrepancy in grain size can be attributed to the incorporation of Er and Sc in Wire-2 and Wire-3. The presence of these elements in the welding wires leads to the formation of Er-containing and Sc-containing phases within the weld zones, which promote nucleation in the joints [22].

The presence of fine equiaxed grains, particularly within the welding zone (WZ), is predominantly attributed to the elevated cooling rates during welding. The rapid cooling process, facilitated by the high thermal conductivity of the base metal (BM) and the plentiful and non-uniform nucleation sites offered by the weld pool's edges, foster grain refinement. These refined grains hinder the growth of dendrites, accounting for the absence of dendritic structures in the WZ.

Figure 5 depict SEM images (back-scattered electron image) of the weld zone for the joints. The EDS results corresponding to points 1 to 9 in Figure 5 are presented in Table 3. Bright regions in Figure 5 indicate an enrichment of higher atomic weight elements, while dark regions represent the α-Al matrix. The results reveal that the black, granular second phase corresponds to the Mg-Si phase, the light gray, blocky second phase corresponds to the Al-Mn-Fe phase, and the bright white granular and lamellar second phases correspond to the Al-Mg phase, which precipitates along grain boundaries. Figure 6 shows the energy spectrum of the SEM image in Figure 5c. From this figure, the enrichment of various elements at the joint of Wire-3 can be observed, which is consistent with the inferred elemental composition in Table 3. Since the compositions of the three welding joints are similar, only the EDS image of Wire-3 is presented.

Figure 4. Secondary electron morphology in WZ of different welded joints: (**a**) Wire-1; (**b**) Wire-2; and (**c**) Wire-3.

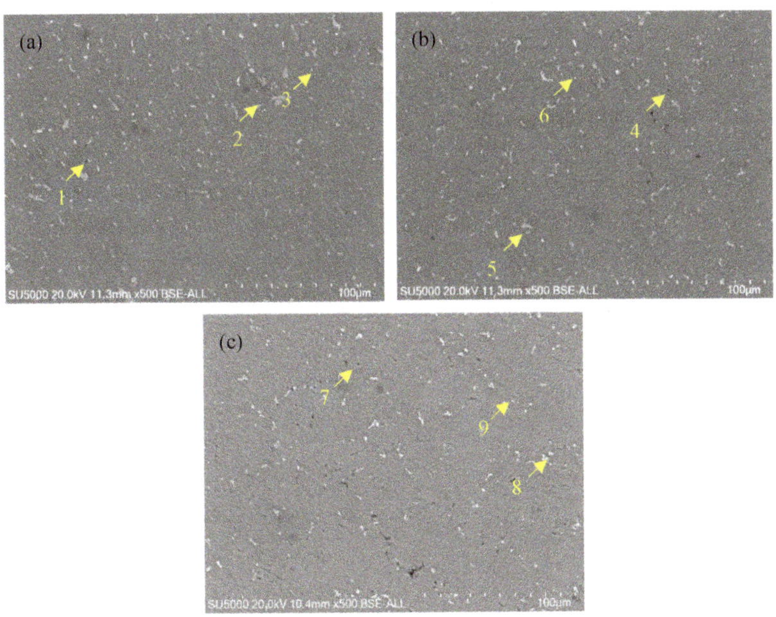

Figure 5. Back-scattered electron images for the WZ of welded joints: (**a**) Wire-1; (**b**) Wire-2; and (**c**) Wire-3.

In the weld zones of the joints, the geometric shapes and distribution of these second phases are largely consistent. Since the three filler wires have almost the same composition and the same welding parameters are used, the phase compositions in the weld zone are identical. Due to the smaller size of the precipitates formed by Er and Sc elements, they could not be observed with SEM microscopy and were further examined in the subsequent TEM analysis.

Table 3. Chemical composition of points in Figure 5 (at.%).

Point	Mg	Al	Si	Mn	Fe	Zr	Inferred Phase
1	48.99	29.00	22.01	0	0	0	Mg_2Si, Si
2	1.17	81.85	0	8.80	8.18	0	$Al_6(Mn,Fe)$, Al_6Mn
3	23.39	76.61	0	0	0	0	Al_3Mg_2, $Al_{12}Mg_{17}$
4	39.63	40.71	19.66	0	0	0	Mg_2Si, Si
5	0.82	82.50	0	9.21	7.47	0	$Al_6(Mn,Fe)$, Al_6Mn
6	33.79	66.21	0	0	0	0	Al_3Mg_2, $Al_{12}Mg_{17}$
7	39.15	43.86	16.99	0	0	0	Mg_2Si, Si
8	1.12	84.85	0	7.43	6.60	0	$Al_6(Mn,Fe)$, Al_6Mn
9	35.33	64.67	0	0	0	0	Al_3Mg_2, $Al_{12}Mg_{17}$

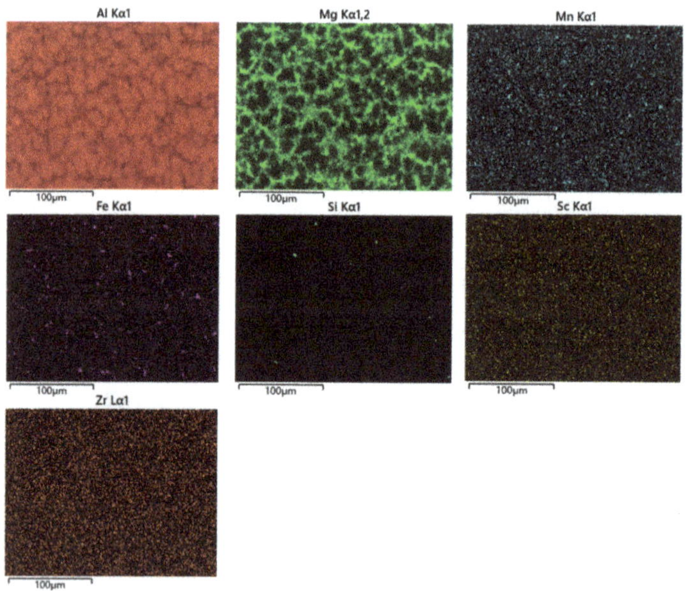

Figure 6. Energy spectrum of the SEM image in Figure 5c.

The existence of Sc and Er triggers eutectic reactions during the solidification of the welding zone, resulting in the formation of submicron-sized Al_3Sc and Al_3Er phases. These phases play a pivotal role in facilitating the precipitation of solute atoms, thereby providing heterogeneous nucleation sites for subsequent solidification processes. This mechanism effectively refines the dendritic structure of the weld while simultaneously inhibiting excessive grain growth, thus achieving fine-grain strengthening in the weld zone.

We utilized an electric discharge wire-cutting machine to slice the test sample into thin sections measuring 5 mm × 5 mm × 1 mm. Subsequently, we polished these thin sections with 320 and 2000-grit sandpaper until they were less than 35 μm thick, ensuring uniformity in thickness. We employed the Gatan 695 C ion-thinning instrument to further refine the transmission sample. Prior to thinning, we used a punch to create and cut out a circular piece with a 3 mm diameter. The ion thinning process involves initially thinning the sample at a large angle (8°) and high voltage (3 eV) for 1–2 h. Once a hole was formed in the sample, we decreased the angle to 6° and the voltage to 3.5 eV, continuing for approximately 0.5 h. Finally, we reduced the angle to 3° while maintaining the voltage at 3.5 eV and thinned this for an additional 10–20 min to achieve a transparent sample. Microstructural changes within the weld zones of joints between Wire-2 and Wire-3, stemming from the introduction of Er and Sc, were meticulously examined via transmission electron microscopy (TEM), as shown in Figures 7 and 8. Although Al_3(Sc,Zr) and Al_3(Er,Zr) particles inherently possess high-temperature stability and relatively elevated melting points (approximately 1320 °C), it is noteworthy that these particles tend to become coarser during the welding process, primarily due to the temperatures reaching up to 1200 °C at the core of the weld [23]. Consequently, the Al_3(Sc,Zr) and Al_3(Er,Zr) particles undergo a transition from the nanoscale to the submicron scale. Detailed observations from Figures 7 and 8 reveal that the size of the Al_3Sc and Al_3Er phases within the weld zones of joints are approximately 200 nm, with no presence of other nanoscale particles [24]. Conventionally, within the context of MIG welding, alloying elements tend to dissolve into the aluminum matrix, and the rapid cooling rates prevalent in this process yield only limited solute precipitation from the matrix, subsequently forming new phases. TEM diffraction patterns unambiguously confirm that the precipitated phases indeed correspond to Al_3Sc and Al_3Er, showcasing a characteristic L12 structural arrangement. Due to the almost identical lattice structure of Al_3Sc, Al_3Er, and α-Al, the discrepancy between the lattice parameters is about 1.5% [25].

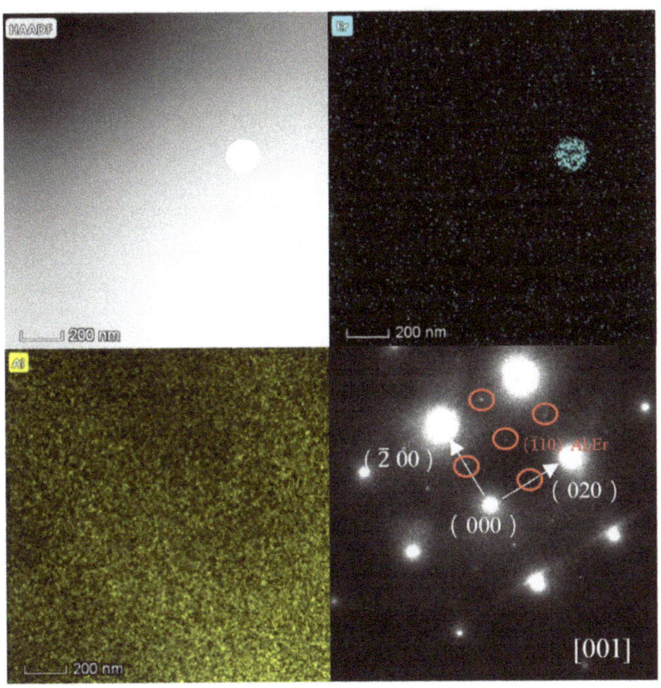

Figure 7. HADDF-STEM image of the Wire-2 weld zone and its element mapping and diffraction pattern of this region with Al_3Er stripes.

In summary, natural air cooling fosters the formation of fine equiaxed grains within the weld zone, and the introduction of Er and Sc elements in the form of submicron Al_3Sc and Al_3Er phases further refines the grains [26]. Furthermore, it is noteworthy that the geometric characteristics of grains within the fusion zone (FZ) remain significantly influenced by the welding thermal cycles.

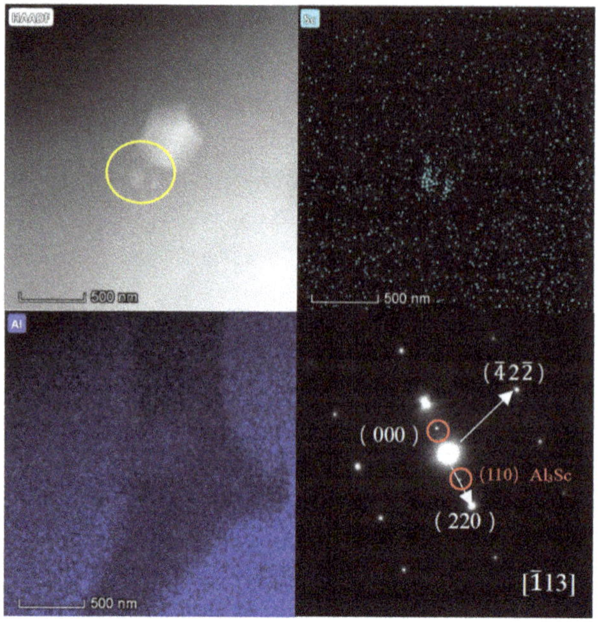

Figure 8. HADDF-STEM image of the Wire-3 weld zone and its element mapping and diffraction pattern of this region with Al_3Sc stripes. The yellow circles highlight the Al_3Sc phases.

3.3. Mechanical Properties

During MIG welding, the three types of welding wires are melted by the electrode and solidified on the adjacent workpiece, resulting in the formation of the weld joint [19]. Figure 9 illustrates the microhardness distribution of MIG-welded joints. The trends in hardness variation were generally consistent across the three groups of joints and exhibited symmetry along the center of the weld. It is apparent that MIG-welded joints comprise the welding zone (WZ), heat-affected zone (HAZ), and base metal (BM). The BM typically exhibits a coarse-grained structure due to the rolling process. The presence of fine $Al_3(Er,Zr)$ precipitates in the BM helps to pin dislocations and sub-grain boundaries, resulting in a high microhardness (approximately 126 HV) and good mechanical strength. The HAZ experiences partial recrystallization due to thermal cycling during welding, leading to a gradient microstructure with coarser grains near the BM and finer grains near the WZ. The fibrous structure in the HAZ is retained due to the suppression of recrystallization by $Al_3(Er,Zr)$ precipitates. The HAZ exhibits a gradual transition in microhardness, with values lower than the BM but higher than the WZ.

In the weld zone, the average microhardness of joints with Wire-2 and Wire-3 is quite similar and significantly higher than that with Wire-1, measuring approximately 82 HV. In contrast, the average microhardness of Wire-1 joints in the weld zone is approximately 78 HV. Within the weld zone of Wire-2 and Wire-3 joints, fine equiaxed grain structures dominate. Furthermore, the precipitated Al_3Er and Al_3Sc phases provide numerous sites for heterogeneous nucleation, refining the grains during the solidification of the welding melt. Consequently, under the same MIG welding parameters, the weld zone hardness of the joint with Wire-1 is lower than that with Wire-2 and Wire-3. As the welding heat input remains consistent under the same welding parameters, the heat-affected zone (HAZ) width is nearly identical among the three welded joints, resulting in similar hardness values in this region.

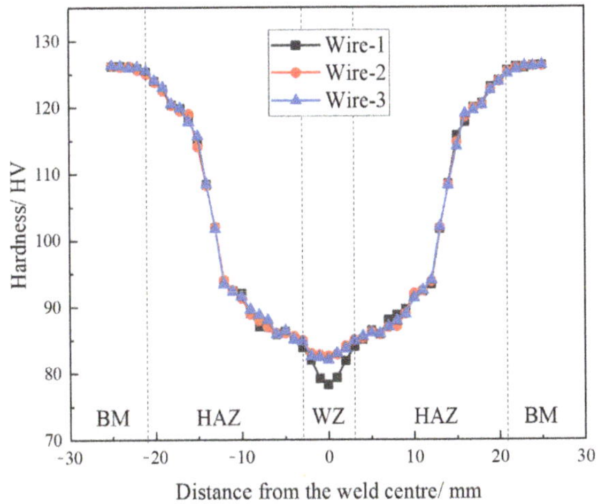

Figure 9. Distribution of microhardness in three kinds of welding wire joints.

In the center of the weld zone, the absence of precipitated $Al_3(Er,Zr)/Al_3(Sc,Zr)$ particles, along with the strengthening effects of substructures and deformation, collectively contribute to the lowest microhardness values observed in this region. As depicted in Figure 3, it is evident that there is a gradual transition from fine equiaxed grains to fibrous structures. Consequently, a recovery in microhardness can be observed, which is attributed to fine grain strengthening and work-hardened states. The high-melting-point $Al_3(Er,Zr)/Al_3(Sc,Zr)$ particles in the base metal do not undergo aggregation or coarsening even when exposed to the elevated temperature of MIG welding. They persist in pinning dislocations and sub-grain boundaries. Consequently, the heat-affected zone is primarily composed of fibrous structures, although some small, recrystallized grains are observed near the fusion line.

The tensile stress–strain curves for joints are presented in Figure 10. To facilitate a more direct comparison of the tensile properties of joints, specific values for ultimate tensile strength and fracture elongation are listed in Table 4.

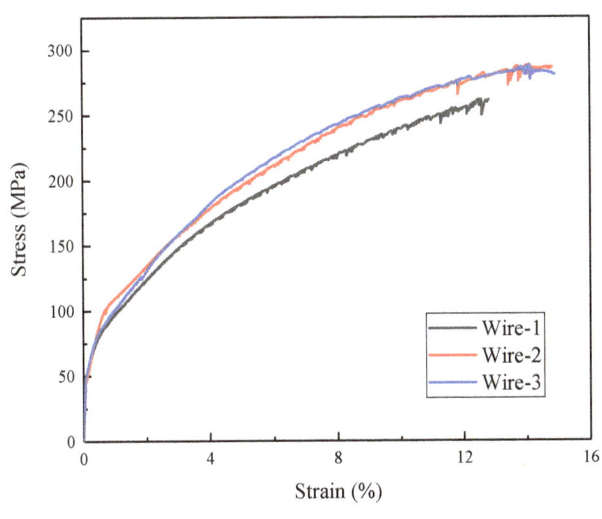

Figure 10. Tensile stress–strain curves of three kinds of wire-welded joints.

Table 4. Mechanical properties of three kinds of welded joints.

Samples	Tensile Strength (MPA)	Elongation (%)
WIRE-1	261 ± 6	12.8 ± 0.3
WIRE-2	287 ± 6	14.8 ± 0.3
WIRE-3	286 ± 6	14.9 ± 0.3

The fracture locations of the tensile specimens are all within the weld zone. In the base metal, coherently distributed $Al_3(Er,Zr)$ particles are uniformly dispersed within the matrix, resulting in fine microstructures and significant strengthening effects. In contrast, in the weld zone, work-hardening effects are absent. Additionally, most of the major alloying elements are dissolved within the supersaturated solid solution of Al, with only small quantities of precipitated Al_3Er and Al_3Sc phases in the joints with Wire-2 and Wire-3. Consequently, considering the strengthening effects of $Al_3(Er,Zr)$ particles through deformation and precipitation, the hardness of all joints is lower than that of the base metal (BM). The tensile properties of the joints are closely related to their grain size. The well-known Hall–Petch equation describes the contribution of grain size to yield strength, with a smaller grain size resulting in higher yield strength. The fine-grained structure of the WZ enhances the mechanical properties through the Hall–Petch effect, where smaller grain sizes lead to higher yield strength. The WZ of Wire-2 and Wire-3 joints exhibits higher microhardness (approximately 82 HV) and superior tensile strength (287 ± 6 MPa and 286 ± 6 MPa, respectively) compared to Wire-1 joints (78 HV and 261 ± 6 MPa). The fine equiaxed grains also contribute to improved ductility, as evidenced by the higher fracture elongation (14.8 ± 0.3% and 14.9 ± 0.3% for Wire-2 and Wire-3, respectively) [27–29]. These results indicate that the mechanical performance of joints with Wire-2 and Wire-3 is quite similar and superior to that of joints with Wire-1.

Figure 11 provides high-magnification micrographs of the fracture surfaces of tensile specimens from the joints. The fracture surfaces of all joints are predominantly composed of numerous dimples without prominent cleavage facets or quasi-cleavage features. This behavior is indicative of a typical ductile fracture, highlighting the good plasticity of the joints, with fracture elongations exceeding 10%.

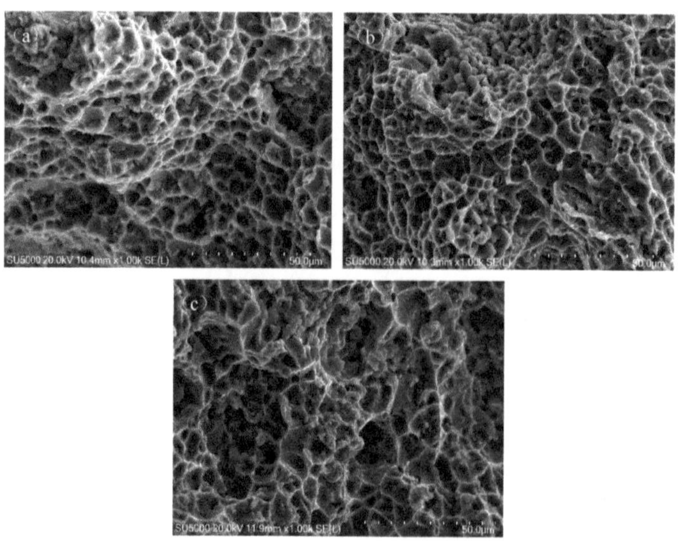

Figure 11. SEM micrographs of fracture surfaces of tensile samples: (**a**) Wire-1-welded joint; (**b**) Wire-2-welded joint; and (**c**) Wire-3-welded joint.

3.4. Corrosion Resistance

The corrosion resistance of aluminum alloys primarily relies on the presence of a passive oxide film on the surface and its capacity for self-repair when it is disrupted. To investigate the influence of the weld zone on the corrosion resistance of joints, the corrosion morphology of the samples in a 3.5% NaCl (by mass) solution and their electrochemical polarization curves were obtained, as illustrated in Figure 12. Employing electrochemical analysis software (CHI760E Electrochemical Testing Software), the self-corrosion potential and corrosion current density of the three welded joints were determined and are presented in Table 5. It is noteworthy that the corrosion current density for joints welded with Wire-1 is 9.266×10^{-7} A/cm^2, accompanied by a corrosion potential of −0.683 V. In contrast, the joints welded with Wire-2 and Wire-3 display corrosion current densities of 8.642×10^{-7} A/cm^2 and 8.775×10^{-7} A/cm^2, coupled with the corrosion potentials of −0.646 V and −0.650 V, respectively. It can be seen from the corrosion morphology that the corrosion area and corrosion depth of the welded joints after adding Er or Sc elements are greatly reduced.

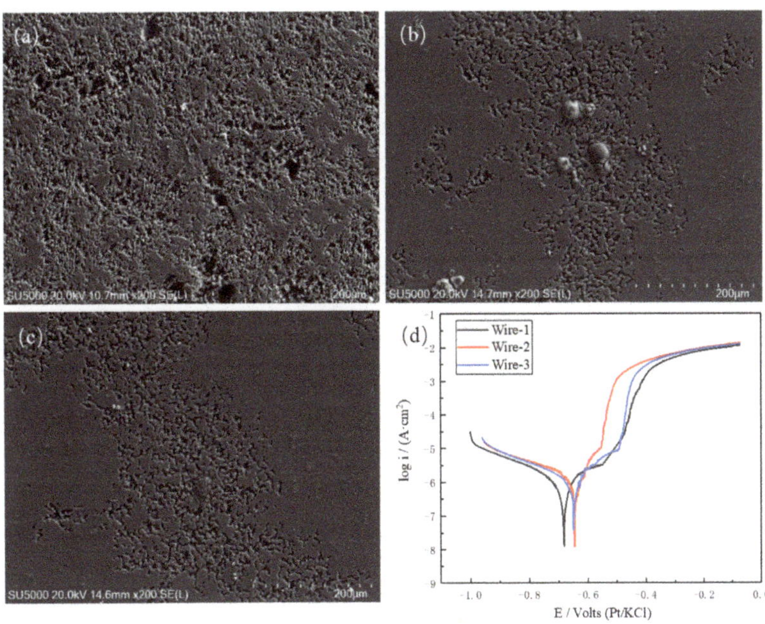

Figure 12. Corrosion morphology of different welded joints: (**a**) Wire-1; (**b**) Wire-2; (**c**) Wire-3; and (**d**) polarization curves of different welded joints.

Table 5. Electrochemical parameters of different WZs in wire joints were obtained from polarization curves immersed in 3.5% NaCl solution (by mass).

Sample	Ecorr (mV vs. Ag/AgCl)	Icorr (A cm^{-2})
Wire-1	−683	9.266×10^{-7}
Wire-2	−646	8.642×10^{-7}
Wire-3	−650	8.775×10^{-7}

Considering that Wire-2 and Wire-3 incorporate Er and Sc elements, the corrosion current density diminishes while the corrosion potential rises. In accordance with the connection between potential and equilibrium constants, an upward shift in the corrosion potential, specifically a reduction in its absolute value, correlates with a decrease in the equilibrium constant of the corrosion reaction. Consequently, corrosion reactions become less favorable. The corrosion current density is associated with the corrosion rate as follows [30]:

$$CR\ (mm/year) = 3.27\frac{M}{n\rho} \cdot i_{corr} \times 10^{-3}$$

The formula demonstrates a general relationship between the metal corrosion rate and corrosion current density (μA/cm^2). In the equation, M represents the atomic weight of the metal atoms (g/mol), n denotes the valency, i represents the current density, and ρ signifies the density of the alloy (g/cm^3). For the aluminum matrix, the numerical value of A/nρ is 3.33 cm^3/mol. Substituting this value into the above equation reveals that the corrosion rate of Wire-1, Wire-2, and Wire-3 is 1.009×10^{-8}, 9.411×10^{-9}, and 9.555×10^{-9} mm/year respectively. The addition of rare earth elements Er and Sc significantly reduces the corrosion rate of the welded joint.

The principal reason for the corrosion resistance of aluminum alloy surfaces is the presence of a protective Al_2O_3 film. The incorporation of Er and Sc leads to the formation of composite Er-Al and Sc-Al oxide films within the interstices of Al_2O_3 [31,32]. The positions occupied by Er and Sc enhance the binding energy for Al-O, resulting in a denser surface oxide film with greater stability, effectively suppressing cathodic reactions. Additionally, the addition of Er and Sc significantly refines the grain size in the weld zone, with the reduction in grain size contributing to a decrease in pinhole density on the oxide film surface. Accordingly, the addition of Er and Sc reduces the corrosion current density and raises the self-corrosion potential in the weld zone. Rare earth elements affect the properties of the oxide film on the surface of aluminum alloy, making the oxide film denser and preventing the dissolution of aluminum ions, so that the anodic dissolution reaction becomes a rate-controlling reaction.

Inclusions, such as oxides and intermetallic phases (e.g., Al_6(Mn,Fe)), act as preferential sites for corrosion initiation due to their higher electrochemical activity compared to the α-Al matrix. These inclusions can lead to localized pitting corrosion, particularly in chloride-containing environments. To mitigate this issue, the addition of Er and Sc in the filler wires (Wire-2 and Wire-3) promotes the formation of Al_3Er and Al_3Sc phases. These phases act as heterogeneous nucleation sites, refining the grain size and reducing the grain boundary area. This refinement lowers the diffusion rate of corrosive media along the grain boundaries, thereby enhancing the intergranular corrosion resistance of the alloy. Additionally, the presence of Sc and Er further stabilizes the microstructure, contributing to improved overall corrosion resistance. [17,18,33].

4. Conclusions

In this study, three different filler wires were employed for the MIG welding of 4 mm thick Al-Mg-Mn-Er-Zr cold-rolled plates, investigating the influence of adding 0.2% Er and 0.2% Sc to the filler wires on the microstructure, mechanical properties, and corrosion performance of the welded joints. The main conclusions are as follows:

(1) All the joints exhibit excellent external appearance without defects. The additions of Er and Sc refine the grains in the WZs from 47 μm to 29 μm and 31 μm, respectively. This refinement is primarily attributed to the presence of submicron-sized Al_3Er and Al_3Sc phases.

(2) The three weld joints possess ultimate tensile strengths of 261 ± 6 MPa, 287 ± 6 MPa, and 286 ± 6 MPa and fracture elongations of 12.8 ± 0.3%, 14.8 ± 0.3%, and 14.9 ± 0.3%, respectively. This improvement, due to the additions of Er and Sc, is mainly attributed to the combined effects of refinement strengthening and dispersion strengthening.

(3) The corrosion resistance of Al-Mg-Mn-Er-Zr alloy-welded joints is significantly influenced by grain size, weld morphology, and the presence of inclusions. In the weld zone (WZ), the fine-grained structure, achieved through the addition of Er and Sc, enhances corrosion resistance by stabilizing the passive oxide film and reducing the susceptibility to intergranular corrosion.

(4) With the addition of 0.2%, wire containing Er can effectively replace wires containing Sc, and has the advantage of low cost.

Author Contributions: Conceptualization, C.Z. and X.Y.; Methodology, C.Z. and X.Y.; Formal analysis, C.Z., X.Y. and Z.M.; Resources, L.H.; Data curation, C.Z. and Z.M.; Writing—original draft, C.Z. and X.Y.; Writing—review and editing, R.W. and L.H.; Visualization, C.Z.; Supervision, R.W. and X.Y.; Project administration, R.W. All authors have read and agreed to the published version of the manuscript.

Funding: This paper was supported by the National Natural Science Foundation of China (52261135538), Fundamental Research Funds for the Central Universities (3072024XX1009), and the Russian Science Foundation (23-49-00098).

Data Availability Statement: The original contributions presented in this study are included in the article. Further inquiries can be directed to the corresponding authors.

Conflicts of Interest: The authors declare no conflicts of interest.

References

1. Yin, Z.; Pan, Q.; Zhang, Y.; Jiang, F. Effect of minor Sc and Zr on the microstructure and mechanical properties of Al–Mg based alloys. *Mater. Sci. Eng. A* **2000**, *280*, 151–155. [CrossRef]
2. Wang, Y.; Zhang, S.; Wu, R.; Turakhodjaev, N.; Hou, L.; Zhang, J.; Betsofen, S. Coarsening kinetics and strengthening mechanisms of core-shell nanoscale precipitates in Al-Li-Yb-Er-Sc-Zr alloy. *J. Mater. Sci. Technol.* **2021**, *61*, 197–203. [CrossRef]
3. Wang, Y.; Zhang, Z.; Wu, R.; Sun, J.; Jiao, Y.; Hou, L.; Zhang, J.; Li, X.; Zhang, M. Ambient-temperature mechanical properties of isochronally aged 1420-Sc-Zr aluminum alloy. *Mater. Sci. Eng. A* **2019**, *745*, 411–419. [CrossRef]
4. Wang, M.; Wei, W.; Shi, W.; Zhou, X.; Wen, S.; Wu, X.; Gao, K.; Rong, L.; Qi, P.; Huang, H.; et al. Synergistic effect of Al_3(Er, Zr) precipitation and hot extrusion on the microstructural evolution of a novel Al–Mg–Si–Er–Zr alloy. *J. Mater. Res. Technol.* **2023**, *22*, 947–957. [CrossRef]
5. Kotov, A.D.; Mochugovskiy, A.G.; Mosleh, A.O.; Kishchik, A.A.; Rofman, O.V.; Mikhaylovskaya, A.V. Microstructure, superplasticity, and mechanical properties of Al–Mg–Er–Zr alloys. *Mater. Char.* **2022**, *186*, 111825. [CrossRef]
6. Nie, Z.R.; Jin, T.; Fu, J.; Xu, G.; Yang, J.; Zhou, J.X.; Zuo, T.Y. Research on rare earth in aluminum. *Mater. Sci. Forum* **2002**, *396*, 1731. [CrossRef]
7. Xue, D.; Wei, W.; Shi, W.; Guo, Y.W.; Wen, S.P.; Wu, X.L.; Huang, H.; Nie, Z.R. Effect of cold rolling on mechanical and corrosion properties of stabilized A-Mg-Mn-Er-Zr alloy. *J. Mater. Res. Technol.* **2021**, *15*, 6329–6339. [CrossRef]
8. Fu, L.; Li, Y.; Jiang, F.; Huang, J.; Xu, G.; Yin, Z. On the role of Sc or Er micro-alloying in the microstructure evolution of Al-Mg alloy sheets during annealing. *Mater. Char.* **2019**, *157*, 109918. [CrossRef]
9. Liu, Y.; Wang, W.; Xie, J.; Sun, S.; Wang, L.; Qian, Y.; Meng, Y.; Wei, Y. Microstructure and mechanical properties of aluminum 5083 weldments by gas tungsten arc and gas metal arc welding. *Mater. Sci. Eng. A* **2012**, *549*, 7–13. [CrossRef]
10. Gungor, B.; Kaluc, E.; Taban, E. Mechanical and microstructural properties of robotic cold metal transfer (CMT) welded 5083-H111 and 6082-T651 aluminum alloys. *Mater. Des.* **2014**, *54*, 207–211. [CrossRef]
11. Madavi, K.R.; Jogi, B.F.; Lohar, G.S. Metal inert gas (MIG) welding process: A study of effect of welding parameters. *Mater. Today: Proc.* **2022**, *51*, 690–698. [CrossRef]
12. Sharma, A.; Verma, A.; Vashisth, D.; Khanna, P. Prediction of bead geometry parameters in MIG welded aluminium alloy 8011 plates. *Mater. Today Proc.* **2022**, *62*, 2787–2793. [CrossRef]
13. Sareekumtorn, P.; Chaideesungnoen, S.; Muangjunburee, P.; Oo, H.Z. The electrochemical corrosion performance of aluminum alloys grade 6082-T6 weld repair. *Mater. Res. Express* **2024**, *11*, 086512. [CrossRef]
14. Zhao, H.; Li, M.; Wang, X.; Shi, Y.; Chang, Y.; Zhang, G.; Cai, B. Intergranular corrosion effect on fatigue behavior of dissimilar 6005A-5083 aluminum alloys weld joints. *Mater. Corros.* **2022**, *73*, 1505–1518. [CrossRef]
15. Rosalbino, F.; Angelini, E.M.M.A.; De Negri, S.; Saccone, A.; Delfino, S. Influence of the rare earth content on the electrochemical behaviour of Al–Mg–Er alloys. *Intermetallics* **2003**, *11*, 435–441. [CrossRef]

16. Yang, Z.; Ji, P.; Wu, R.; Wang, Y.; Turakhodjaev, N.; Kudratkhon, B. Microstructure, mechanical properties and corrosion resistance of friction stir welded joint of Al–Mg–Mn–Zr–Er alloy. *Int. J. Mater. Res.* **2022**, *114*, 65–76. [CrossRef]
17. Algendy, A.Y.; Rometsch, P.; Chen, X.G. Mechanical and corrosion performances of Al-Mg-Mn 5083 rolled alloys microalloyed with Sc and Zr in different thermomechanical processing conditions. *J. Mater. Sci.* **2024**, *59*, 14692–14715. [CrossRef]
18. Shi, J.; Hu, Q.; Zhao, X.; Liu, J.; Zhou, J.; Xu, W.; Chen, Y. Densification, Microstructure and Anisotropic Corrosion Behavior of Al-Mg-Mn-Sc-Er-Zr Alloy Processed by Selective Laser Melting. *Coatings* **2023**, *13*, 337. [CrossRef]
19. Shih, J.S.; Tzeng, Y.F.; Yang, J.B. Principal component analysis for multiple quality characteristics optimization of metal inert gas welding aluminum foam plate. *Mater. Des.* **2011**, *32*, 1253–1261. [CrossRef]
20. Deng, Y.; Xu, G.; Yin, Z.; Lei, X.; Huang, J. Effects of Sc and Zr microalloying additions on the recrystallization texture and mechanism of Al–Zn–Mg alloys. *J. Alloys Compd.* **2013**, *580*, 412–426. [CrossRef]
21. Geng, S.; Jiang, P.; Shao, X.; Guo, L.; Gao, X. Heat transfer and fluid flow and their effects on the solidification microstructure in full-penetration laser welding of aluminum sheet. *J. Mater. Sci. Technol.* **2020**, *46*, 50–63. [CrossRef]
22. Han, C.; Jiang, P.; Geng, S.; Mi, G.; Wang, C.; Li, Y. Nucleation mechanisms of equiaxed grains in the fusion zone of aluminum-lithium alloys by laser welding. *J. Mater. Res. Technol.* **2021**, *14*, 2219–2232. [CrossRef]
23. Samiuddin, M.; Li, J.L.; Taimoor, M.; Siddiqui, M.N.; Siddiqui, S.U.; Xiong, J.T. Investigation on the process parameters of TIG-welded aluminum alloy through mechanical and microstructural characterization. *Def. Technol.* **2020**, *17*, 1234–1248. [CrossRef]
24. Tong, X.; Wu, G.; Zhang, L.; Wang, Y.; Liu, W.; Ding, W. Microstructure and mechanical properties of repair welds of low-pressure sand-cast Mg-Y-RE-Zr alloy by tungsten inert gas welding. *J. Magnesium Alloys* **2022**, *10*, 180–194. [CrossRef]
25. Zhang, X.; Wang, H.; Yan, B.; Zou, C.; Wei, Z. The effect of grain refinement and precipitation strengthening induced by Sc or Er alloying on the mechanical properties of cast Al-Li-Cu-Mg alloys at elevated temperatures. *Mater. Sci. Eng. A* **2021**, *822*, 141641. [CrossRef]
26. Xiang, H.; Liu, P.L.; Huang, Y.; Liu, Z.H.; Deng, S.X.; Liu, T.L.; Li, J.F.; Liu, D.Y. Effects of the filler wire composition on the structures and mechanical performance of 2195 AlLi alloy TIG joints. *Mater. Charact.* **2023**, *198*, 112748. [CrossRef]
27. Zhang, D.K.; Wu, A.; Zhao, Y.; Shan, J.; Wan, Z.; Wang, G.; Song, J.; Zhang, Z.; Liu, X. Microstructural evolution and its effect on mechanical properties in different regions of 2219-C10S aluminum alloy TIG-welded joint. *Trans. Nonferrous Metals Soc. China* **2020**, *30*, 2625–2638. [CrossRef]
28. Jia, Q.; Rometsch, P.; Kürnsteiner, P.; Chao, Q.; Huang, A.; Weyland, M.; Bourgeois, L.; Wu, X. Selective laser melting of a high strength Al-Mn-Sc alloy:alloy design and strengthening mechanisms. *Acta Mater.* **2019**, *171*, 108–118. [CrossRef]
29. Yang, Z.; Liao, Y.; Wu, R.; Sun, D.; Liu, M.; Wang, Y. Effect of Single-Pass Large-Strain Rolling on Microstructure and Mechanical Properties of Al-3Li-1Cu-0.2Er-0.1Zr Alloy. *J. Mater. Eng. Perform.* **2022**, *31*, 3287–3298. [CrossRef]
30. *ASTM G102-89*; Standard Practice for Calculation of Corrosion Rates from Electrochemical Measurements. ASTM International: West Conshohocken, PA, USA, 2015.
31. Ahmad, Z.; UI-Hamid, A.; Abdul-Aleem, B.J. The corrosion behavior of scandium alloyed Al 5052 in neutral sodium chloride solution. *Corros. Sci.* **2001**, *43*, 1227–1243. [CrossRef]
32. Chen, H.; Yu, T.; Qi, Z.; Wu, R.; Wang, G.; Lv, X.; Cong, F.; Hou, L.; Zhang, J.; Zhang, M. Effect of Minor Er on the Microstructure and Properties of Al-6.0Mg-0.4Mn-0.1Cr-0.1Zr Alloys. *J. Mater. Eng. Perform.* **2018**, *27*, 5709–5717. [CrossRef]
33. Xing, Q.; Wu, X.; Zang, J.; Meng, L.; Zhang, X. Effect of Er on Microstructure and Corrosion Behavior of Al–Zn–Mg–Cu–Sc–Zr Aluminum Alloys. *Materials* **2022**, *15*, 1040. [CrossRef] [PubMed]

metals

Article

Influence of Post-Weld Heat Treatment on S960QL High-Strength Structural Steel Electron-Beam-Welded Joint

Raghawendra Pratap Singh Sisodia [1,*], Piotr Sliwinski [2], Dániel Koncz-Horváth [3] and Marek St. Węglowski [2]

1 Institute of Materials Science and Technology, Faculty of Mechanical Engineering and Informatics, University of Miskolc, 3515 Miskolc, Hungary
2 Łukasiewicz—Upper Silesian Institute of Technology, Karola Miarki Str. 12-14, 44-100 Gliwice, Poland; piotr.sliwinski@git.lukasiewicz.gov.pl (P.S.); marek.weglowski@git.lukasiewicz.gov.pl (M.S.W.)
3 Institute of Physical Metallurgy, Metal Forming and Nanotechnology, University of Miskolc, 3515 Miskolc, Hungary; daniel.koncz-horvath@uni-miskolc.hu
* Correspondence: raghawendra.sisodia@uni-miskolc.hu

Abstract: Electron beam welding (EBW) is one of the most highly precise methods that is gaining more importance in high-strength structural steel (HSSS) thicker plate application in various vehicles, construction industries, etc. Since it offers particular advantages over arc welding processes like narrow welds, reduced heat-affected zone (HAZ), and low distortion, it inherits lower linear heat input characteristics. The main purpose of this study is to analyze and compare the effect of localized electron beam–post-weld heat treatment (LEB-PWHT) with that of an as-welded EB-welded S960QL joint of a thickness of 12 mm for various joint and HAZ properties. LEB-PWHT can be beneficial in terms of time saving, more local treatment, higher flexibility, energy saving, greater efficiency, increased productivity, etc. In this study, LEB-PWHT was applied to an autogenous EB-welded S960QL joint using a defocused beam. Microstructural characteristics were observed through light optical and scanning electron microscopy (SEM) while mechanical properties, including microhardness, tensile strength, bending, and Charpy V-notch (CVN) impact test, are compared in as-welded and LEB-PWHT joints. The microstructural results showed that the EBW coarse-grain heat-affected zone (CGHAZ) consists of martensite, while the PWHT weld metal contains tempered martensite with carbide precipitates. The fine-grain heat-affected zone (FGHAZ) of EBW exhibits a martensitic and bainitic microstructure, whereas the FGHAZ of the PWHT joint exhibits equiaxed grain with finely dispersed carbides. The hardness decrease after LEB-PWHT in the weld metal and HAZ was approximately 23% and 21%, respectively. An increase in tensile strength (3%) was observed in the LEB-PWHT joints (1082 MPa) compared to the EBW joint (1051 MPa). Both tensile and bending tests demonstrated improved ductility behavior after PWHT. However, the impact test at −40 °C indicated a reduction in toughness in the weld metal of LEB-PWHT (27 J) compared to EBW (63 J).

Keywords: electron beam welding; S960QL; local electron beam–post-weld heat treatment; microstructural characterization; mechanical properties

1. Introduction

High-strength steels (HSSs) are widely used in today's steel structural applications due to various factors such as sustainability, climate change, and higher weight reduction with increased strength [1,2]. To meet these requirements, several researchers are looking for an efficient method of producing HSSs with increasing strength and toughness using new manufacturing methods, the careful selection of alloy content, or a combination of both. For various delivery conditions of high-strength structural steels (HSSSs), the following technological processes are applied in their manufacturing: rolled (AR), normalized (N), quenched and tempered (Q + T), and thermomechanically rolled (TM) [3]. Since the introduction of the Q + T steel group in the 1970s, technological breakthroughs and rising demand for HSSSs have resulted in considerable increases in yield strength. Today, the maximum yield strength has been increased to 1300 MPa by combining various alloying materials and optimizing Q + T heat treatment methods [4]. The addition of alloying elements to steels increases their strength and hardenability, which, in turn, increases the risk of brittle fracture, especially if the optimal welding parameters are not used [5]. However, it should be noted that filler materials (FMs) are currently available in the market with yield strengths of up to 1100 MPa, so for these extreme strength HSSSs, only undermatching (UM) (approximately 15–20 percent) is available if it is permitted. S960QL is more widespread in industrial applications like cranes [6], bridges, and constructional equipment industries like bulldozers [4,7,8]. Until now, EN 10025-6 [9] has standardized Q + T steel grades with yield strength grades up to 960 MPa, but the S1100 and S1300 grades are not included.

The main goal in the development of various HSSS is an overall "property package" of very appealing mechanical properties, which has resulted in widespread adoption and piqued the interest of several industries. However, it is important to use appropriate welding processes that can meet the weldability criteria, welded joint strength requirements, etc. [10]. The wider applications of these steels would necessitate advancements in welding and joining technology, as well as pre- and post-weld heat treatment processes. As a result, the weldability study is critical, as it is one of the key technological characteristics of structural steels that is influenced by processing, carbon concentration, and alloying element composition [11]. Difficulties arise in the welding of HSSSs due to their inhomogeneous grain structures resulting from the refining method used in the production of the steels. One of the most important points to consider for weldability in the case of these HSSSs is the heat input (Q), which can be described with linear energy. If this value is too low, the cooling rate of the welded joint may be too fast, resulting in cold cracks [12,13]. In the opposite case, a strong coarse-grain microstructure can form in the heat-affected zone (HAZ), which decreases the strength and toughness features [14,15]. As a result, the welded joint's HAZ has a highly inhomogeneous microstructure and different mechanical properties than the base material (BM). Higher hardness peaks and brittle parts formed locally in the HAZ as a result of the thermal cycle can be sites of crack formation, especially when combined with a sufficient diffusible hydrogen content [16]. Furthermore, the additional internal stresses caused by welding, combined with the low deformability of HSSs, increase the tendency of cracks to form.

So, the appropriate low linear heat input will be the best option, and the electron beam welding (EBW) process is the most appropriate and unique technology under the category of high-energy beam welding processes which encompasses several unparallel benefits due to its owing characteristic features. Due to its high energy density, thicker plates can be welded by the keyhole technique with full penetration without using filler materials in a single pass. The use of this high-energy-density welding process not only produces quality joints that perform their intended function, but it also reduces the area influenced by heat, i.e., the HAZ, to achieve mechanical properties similar to the BM. It has the ability to weld with extremely narrow HAZ, deep penetrated welds, and low heat distortions [17–19]. This is a significant difference from the way arc welds fuse together and is caused by the unique penetration mechanism of beam welding processes [20]. In EBW, the cooling time from 800 °C to 500 °C ($t_{8/5}$) is usually between 2 and 5 s due to the high energy density, and the hardness peaks in the HAZ are even higher due to the brittle martensite microstructure in the HAZ of HSSSs with yield strengths between 690 MPa and 1300 MPa and high carbon equivalent (CE) values [21]. Therefore, preheating or post-weld heat treatment may be necessary to reduce HAZ hardness. Localized electron beam–post-weld heat treatment (LEB-PWHT) can be effective in reducing the hardness peaks across the HAZ sub-zones (coarse-grained heat-affected zone (CGHAZ), fine-grained heat-affected zone (FGHAZ), and inter-critical heat-affected zone (ICHAZ)) [22]. LEB-PWHT also induces microstructural change in the welded joint, which improves ductility and reduces the risk of brittle fracture, increases the fatigue resistance [23], and improves the impact toughness and tensile residual stresses (TRSs) [24]. However, its primary benefits include high purity, local PWHT, higher precision, energy saving, time saving, flexibility, greater efficiency, and higher productivity [25]. The scientific novelty of this work lies in its specialized investigation of LEB-PWHT on EB-welded S960QL steel, an HSSS with limited prior research on this topic. This study uniquely examines how localized EB-PWHT affects the microstructure and mechanical properties of S960QL, providing new insights through microstructural characterization including hardness, tensile, bending, and toughness tests. These findings fill a significant gap in the literature, contributing valuable data and methodologies for optimizing localized EB-PWHT procedures and establishing standards for HSSS applications.

In this study, an autogenous S960QL joint was produced by the EBW process, followed by LEB-PWHT with a defocused electron beam. Mechanical and microstructural analyses were conducted, and the results of PWHT-welded joints were compared with those of EB-welded S960QL joints to evaluate the effects of localized heat treatment.

2. Materials and Methods

2.1. The Investigated Base Material

The base material DILLIMAX S965T, Dillinger, Dillingen, Germany (S960QL in standard EN 10025-6, quenched and tempered steel) with a plate thickness of 12 mm was used for the experiments. The sample was tack-welded at a few positions along the weld length by manual tungsten inert gas (TIG) welding. The chemical composition and mechanical properties of the base material are presented in Table 1 and Table 2, respectively. The mentioned details are from the material's data sheet [26].

Table 1. Chemical composition of the investigated S960QL steel in wt%. Reprinted from Ref. [26].

C	Si	Mn	P	Cr	Ni	Mo	V + Nb	CEV/CET
≤0.20	≤0.50	≤1.40	≤0.018	≤0.90	≤2.0	≤0.70	≤0.10	0.70/0.43

Table 2. Mechanical properties of the investigated S960QL steel. Reprinted from Ref. [26].

Material	$R_{p0.2}$ MPa	R_m MPa	A_5 %	CVN (at −40 °C) J	HV0.1
S960QL	≥960	980–1150	12	≥27	* 380

* Lab measurement data.

The carbon equivalent values for the S960QL base material are CEV = 0.70 and CET = 0.43 (material certificate [26]). Figure 1a, 1b and 1c show scanning electron microscopy (SEM), inverse pole figure (IPF), and image quality (IQ) photos of the S960QL base material (BM), respectively. The microstructure of the S960QL base material, as received, is composed of tempered martensite (TM) and bainite (B).

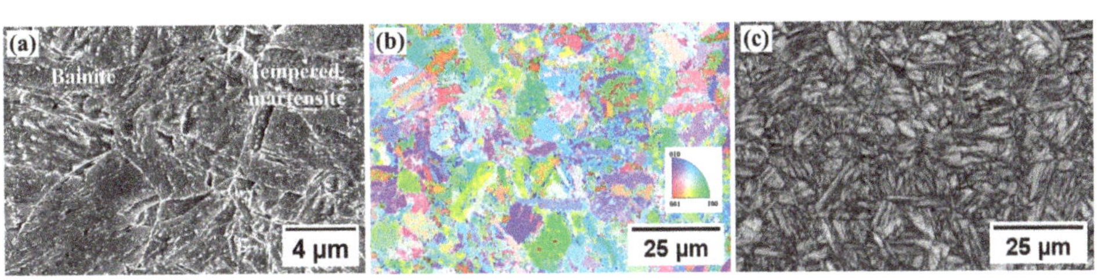

Figure 1. Base material (S960QL) microstructure: (**a**) SEM (M = 10,000×), (**b**) IPF, and (**c**) IQ micrographs (after 2% NITAL chemical etching).

2.2. Welding Experiments

2.2.1. Electron Beam Welding

EBW and LEB-PWHT were performed in situ on a 12 mm thick plate by means of an electron beam welding machine at the Lukasiewicz—Upper Silesian Institute of Technology, Gliwice, Poland. The S960QL HSSs (compliance to EN 10025-6) plates with dimensions of 350 mm × 150 mm × 12 mm in two pieces for a butt-welded (BW) joint were used. EBW was performed in a vacuum with a chamber pressure of 1×10^{-4} mbar and a gun pressure of 1×10^{-5} mbar. Before welding, joints were prepared carefully to avoid any defects or misalignments. The joining surface was initially cleaned mechanically using sandpaper and a steel wire brush to remove impurities, followed by cleaning with acetone to remove any remaining residual particles. The EB-welded and LEB-PWHT joints are shown in Figure 2a,b. Figure 2c,d show the thermal cycle measurement set-up and the thermocouple arrangement on both the top and bottom surfaces.

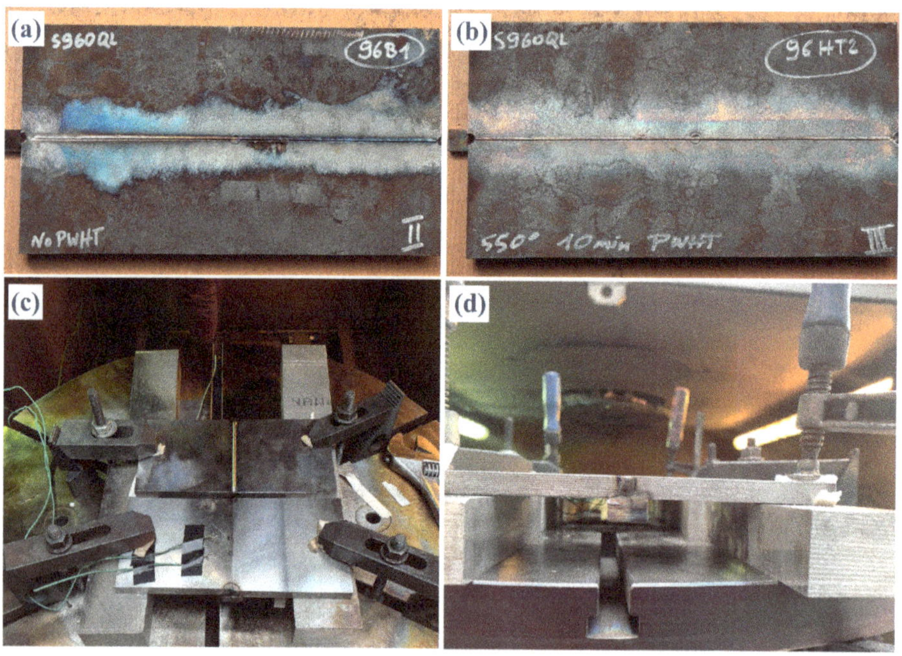

Figure 2. (**a**) Electron beam welded joint; (**b**) LEB-PWHT joint; (**c**) clamping and thermocouples for the thermal cycle on the top of the joint; (**d**) thermocouple arrangement on the bottom of the joint for temperature measurement during the PWHT process.

The optimal parameters for EBW are presented in Table 3.

Table 3. EBW optimal parameters.

Steel	Process	V_a (kV)	I_b (mA)	v (mm/s)	d_b (mm)
S960QL	EBW	140	28	10	0.3

V_a = accelerating voltage; I_b = beam current; v = welding speed; d_b = beam diameter.

The measured cooling time (with thermocouple arrangement) at a distance of 1.55 mm from the weld center was found to be 2.24 s (Figure 3a). Welding was performed at a speed (v) of 10 mm/s, an accelerating voltage (V_a) of 140 kV, a beam diameter (d_b) of 0.3

mm, and a beam current (I_b) of 28 mA. The linear heat input was calculated using Equation (1) [27], with parameters provided in Table 3 and efficiency (η = 0.9) [28] as 0.353 kJ/mm.

$$Q = \eta \frac{V_a I_b}{v} \quad (1)$$

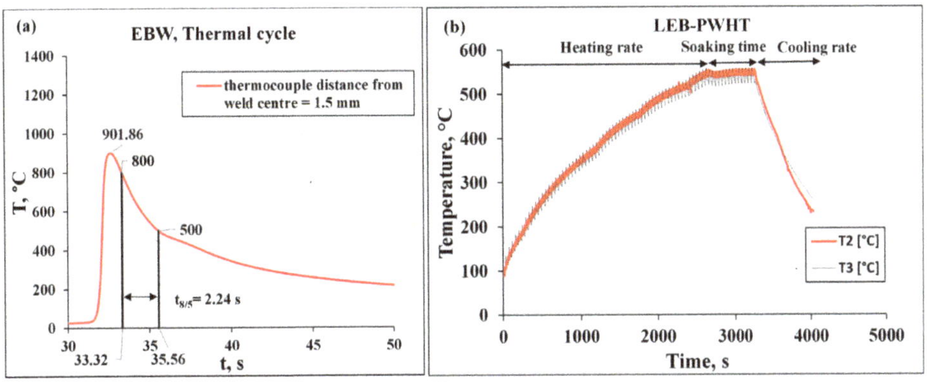

Figure 3. Thermal cycle: (**a**) EBW and (**b**) LEB-PWHT.

2.2.2. Local Electron Beam–Post-Weld Heat Treatment (LEB-PWHT)

LEB-PWHT was carried out with great care and special arrangements. In performing LEB-PWHT, a temperature of 550 °C was selected. According to the AWS [29], the maximum allowable heat treatment temperature for quenched and tempered steels is 600 °C. Zhao et al. [30] mentioned in their study that normal-strength steel heat treatment temperatures range from 600 to 650 °C. During the heat treatment of steels, the maximum soaking or holding temperature and the soaking or holding time at the maximum soaking temperature are the two most important factors that would influence the final mechanical properties of the steel under heat treatment [31]. In this study, a soaking time of 10 min was selected for LEB-PWHT. The heat treatment cycle for LEB-PWHT is shown in Figure 3b. To measure the heating temperature at full penetration, thermocouples T2 and T3 were welded at two different locations along the weld joint on the bottom of the plate. The sample travel speed during LEB-PWHT (which corresponds to the welding speed during the EBW process) was 12.5 mm/s. The detailed parameters for LEB-PWHT are presented in Table 4.

Table 4. The local electron beam–post-weld heat treatment (LEB-PWHT).

Steel	Process	V_a, (kV)	I_b, (mA)	v, (mm/s)	S_t, (min)	S_{temp}, (°C)
S960QL	LEB-PWHT	140	8	12.5	10	550

V_a = accelerating voltage; I_b = beam current; v = sample travel speed; S_t = soaking time; S_{temp} = soaking temperature.

For LEB-PWHT, an oscillating defocused beam (100 mA on focusing coil in relation to the focus point above the surface) with a frequency of 20 Hz and a heating field of 40 mm × 40 mm with 200 × 200 points was used. The defocused beam current (I_{bf}) for the LEB-PWHT process was 640 mA. During the PWHT process, a beam current of 12 mA (with a 140 kV accelerating voltage) was applied for the first 1200 s. The current was then increased to 14 mA for the next 600 s. From 1800 s to 2600 s, the current was further raised to 18 mA, which helped the material to gain the required temperature (550 °C) faster. Following this, the beam current was reduced to 16 mA to maintain a temperature of approximately 550 °C from 2600 s until the end, ensuring that the welded joint remained at this temperature for the entire heat treatment process (with a soaking time of 10 min). Therefore, during the process, the sequence for the holding time and beam currents is as follows: 0–1200 s → 12 mA; 1200–1800 s → 14 mA; 1800–2600 s → 18 mA; and 2600 s–end → 16 mA.

After welding and PWHT, the microstructural and mechanical properties of the EBW joints were investigated. The samples for light optical microscopy (LOM) and scanning electron microscopy (SEM) observations were sectioned transversely through the weld. The sectioned samples were polished with SiC waterproof papers of 120, 400, 800, and 2000 ANSI grits, followed by a disk of 1 μm diamond paste. The specimens were then etched for 10 s with 2% Nital (2% HNO_3). The microstructural examinations were carried out using an LOM (Axio Observer D1m, Zeiss, Oberkochen, Germany) and SEM (ThermoScientific Helios G4 PFIB CXe, ThermoScientific, Walthum, MA, USA; Zeiss Evo MA10 Carl Zeiss Microscopy GmbH, Jena, Germany). The microhardness tests were performed using KB Prüftechnik KB50BVZ-FA (KB Prüftechnik, Hochdorf-Assenheim, Germany) automated digital Vickers tester with a 100 gm (HV0.1) load and a 10 s dwell time in accordance with EN ISO 9015-2 [32]. The hardness measurement was performed 5 mm below the top edge of the specimen. Transverse (perpendicular to the welding direction) tensile tests on EB welded and LEB-PWHT joints were conducted in accordance with the EN ISO 4136 [33] standard. An MTS Criterion Model 64 (MTS Systems, Eden Prairie, MN, USA) machine with a maximum capacity of 600 kN was utilized to load these specimens. The bending tests were performed according to EN ISO 5173 [34]. The impact tests were performed with a LabTest CHK 450J-I machine (Labortech, s.r.o., Rolnická, Opava, Czech Republic) in accordance with the EN ISO 148-1 [35] and EN ISO 9016 [36] standards. Three samples each of BM and weld metal were used to perform Charpy V-notch (CVN) impact tests with dimensions of 10 mm × 10 mm × 55 mm, and the test was performed at −40 °C. Specimens were incised in the weld metal.

3. Results and Discussion

3.1. Microstructural Tests

Figure 4a,b depict a typical light optical panoramic cross-sectional micrograph of the S960QL EB-welded and PWHT joints, respectively, demonstrating the changes in microstructure from the base material to the weld center throughout the weld (roughly in the mid-thickness of the welded cross-section).

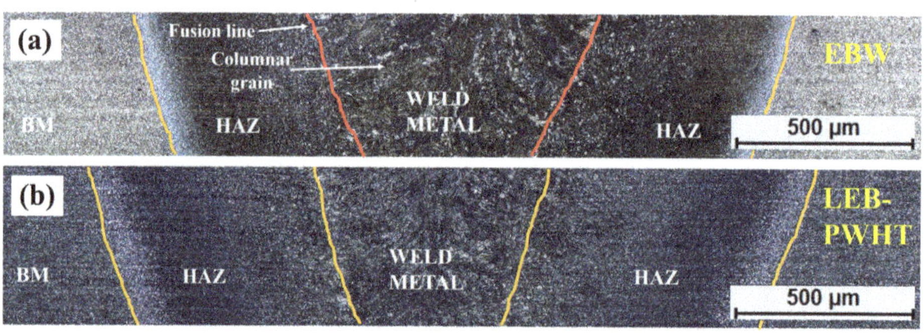

Figure 4. Light optical microscopic panoramic photo of BM, weld metal cross-section, and HAZ: (**a**) EBW and (**b**) LEB-PWHT.

The microstructure of the weld metal for EBW exhibits finer and columnar characteristics (Figure 4a), as evidenced by the faster cooling rate and subsequently shorter cooling time (2.24 s) associated with the process involved. Marcell and András [37] investigated the gas metal arc welding (GMAW) of Q + T steel (S960QL) and found that a linear energy setting of 0.7 kJ/mm for filler passes leads to a longer cooling time ($t_{8/5}$) limit of 5–6 s. The columnar grains grow from the fusion boundary in a direction parallel to and opposite to the heat flow (Figure 4a). The solidification begins from the fusion boundary, adjacent to BM, which acts as a heat sink and proceeds toward the weld center due to a preferential growth mechanism, resulting in a columnar structure.

The light optical micrograph location taken on the transverse cross-section of the welded joint of the EBW and LEB-PWHT processes is shown in Figure 5.

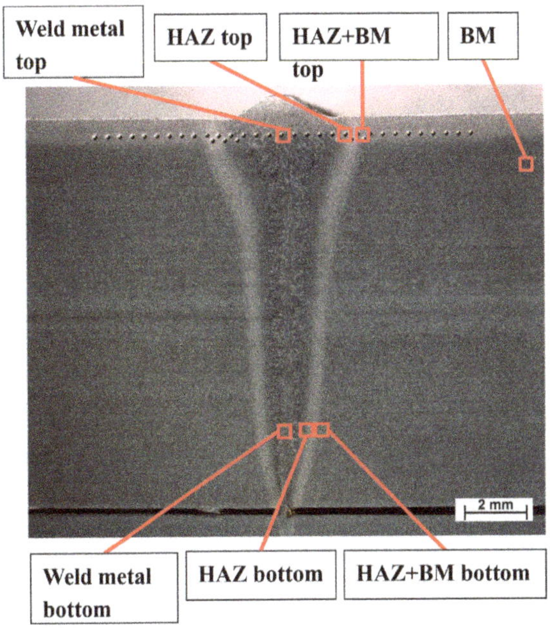

Figure 5. EBW and LEB-PWHT light optical micrograph location on the top and bottom face of the transverse weld cross-section.

Figures 6a–f and 7a–f depict the light optical microstructure of the weld metal, HAZ, and HAZ-BM interface for the top and bottom faces of the EBW and LEB-PWHT process transverse cross-sections of the weld.

Figure 6a,d show that the weld metal in EBW exhibits a martensitic microstructure. However, in LEB-PWHT, Figure 7a,d shows that the weld metal undergoes tempering of the microstructure. In the HAZ, Figure 6b,e show a coarse, lath-like martensitic microstructure with visible boundaries, while Figure 7b,e demonstrate that after LEB-PWHT, the HAZ appears more homogenous and refined due to the tempering effect on the martensite. The interface between the HAZ and BM under LEB-PWHT conditions, as seen in Figure 7c,f), shows a smoother, less pronounced transition than in EBW, where Figure 6c,f, shows a distinctive boundary between the HAZ and the BM.

The HAZ consists of four sub-zones, namely, coarse-grained HAZ (CGHAZ), fine-grained HAZ (FGHAZ), inter-critical HAZ (ICHAZ), and sub-critical heat-affected zone (SCHAZ) [38]. Figure 8 shows the SEM microstructure of the weld metal and CGHAZ, while Figures 9 and 10 present SEM images illustrating the spot energy spectrum analysis at the reference point (RP) and the carbide locations (C1 and C2) in the weld metal and CGHAZ of the S960QL LEB-PWHT joint, respectively. In Figure 8c, the weld metal of the EBW joint shows a martensitic structure, while the PWHT weld metal (Figure 8d) shows a more equiaxed and uniform microstructure compared to the as-welded microstructure. Also, carbide precipitation in the weld metal of LEB-PWHT is visible in Figure 9. Figure 8g shows the martensitic microstructure of the CGHAZ of EBW and Figure 8h shows the tempered microstructure of the CGHAZ after heat treatment, which consists mainly of carbides. The carbides appear very fine and evenly dispersed within the tempered martensite; however, coarser carbide precipitation can also be observed along the lath boundaries (Figures 8 and 10). As a result, the coarse carbides pose a risk to the material's impact toughness.

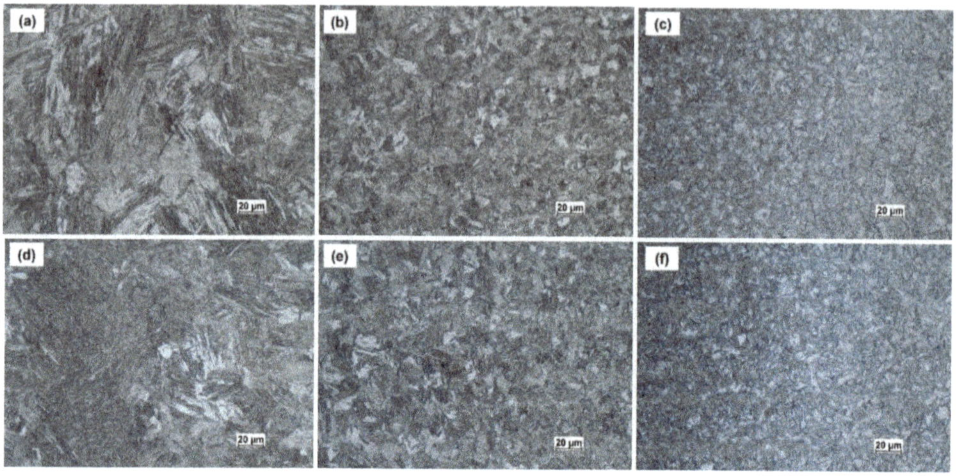

Figure 6. S960QL, EBW, optical micrograph. Top: (**a**) Weld metal, (**b**) HAZ, (**c**) HAZ-BM. Bottom: (**d**) Weld metal, (**e**) HAZ, (**f**) HAZ-BM. M = 500× (after 2% NITAL chemical etching).

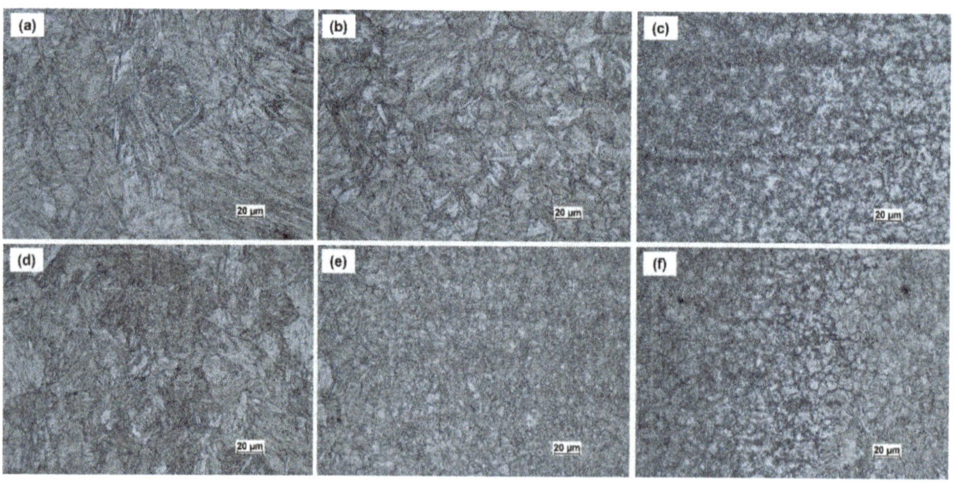

Figure 7. S960QL, LEB-PWHT, optical micrograph. Top: (**a**) Weld metal, (**b**) HAZ, (**c**) HAZ-BM. Bottom: (**d**) Weld metal, (**e**) HAZ, (**f**) HAZ-BM. M = 500× (after 2% NITAL chemical etching).

Figure 11 illustrates the SEM images of the FGHAZ and SCHAZ for both the EBW and LEB-PWHT joints, while Figure 12 presents SEM images with spot energy spectrum analysis at the reference point (RP) and carbide locations (C1 and C2) in the SCHAZ of the S960QL LEB-PWHT joint. From Figures 8 and 11, it is evident that the grain sizes in the CGHAZ are bigger than those in the FGHAZ of the EBW joint. In Figure 11, the FGHAZ of EBW clearly shows a martensitic and bainitic microstructure, while the FGHAZ of PWHT exhibits a martensitic structure with finely dispersed carbides and equiaxed grains. In the case of EBW, the SCHAZ microstructure morphology appears more like spheroidal, which can be attributed to the lower temperature of sub-critical range. However, after PWHT, more finely dispersed carbide precipitates are observed alongside coarse carbides, as shown in Figure 12.

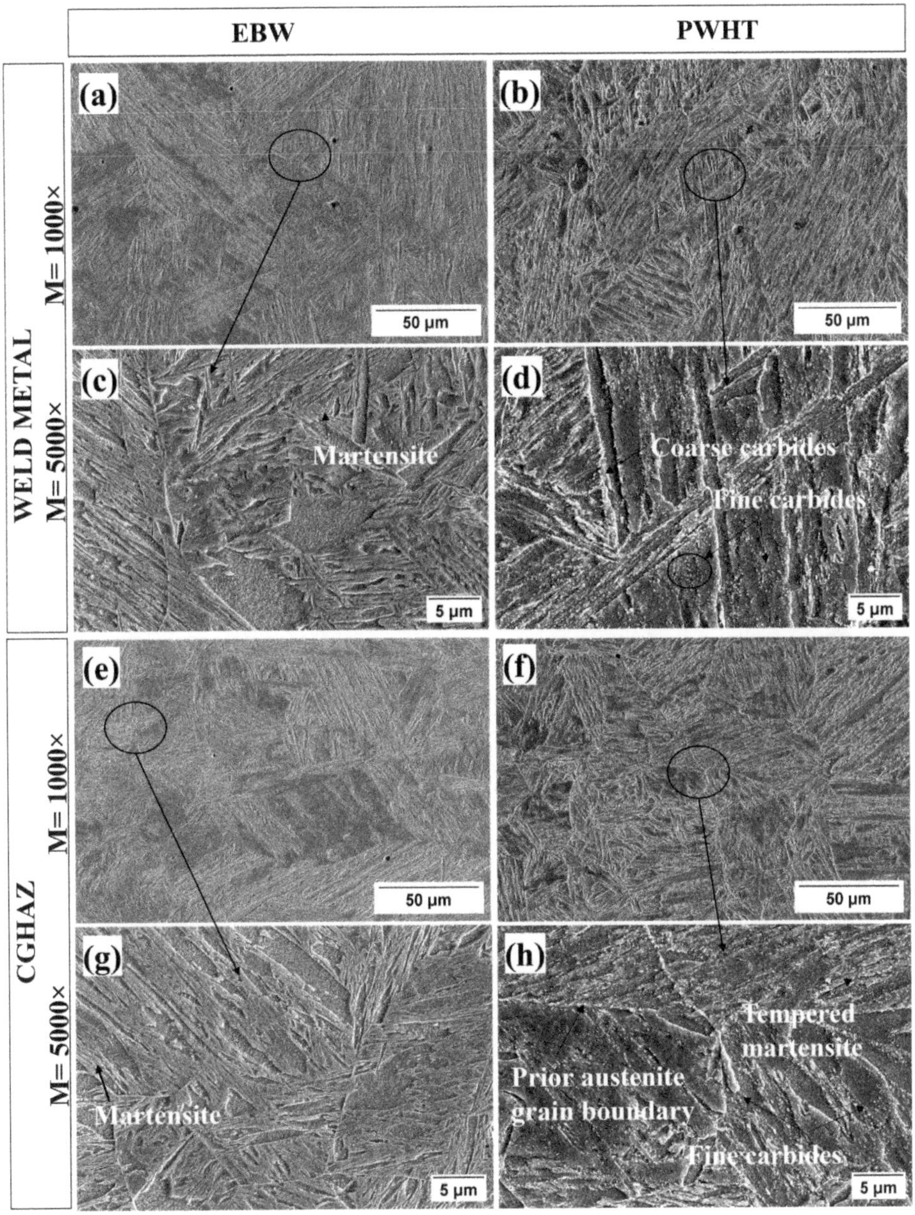

Figure 8. S960QL SEM micrographs of normal EBW and LEB-PWHT specimens: (**a**,**b**) weld metal, M = 1000×; (**c**,**d**) weld metal, M = 5000×; (**e**,**f**) CGHAZ, M = 1000×; and (**g**,**h**) CGHAZ, M = 5000×.

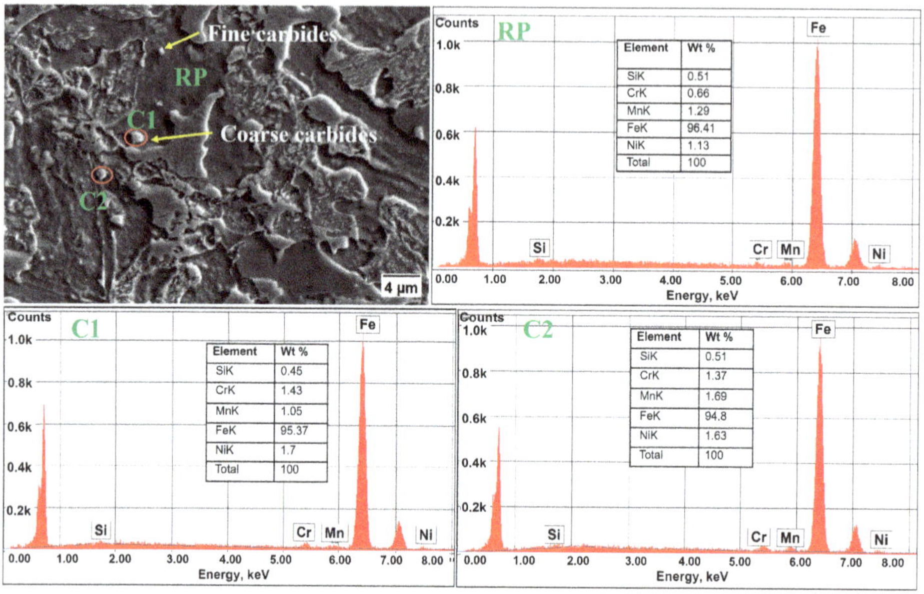

Figure 9. Energy spectrum analysis of S960QL LEB-PWHT joint. Weld metal SEM micrograph, M = 2500×; reference point (RP) = EDS spot RP; carbide (C1) = EDS spot C1; carbide (C2) = EDS spot C2.

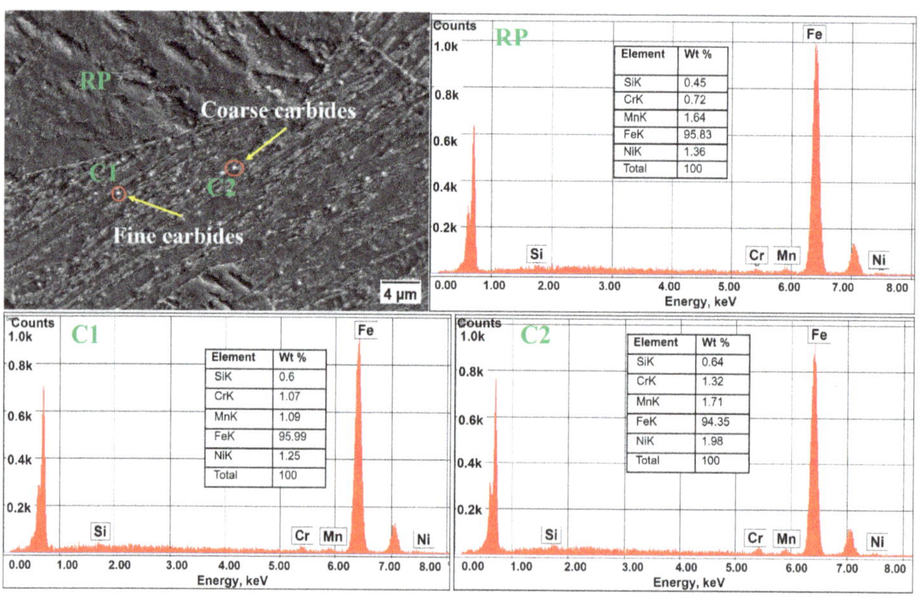

Figure 10. Energy spectrum analysis of S960QL LEB-PWHT joint. CGHAZ SEM micrograph, M = 2500×; reference point (RP) = EDS spot RP; carbide (C1) = EDS spot C1; carbide (C2) = EDS spot C2.

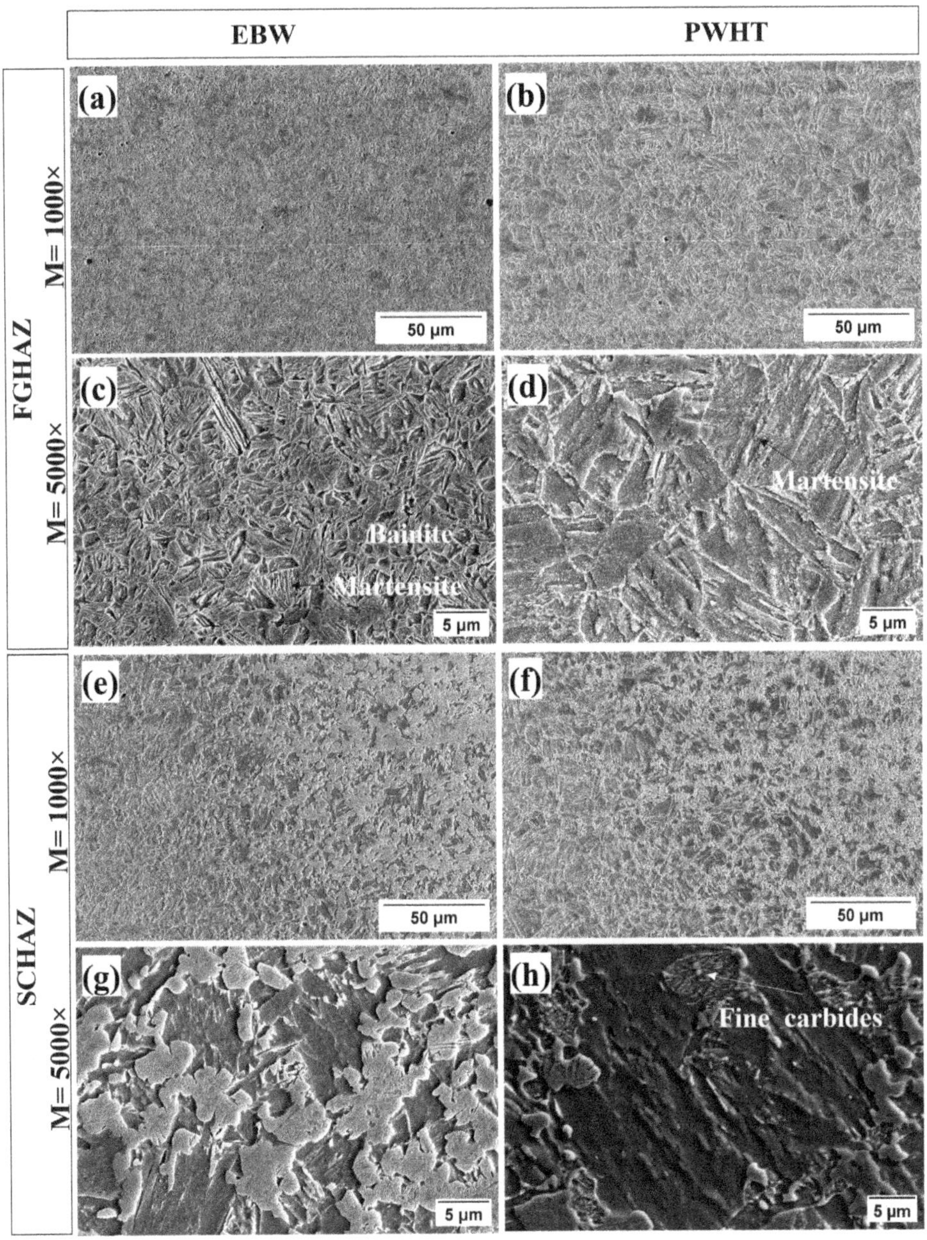

Figure 11. S960QL SEM micrographs of normal EBW and PWHT specimens: (**a**,**b**) FGHAZ, M = 1000×; (**c**,**d**) FGHAZ, M = 5000×; (**e**,**f**) SCHAZ, M = 1000×; and (**g**,**h**) SCHAZ, M = 5000×.

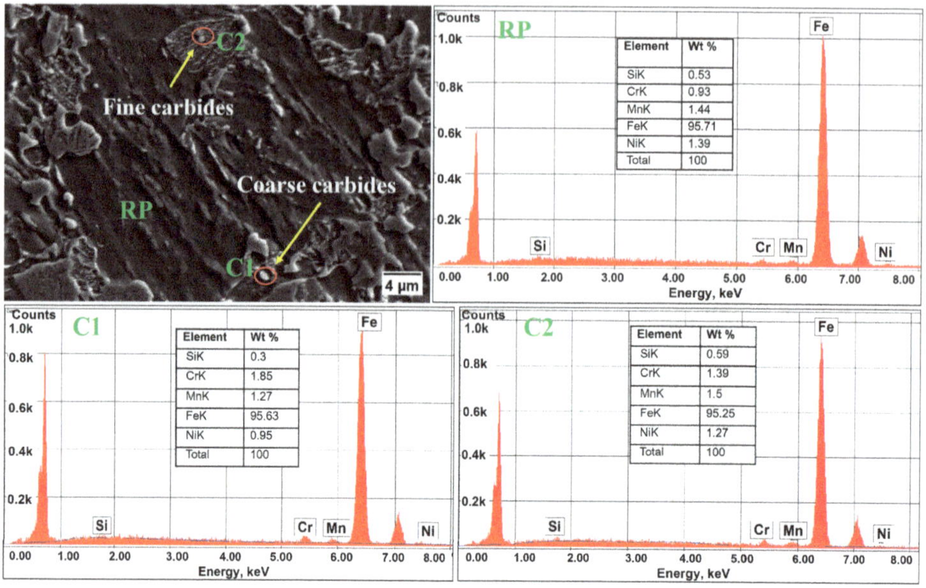

Figure 12. Energy spectrum analysis of S960QL LEB-PWHT joint. SCHAZ SEM micrograph, M = 2500×; reference point (RP) = EDS spot RP; carbide (C1) = EDS spot C1; carbide (C2) = EDS spot C2.

The energy spectra of the S960QL LEB-PWHT joint were evaluated using energy-dispersive spectroscopy (EDS) for the weld metal, CGHAZ, and SCHAZ, as shown in Figures 9, 10 and 12, respectively. EDS spot analyses were conducted at the reference points (RP) in each zone and at two locations (C1 and C2) of carbide precipitates.

3.2. Hardness Test

The distribution of Vicker's microhardness across the weld cross-section was examined for both EBW and LEB-PWHT joints. The microhardness of the base material (S960QL) was measured at 380 ± 10 HV0.1. The average measured hardness for the weld metal along the width of the EBW and LEB-PWHT specimens (Figure 13) is 519 ± 35 HV0.1 and 399 ± 24 HV0.1, respectively. Figure 13a clearly depicts the decrease in the hardness values after LEB-PWHT, and the overall average decrease is about 23%. The average measured hardness for HAZ along the width of the EBW and LEB-PWHT specimens (Figure 13) is 473 ± 77 HV0.1 and 376 ± 28 HV0.1, respectively. This indicates a 21% decrease in hardness for the LEB-PWHT specimen compared to the EBW specimen. In the case of EBW, higher hardness was observed in the CGHAZ, i.e., 589 HV0.1, than in the BM because of the martensitic microstructure obtained after welding thermal cycles. In the case of LEB-PWHT, the maximum hardness was observed in the FGHAZ, i.e., 434 HV0.1. Figure 13b shows the average hardness comparison between EBW and LEB-PWHT in the weld.

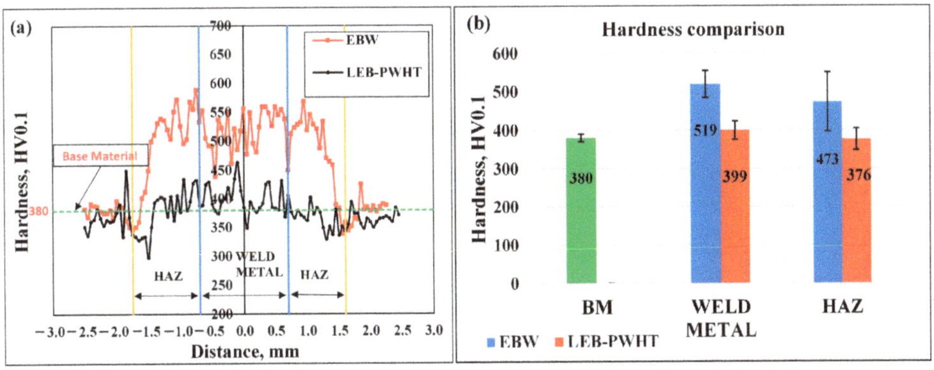

Figure 13. S960QL, microhardness. (**a**) EBW and LEB-PWHT along the width, and (**b**) base material average hardness, average hardness comparison between EBW and LEB-PWHT in the weld.

3.3. Transverse Tensile Test

In accordance with the minimum requirements specified in EN ISO 15614-11 [39], two tensile test specimens were taken from both the EBW and LEB-PWHT joints. The load was applied with a displacement rate of 0.2 mm/min. The specimens were milled and etched in order to see the fracture along the welded specimens by viewing the weld metal and HAZ from the welded plates. The tensile results for the EB-welded and LEB-PWHT joints, along with their standard deviations, are summarized in Table 5.

Table 5. Tensile properties of EB-welded and LEB-PWHT joints.

Process	Sample No.	Tensile Strength (MPa)	Average Tensile Strength, MPa	Std. Dev.	Fracture Location
EBW	1	1048	1051	3.5	Rupture out of BM
	2	1053			Rupture out of BM
LEB-PWHT	1	1086	1082	5.7	Rupture in weld metal
	2	1078			Rupture in weld metal

The average tensile strength of the EBW and LEB-PWHT joints was 1051 MPa and 1082 MPa, respectively. A three percent increase in tensile strength was observed in the LEB-PWHT joint compared to the EBW joint, as shown in Table 5. This slight rise in tensile strength may be due to the refinement of the microstructures in the FGHAZ and the carbide precipitation that occurred during the PWHT, as shown in Figures 8 and 11. It is also important to note that the tensile strength of LEB-PWHT reached the higher end of the strength range provided in the material's certificate (Table 2).

3.4. Bending Test

The bending tests were performed according to EN ISO 5173 [34]. The 2-2 samples were tested for the weld-face tensile bend test (FBB) and root-face tensile bend test (RBB) for both EBW and LEB-PWHT. The diameter (100 mm) of bending mandrel was chosen for A_5 = 10% and L = 180 mm. The bending test results are presented in Table 6.

The required bending angle of 180° was achieved. Based on the experimental observations, no fracture or cracks were reported on test specimens.

Table 6. Bend tests for the EBW and LEB-PWHT joints.

BM	Process	Sample No.	Bending Mandrel Diameter (mm)	Fracture Location
S960QL	EBW	96B1/FBB/1	100	No crack
	EBW	96B1/FBB/2		No crack
	EBW	96B1/RBB/1		No crack
	EBW	96B1/RBB/2		No crack
	LEB-PWHT	96HT/FBB/1		No crack
	LEB-PWHT	96HT/FBB/2		No crack
	LEB-PWHT	96HT/RBB/1		No crack
	LEB-PWHT	96HT/RBB/2		No crack

3.5. Charpy V-Notch (CVN) Impact Test

According to EN 10025-6, the required minimum impact energy for S960QL steel is 27 J at −40 °C. The Charpy V-notch test was used to measure the toughness of the welded joints at the base material's guaranteed operating temperature [40]. The investigated BM steel has a CVN absorb energy of 44 J, while for the EBW weld metal, it was measured as 63 J. The comparison of the CVN results for the EBW and LEB-PWHT joints from the investigated S960QL material is presented in Table 7.

Table 7. Measured CVN values of EB-welded and LEB-PWHT S960QL joints.

Zone	S. No.	CVN, J		CVN, Avg.		Toughness, J/cm^2		Toughness (Avg.), J/cm^2		Remark	
BM	1	46		44		61		55		-	
	2	36				46				-	
	3	46				58				-	
	S. No.	EBW	PWHT	EBW	PWHT	EBW	PWHT	EBW	PWHT	EBW	PWHT
Weld metal	1	51	24	63	27	64	30	78	34	B *	M *
	2	68	26			85	33			B	M
	3	69	30			86	38			B	M

* B = Brittle failure; M = Mixed mode failure.

The Charpy V-notch impact test results show that LEB-PWHT leads to a decrease in the average toughness of the weld metal (27 J), which is approximately half the toughness of the EB-welded joint (63 J). This decrease in toughness is attributed to the tempering of the microstructure and carbide precipitation in the weld metal (Figures 8 and 9). Following heat treatment, the CGHAZ exhibits a tempered microstructure, primarily composed of fine, evenly dispersed carbides within the tempered martensite. However, coarser carbide precipitates can also be observed along the lath boundaries (Figures 8 and 10), which may pose a risk to the material's impact toughness. Also, the weld metal impact test of EBW indicates brittle fracture behavior in all the tested samples, whereas the weld metal impact test of PWHT demonstrated mixed-mode failure.

4. Conclusions

The following conclusions can be drawn from our investigations and their findings:

(a) EBW results in a martensite microstructure in the weld metal, while LEB-PWHT produces equiaxed and uniform grains with carbide formation in the weld metal. The EBW CGHAZ predominantly consists of martensite, whereas the tempered martensite in the PWHT CGHAZ contains both fine and coarse carbides. The EBW FGHAZ exhibits martensitic and bainitic microstructures, while the PWHT FGHAZ shows equiaxed grains with finely dispersed carbides.

(b) After LEB-PWHT, the average hardness of the weld metal and HAZ in the LEB-PWHT joint decreased compared to the EBW joint, with values of 399 ± 24 HV0.1 and

376 ± 28 HV0.1, respectively, compared to 519 ± 35 HV0.1 and 473 ± 77 HV0.1 in the EBW joint.

(c) The tensile strength of the LEB-PWHT joint increased by 3% compared to the EBW joint, and the bending test successfully met the required 180° bending angle without fracture or cracks.

(d) After LEB-PWHT, a decrease in impact toughness was observed in the welded joint, attributed to the tempering of the microstructure and carbide precipitation.

This study provides important direction for the future applications of the LEB-PWHT technique across various fields that use high-strength structural steels. The LEB-PWHT results demonstrated a reduction in hardness peaks in both the weld metal and HAZ, along with an increase in tensile strength and good bendability. However, even if the CVN results are under the as-welded values, these results are relevant because they were obtained with an economical process and represent a robust property assessment for structural applications. Additionally, these results may be further enhanced with the use of standard post-weld heat treatment, along with optimized soaking temperature, soaking time, and the proper size of the heating field.

Author Contributions: Conceptualization, Methodology, Investigation, Data curation, Formal analysis, Validation Writing—Original draft preparation, and Supervision, R.P.S.S.; Methodology, Investigation, Data Curation, Validation, and Writing—review and editing, P.S.; Investigation, Visualization, and Data Curation, D.K.-H.; Resources, Writing—Reviewing and Editing, and Supervision, M.S.W. All authors have read and agreed to the published version of the manuscript.

Funding: This research received no external funding.

Data Availability Statement: The original contributions presented in this study are included in the article. Further inquiries can be directed to the corresponding author.

Acknowledgments: The authors express their deep appreciation to the staff at Lukasiewicz—Upper Silesian Institute of Technology, Gliwice, Poland, for their generous cooperation in carrying out the experiments for this research. The authors are grateful to the Institute of Physical Metallurgy, Metal forming and Nanotechnology at the University of Miskolc for facilitating the SEM examination.

Conflicts of Interest: The authors declare no conflicts of interest.

References

1. Lukács, J.; Dobosy, Á. Matching effect on fatigue crack growth behaviour of high-strength steels GMA welded joints. *Weld. World* **2019**, *63*, 1315–1327. [CrossRef]
2. Koncsik, Z.; Lukács, J.; Nagy, G. Fracture Mechanical Analysis of Gleeble Simulated Heat Affected Zones in High Strength Steels. *Period. Polytech. Mech. Eng.* **2022**, *66*, 83–89. [CrossRef]
3. Mičian, M.; Harmaniak, D.; Nový, F.; Winczek, J.; Moravec, J.; Trško, L. Effect of the t8/5 cooling time on the properties of s960mc steel in the HAZ of welded joints evaluated by thermal physical simulation. *Metals* **2020**, *10*, 229. [CrossRef]
4. Sisodia, R.P.S. High Energy Beam Welding of Advanced High Strength Steels. Ph.D. Thesis, University of Miskolc, Miskolc, Hungary, 2021.
5. Kovács, J.; Lukács, J. Influence of Filler Metals on Microstructure and Mechanical Properties of Gas Metal Arc Welded High Strength Steel. In *Vehicle and Automotive Engineering 4*; Jármai, K., Cservenák, Á., Eds.; Springer International Publishing: Cham, Switzerland, 2023; pp. 995–1005.
6. Mert, T.; Gürol, U.; Tümer, M. The effect of weaving and non-weaving multi-pass procedure on microstructure and mechanical properties in GMAW of S960QL. *Int. J. Adv. Manuf. Technol.* **2023**, *129*, 4731–4742. [CrossRef]
7. Kovács, J.; Gáspár, M.; Lukács, J.; Tervo, H.; Kaijalainen, A. Comparative study about the results of HAZ physical simulations on different high-strength steel grades. *Weld. World* **2024**, *68*, 1965–1980. [CrossRef]
8. Lukács, J. Fatigue crack propagation limit curves for high strength steels based on two-stage relationship. *Eng. Fail. Anal.* **2019**, *103*, 431–442. [CrossRef]
9. *EN 10025-6*; Hot Rolled Products of Structural Steels—Part 6: Technical Delivery Conditions for Flat Products of High Yield Strength Structural Steels in the Quenched and Tempered Condition. European Committee for Standardization: Brussel, Belgium, 2019.
10. Commission of the European Communities. *Properties and Service Performance: High Strength Structural Steels: A European Review*; Commission of the European Communities: Brussels, Belgium, 1988; Volume 1, p. 218.
11. Błacha, S.; Węglowski, M.S.; Dymek, S.; Kopuściański, M. Microstructural and mechanical characterization of electron beam welded joints of high strength S960QL and WELDOX 1300 steel grades. *Arch. Met. Mater.* **2017**, *62*, 627–634. [CrossRef]

12. Gáspár, M. Effect of welding heat input on simulated haz areas in s960ql high strength steel. *Metals* **2019**, *9*, 1226. [CrossRef]
13. Gáspár, M.; Dobosy, Á.; Meilinger, Á.; Nagy, G. Low-cycle fatigue behavior of differently matched welded joints of quenched and tempered steel. *Weld. World* **2023**, *67*, 2247–2259. [CrossRef]
14. Sága, M.; Blatnická, M.; Blatnický, M.; Dižo, J.; Gerlici, J. Research of the fatigue life of welded joints of high strength steel S960 QL created using laser and electron beams. *Materials* **2020**, *13*, 2539. [CrossRef]
15. Wang, Z.; Wang, X.; Shang, C. Effect of Pre-Weld Heat Treatment on the Microstructure and Properties of Coarse-Grained Heat-Affected Zone of a Wind Power Steel after Simulated Welding. *Metals* **2024**, *14*, 587. [CrossRef]
16. Tomków, J.; Landowski, M.; Rogalski, G. Application possibilities of the S960 steel in underwater welded structures. *FACTA Univ. Ser. Mech. Eng.* **2022**, *20*, 199–209. [CrossRef]
17. Schultz, H. *Electron Beam Welding*; Abington Publishing: Cambridge, UK, 1993.
18. Węglowski, M.S.; Błacha, S.; Phillips, A. Electron beam welding—Techniques and trends—Review. *Vacuum* **2016**, *130*, 72–92. [CrossRef]
19. Lippold, J.C. *Welding Metallurgy and Weldability*; John Wiley & Sons, Inc.: Hoboken, NJ, USA, 2015; ISBN 9781118230701.
20. Abstract, N. science Air space frontiers challenge welding. *Weld. Eng.* **1969**, *54*, 39–44.
21. Tümer, M.; Schneider-Bröskamp, C.; Enzinger, N. Fusion welding of ultra-high strength structural steels—A review. *J. Manuf. Process.* **2022**, *82*, 203–229. [CrossRef]
22. Hirohata, M.; Nozawa, S.; Jármai, K. An Economical and Mechanical Investigation on Local Post-Weld Heat Treatment for Stiffened Steel Plates in Bridge Structures. *Appl. Mech.* **2021**, *2*, 714–727. [CrossRef]
23. Schönbauer, B.M.; Ghosh, S.; Karr, U.; Pallaspuro, S.; Kömi, J.; Frondelius, T. Mean-stress sensitivity of an ultrahigh-strength steel under uniaxial and torsional high and very high cycle fatigue loading. *Fatigue Fract. Eng. Mater. Struct.* **2022**, *45*, 3361–3377. [CrossRef]
24. Hu, M.J.; Liu, J.H. Effects of zonal heat treatment on residual stresses and mechanical properties of electron beam welded TC4 alloy plates. *Trans. Nonferrous Met. Soc. China (Engl. Ed.)* **2009**, *19*, 324–329. [CrossRef]
25. Chen, F.R.; Huo, L.X.; Zhang, Y.F.; Zhang, L.; Liu, F.J.; Chen, G. Effects of electron beam local post-weld heat-treatment on the microstructure and properties of 30CrMnSiNi2A steel welded joints. *J. Mater. Process. Technol.* **2002**, *129*, 412–417. [CrossRef]
26. DILLIMAX 965 High Strength Fine Grained Structural Steel Quenched and Tempered. Available online: https://www.dillinger.de/app/uploads/2024/06/DILLIMAX_965_01_2024_D.pdf (accessed on 7 November 2024).
27. Kou, S. *Welding Metallurgy*, 2nd ed.; John Wiley & Sons, Inc.: Hoboken, NJ, USA, 2002; ISBN 0-471-43491-4.
28. Maurer, W.; Ernst, W.; Rauch, R.; Kapl, S.; Pohl, A.; KrÜssel, T.; Vallant, R.; Enzinger, N. Electron Beam Welding Of A TMCP Steel with 700 Mpa Yield Strength. *Weld. World* **2012**, *56*, 85–94. [CrossRef]
29. *ANSI/AWS D1.1*; Structural Welding Code–Steel; American National Standards Institue: Washington, DC, USA, 2000.
30. Zhao, M.S.; Chiew, S.P.; Lee, C.K. Post weld heat treatment for high strength steel welded connections. *J. Constr. Steel Res.* **2016**, *122*, 167–177. [CrossRef]
31. Connor, L.P. *Welding Handbook—Volume 1: Welding Technology*, 8th ed.; Amer Welding Society: Doral, FL, USA, 1987.
32. *EN ISO 9015-2*; Destructive Tests on Welds in Metallic Materials—Hardness Testing, Part 2: Microhardness Testing of Welded Joints. International Organization for Standardization: Geneva, Switzerland, 2016.
33. *EN ISO 4136*; Destructive Tests on Welds in Metallic Materials—Transverse Tensile Test. International Organization for Standardization: Geneva, Switzerland, 2022.
34. *EN ISO 5173*; Destructive Tests on Welds in Metallic Materials—Bend Tests. International Organization for Standardization: Geneva, Switzerland, 2023.
35. *EN ISO 148-1*; Metallic Materials—Charpy Pendulum Impact Test. International Organization for Standardization: Geneva, Switzerland, 2016.
36. *EN ISO 9016*; Destructive Tests on Welds in Metallic Materials—Impact Tests—Test Specimen Location, Notch Orientation and Examination. International Organization for Standardization: Geneva, Switzerland, 2022.
37. Gaspar, M.; Balogh, A. Gmaw Experiments for Advanced (Q + T) High Strength Steels. *Prod. Process Syst.* **2013**, *6*, 9–24.
38. Tomerlin, D.; Kozak, D.; Ferlič, L.; Gubeljak, N. Experimental and Numerical Analysis of Fracture Mechanics Behavior of Heterogeneous Zones in S690QL1 Grade High Strength Steel (HSS) Welded Joint. *Materials* **2023**, *16*, 6929. [CrossRef] [PubMed]
39. *EN ISO 15614-11*; Specification and Qualification of Welding Procedures for Metallic Materials—Welding Procedure Test—Part 11: Electron and Laser Beam Welding. International Organization for Standardization: Geneva, Switzerland, 2002.
40. Yang, Z.; Liu, Z.; He, X.; Qiao, S.; Xie, C. Effect of microstructure on the impact toughness and temper embrittlement of SA508Gr.4N steel for advanced pressure vessel materials. *Sci. Rep.* **2018**, *8*, 207. [CrossRef]

metals

MDPI

Article

Direct In Situ Fabrication of Strong Bonding ZIF-8 Film on Zinc Substrate and Its Formation Mechanism

Haidong Wang [1], Jie Liu [2], Baosheng Liu [2], Zhechao Zhang [2], Xiaoxia Ren [2], Xitao Wang [2,*], Pengpeng Wu [2] and Yuezhong Zhang [2,*]

1 School of Applied Technology, Changchun Institute of Technology, Changchun 130012, China; jd_whd@ccit.edu.cn

2 School of Materials Science and Engineering, Taiyuan University of Science and Technology, Taiyuan 030024, China; s202214210390@stu.tyust.edu.cn (J.L.); liubaosheng@tyust.edu.cn (B.L.); b202414110029@stu.tyust.edu.cn (Z.Z.); 2019012@tyust.edu.cn (X.R.); 2022041@tyust.edu.cn (P.W.)

* Correspondence: xitao.wang@gmail.com (X.W.); yuezhongzhang@tyust.edu.cn (Y.Z.)

Abstract: There is much promise for creating metal organic framework (MOF) films on metal substrates in fields including sensing and electrical conduction. For these applications, direct production of MOF films with strong bonding on metal substrates is extremely desirable. In this study, a simple one-step method without the need for additives or pre-modification is used to directly create zeolitic imidazolate framework-8 (ZIF-8) films with strong bonding on zinc substrate. The formation mechanisms of ZIF-8 film are analyzed. The strong bonding ZIF-8 film can be attributed to an in-situ grown ZnO interlayer between the ZIF-8 and substrate. The growth process shows the formation time of zinc oxide on the substrate, which is subsequently covered by ZIF-8 crystals. The ZnO interlayer results from a combination of decomposition products of the solvent and the zinc ions. Furthermore, the ZnO interlayer serves as a sacrificial precursor for the in-situ nucleation and continuous growth of ZIF-8 film. It serves as an anchoring site between ZIF-8 film and substrate, resulting in strong adhesion. This paper describes a simple and straightforward production process that is expected to provide a theoretical basis for the laboratory preparation of ZIF films.

Keywords: ZIF-8 film; strong bonding; in situ fabrication; zinc substrate; ZnO interlayer

1. Introduction

A subclass of metal organic frameworks (MOFs) known as zeolitic imidazolate frameworks (ZIFs) combines the special qualities of both zeolites and MOFs, including large internal surface areas, high grades of crystallinity, permanent porosity, adjustable pore sizes, and chemical and thermal stability [1]. More recently, ZIFs have demonstrated their intriguing potential as novel functional materials by being used in energy storage [2], sensors [3], optoelectronics [4], membrane separations [5], and catalysis [6]. It is more advantageous to fabricate ZIF materials as films or membranes (rather than powders) for use in these applications, as the former can help with the direct fabrication of ZIFs into gas sensors [7–9], electronic devices [10], and other microelectronic devices, while the latter can be used for gas separations [11–13].

Various MOF films or membranes that correspond to thin layers have been created thus far using either in situ or secondary growth techniques on a variety of support materials [14], including aluminum oxide [15], titanium dioxide [16], glass [17], organic polymers, etc. [18–20]. Less research has been done on ZIF films or membranes synthesized directly on metal substrates [21,22], which could be employed as sensors or electrodes in electrochemical and electrical applications. Furthermore, it is still a challenge to directly fabricate strong-bonding MOF films on metal substrates. The main handicap for directly growing MOF films or membranes on metal substrates is weak interface adhesion and poor heterogeneous nucleation sites [23–25]. Therefore, modification of the support surface with suitable functional groups like organosilanes, imidazole derivatives, dopamine, and metal oxides became necessary to improve binding affinity or/and nucleation density. Compared with organic functional, the inorganic modification layer is more favored because it not only can enhance affinity and compatibility but also has strong thermal stability and environmental friendliness. To enhance the binding strength of the MOF layer with the substrate, modifying a metal substrate with metal oxide or hydroxide micro-nanostructures is employed to promote MOF crystal nucleation because these structures can act as both nucleation centers and anchor bars for film growth. For example, Falcaro and co-workers fabricated HKUST-1 ($Cu_3(BTC)_2$) film on Cu substrate by converting pre-prepared copper hydroxide coating into HKUST-1 [26]. This approach may be applied to four different MOF films (Cu-BTC, Cu-BDC, ZIF-8, and MOF-5) on either a Cu or Zn substrate, as Kim et al. found [27]. Li et al. also fabricated $Co_3(HCOO)_6$ membranes on a Co_3O_4-modified Ni foam substrate [28]. Nevertheless, these strategies always involved a preceding oxidation step of the metal substrate.

Recently, modified ZnO micro/nanostructures on substrates have been proven to be an effective method for promoting ZIF film or membrane growth and enhancing binding strength to the substrate. For example, a series of ZIFs (ZIF-8, ZIF-68, ZIF-71, and ZIF-78) membranes were fabricated on porous ZnO or ZnO-modified α-Al_2O_3 substrates [29–36]. In these cases, the ZnO layer acts as a multifunctional role, inducing the nucleation for ZIF crystals, providing the zinc source for Zn-based ZIFs films, and the anchorages between the film and substrate. However, in most of the previous research, ZnO layer deposition was completed via sol-gel and heating methods, resulting in an unstable ZnO layer and complicated operation. Furthermore, activation of the ZnO layer is a non-negligible step to create nucleated species for homogeneous nucleation and induce the growth of ZIF crystals on the ZnO layer surface [37,38]. Therefore, to develop a facile and direct method to fabricate strong-bonding ZIF films derived from ZnO layers on metal substrates is highly desirable.

Here, a one-step solvothermal method was used to directly create a dense zeolitic imidazolate framework-8 (ZIF-8) layer on a zinc substrate. In our strategy, DMF solution only containing zinc ions and 2-methyl imidazolate (mim) ligands were used to synthesize ZIF-8 film (schematically shown in Figure 1). During this one-step process, a hexagonal plate-like ZnO interlayer was in situ constructed between the ZIF-8 crystal layer and substrate, resulting in a separate ZnO preparation or/and activation process that was omitted in previous reports. For the ZIF-8 film to nucleate and grow continuously, the ZnO interlayer acted as a sacrificial precursor. It also acted as an anchoring site between the ZIF-8 film and substrate, strengthening the stability of the film.

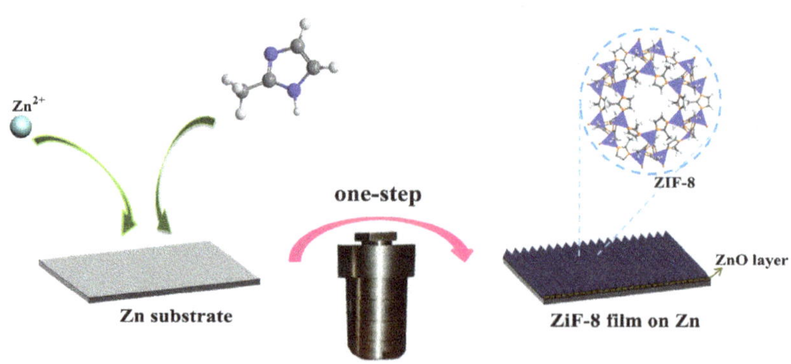

Figure 1. Schematic diagram of the in situ synthesis process of ZIF-8 film on zinc substrate. (ZIF-8: gray means C, orange means N, blue means Zn, and white means H).

2. Experimental

2.1. Materials and Methods

We bought 99.9% zinc foil with a thickness of 0.2 mm from Yi Tian Metal Products Co., Ltd., Shanghai, China. The chemicals that were employed were N, N′-dimethylformamide (DMF, ≥99.9%), 2-methylimidazole (mim), ethanol, and zinc nitrate hexahydrate. (Sinopharm Chemical Reagent Co., Ltd., Shanghai, China, analytical grade).

2.2. Preparation of the ZIF-8 Film

As the substrate, 20 mm by 30 mm zinc foil was utilized. Before the synthesis began, the zinc foil was cleaned by ultrasonically degreasing it in absolute ethanol for five minutes, rinsing it with ultrapure water, and abrading it with emery papers rated 200, 400, and 800#. In order to prepare the growth solution of ZIF-8 film grown in situ, 2-methylimidazole and $Zn(NO_3)_2$ were dissolved in DMF solution and stirred magnetically for 10 min to obtain mixed solution concentrations of 0.024 mol/L mim and 0.030 mol/L $Zn(NO_3)_2$. A growth solution was made for the in-situ growth of ZIF-8 film by dissolving 0.0457 g of 2-methylimidazole and 0.1812 g of $Zn(NO_3)_2 \cdot 6H_2O$ in 20 milliliters of DMF solution. First, clean Zn foil was positioned vertically in a stainless steel autoclave lined with Teflon, and 20 milliliters of growth fluid were added. After that, the autoclave was covered with a lid, kept at 150 °C for a predetermined amount of time, and then allowed to cool to room temperature. Then, the Zn foil was removed and dried in air. For comparison, the zinc foil was treated in a quite similar procedure, with methanol instead of DMF or without 2-methylimidazole.

2.3. Characterization

The treated Zn foil surface's morphological structures were investigated using scanning electron microscopy (SEM, Hitachi S-4800, Tokyo, Japan). Utilizing an X-ray diffraction machine (XRD, Rigaku D/MAX-2500PC, Tokyo, Japan), the phase constitutions were examined. Using Fourier transform infrared spectroscopy (FTIR, Nicolet iN10, Waltham, MA, USA), the samples' infrared spectra were examined. A Thermoelectron ESCALAB 250 X-ray photoelectron spectrometer (Thermo Fisher Scientific, Waltham, MA, USA) was used to analyze the samples' surface chemical composition.

3. Results and Discussion

3.1. Synthesis and Characterization of ZIF-8 Film

Typical SEM pictures of the zinc foil following solvothermal treatment for 24 h are given in Figure 2. The SEM image (Figure 2a) at low magnification shows that zinc foil is growing a well-intergrown crystal layer. A higher magnification image of the treated Zn foil (Figure 2b) shows well-formed crystal facets that some small polyhedral crystals randomly distributed with a diameter of 1–3 μm. The cross-sectional view in Figure 2c demonstrates that the film consists of intergrown crystals. The crystals penetrate inside the substrate, making the boundary of the crystals and support very undistinguishable, but a careful inspection still illustrates that the film is of thickness of about 40–50 μm. The treated zinc foil's typical XRD pattern is displayed in Figure 2d. In addition to the prominent zinc substrate diffraction peaks, a portion of the remaining diffraction peaks ($2\theta > 30°$) can be satisfactorily indexed to the wurtzite hexagonal ZnO structure, indicating that the as-prepared film is primarily composed of ZnO and ZIF-8.

The XPS detection was used to look into the chemical condition and surface composition of ZIF-8 film. The O element and all ZIF-8 ($C_8H_{10}N_4Zn$) characteristic elements, such as zinc (coordinating metal), nitrogen, and carbon (imidazole linker), are visible in the XPS survey spectrum (Figure 3a). Three components originating from adventitious carbon (284.6 eV), C-N-C (286.2 eV), and C=C (288.6 eV) could be deconvoluted from the C 1s high-resolution spectrum (Figure 3b) [39]. Two peaks can be identified in the N 1s high-resolution spectrum (Figure 3c) for ZIF-8 film: a larger peak at 399.0 eV and a smaller peak at 400.5 eV, which are attributed to the imidazole and secondary amine groups [22].

DMF breakdown and a trace amount of uncoordinated methylimidazole linkers can produce secondary amines. The autoclaves of the former emit a distinct and powerful fishy odor upon opening for each synthesis. Moreover, Zn^{2+} of the ZIF-8 is responsible for the Zn 2p 3/2 and Zn 2p 1/2 peaks, which are located at approximately 1021.9 and 1044.9 eV, respectively (Figure 3d) [40].

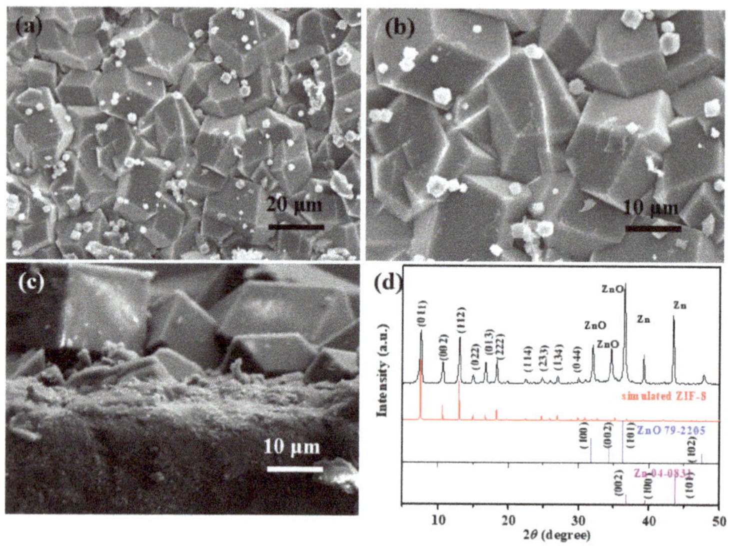

Figure 2. SEM of the zinc foil after solvothermal treatment for 24 h: (**a**,**b**) Top view and (**c**) cross-section, (**d**) XRD patterns of the treated zinc foil.

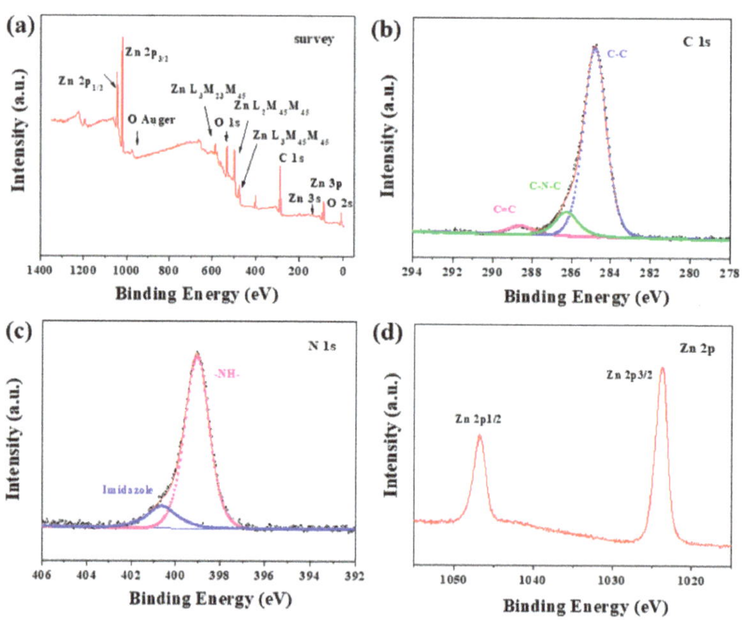

Figure 3. XPS (Al Kα) spectra of the zinc foil after solvothermal treatment for 24 h: (**a**) survey, (**b**) C 1s, (**c**) N 1s, and (**d**) Zn 2p.

Furthermore, FTIR spectroscopy was used to investigate the ZIF-8 film's distinctive functional groups. Zn-N stretching bands exhibit a distinctive absorption peak at 418 cm^{-1}, as depicted in Figure 4. The mim ring's out-of-plane and in-plane bending are attributed to the bands at 600–800 cm^{-1} and 900–1350 cm^{-1}, respectively [41,42]. The imidazole's whole ring stretching is situated between 1350 and 1500 cm^{-1} [41,42]. The methyl groups from the mim linker's aliphatic C-H stretching bands may be identified as the peak at 1310 cm^{-1}, while the N-H bending vibration of mim is linked to the peak at 1543 cm^{-1} [43]. The spectral area of 2900–3200 cm^{-1} derives from the aliphatic and aromatic C-H stretching of mim. The XRD and XPS results and this result match one another quite well. Therefore, it can be concluded that ZIF-8 completely covers the ZnO layer.

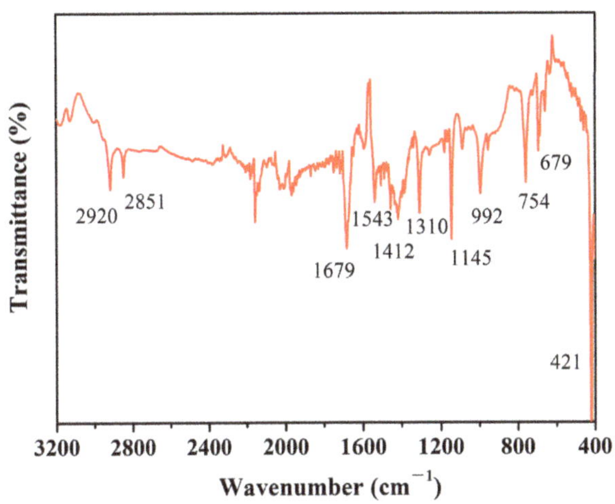

Figure 4. FTIR spectra of the zinc foil after solvothermal treatment for 24 h.

3.2. Evolution of the ZIF-8 Film Formation

A number of operations with varying synthesis times (1 h, 3 h, 6 h, and 12 h) were carried out in order to investigate the formation process of ZIF-8 film in this in-situ approach. After the reaction ran for one hour, the majority of the substrate was covered in randomly distributed tiny protrusions on the zinc foil, as shown in the low-magnification SEM image of the foil (Figure 5a). From the high-resolution SEM image (Figure 5b), the protrusions show irregular polyhedron structure. After just 3 h, some cubes with truncated edges with a diameter of 20 μm and a large number of hexagonal plate-like structures with a diameter of 3–5 μm and a thickness of 0.50–1.4 μm are observed in Figure 5c and d. As reported, the cube with truncated edges is a typical intermediate shape of ZIF-8 (Figure S1, in ESI) [44,45]. The plate-like structures are highly similar to ZnO prepared without mim (Figure 7c). Furthermore, the elemental composition of different morphology protrusions was tested by EDS, and elemental mapping of the sample confirmed that the polyhedron was ZIF-8 crystal, whereas the hexagonal plate-like structure was ZnO (Figure S2, in ESI). Furthermore, the elemental composition of different morphology protrusions was tested by EDS, and elemental mapping of the sample confirmed that the polyhedron was ZIF-8 crystal, whereas the hexagonal plate-like structure was ZnO (Figure S3, in ESI). After just 6 h, a small amount of well-defined independent rhombic dodecahedral crystal protrusions with an average size of 12–18 μm are obtained (Figure 5e), which is a stable equilibrium morphology of ZIF-8 [46–49]. For the non-protuberant section between the rhombic dodecahedral protrusions (Figure 5f), the small-sized (1–3 μm) interconnected crystals and several hexagonal plate-like structures are obtained. After just 12 h, rhombic dodecahedral crystals become larger and interdigitated with a diameter of 40–60 μm (Figure 5g), and cross-linking small crystals become denser in the non-protuberant section

(Figure 5h). When the reaction time was extended to 24 h, the lamellar structure disappeared completely and the polyhedron ZIF-8 was observed (Figure 5i). In addition, small particles of polyhedral crystals have also been observed (Figure 5j). Based on the above results, we can conclude that the hexagonal plate-like ZnO layer is preferentially grown and a dense and crosslinked ZIF-8 crystal layer is obtained on the zinc substrate through nucleation, evolution, and growth processes.

Figure 5. SEM images of zinc foil after the different treatment times: (**a**,**b**) 1 h, (**c**,**d**) 3 h, (**e**,**f**) 6 h, (**g**,**h**) 12 h, and (**i**,**j**) 24 h.

Additionally, the ZIF-8 film underwent time-dependent XRD testing, the results of which are displayed in Figure 6. As seen in Figure 6 and Figure S3, weak characteristic diffraction peaks of ZnO are detected just after 1 h of treatment. Moreover, the peaks of ZnO are also observed for different synthesis times. These results indicate that zinc oxide grows preferentially and is present throughout the process. It is noted that the diffraction pattern of ZIF-8 is detected for the 6h sample, and the intensity is substantially weaker compared with that of 12 h. This is because zinc ions competitively interacted with hydroxyl groups and ligands to form ZnO and ZIF-8 in the case of an in situ process. This outcome agrees with the SEM findings (Figure 5a). Furthermore, ZIF-8's intensity and degree of crystallinity grow steadily throughout reaction time. Thus, we can conclude that the hexagonal plate-like ZnO is formed prior to the ZIF-8 crystals, then spontaneously converts to ZIF-8, and is finally completely covered by ZIF-8 crystals.

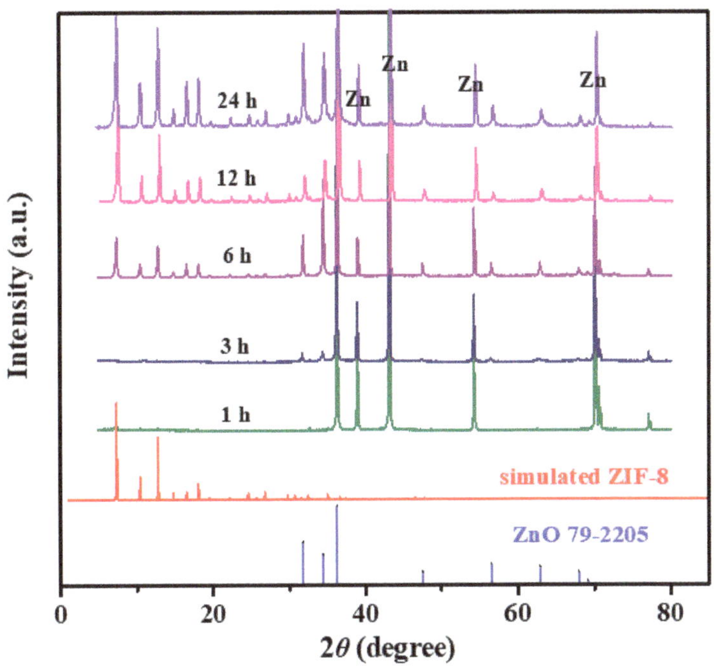

Figure 6. XRD patterns of Zn foil after the different treatment times.

3.3. *Effect of Solvent and Reactives on ZnO Formation*

To probe the formation mechanism of ZnO, some comparative experiments were carried out. When DMF is replaced by methanol alone, spherical ZnO with a diameter of 0.5–5 μm is detected (Figure 7a,b, and Figure S4a in ESI). These findings suggest that the ZnO layer's shape and crystal growth direction are significantly influenced by the solvent. The N 1s XPS high-resolution spectrum (Figure 3c) confirms that the amine was created throughout the preparation process when these syntheses were carried out in DMF without the inclusion of a mim linker. There was also a strong, distinct, fishy smell when the autoclaves were opened. The amine may be attributed to dimethylamine decomposed by DMF at the reaction temperature [50]. Moreover, the ZnO layer is formed on a zinc substrate, which consists of a stacked hexagonal plate (Figure 7c,d, and Figure S4b in ESI). This morphology is highly similar to the sample after treatment for 3 h with ligand. These findings suggest that although the ligand aids in the growth of the ZnO layer, it is not necessary for the ZnO layer to form. Additionally, the ligand (mim) also facilitated this process, as confirmed by Carreon et al. [51] due to its weak alkaline (pKb = 7.75) [52].

According to Carreon et al. [51], the mim might function as an organic amino group catalyst that encourages ZnO production. The morphology of the synthesis gel takes on the shape of a hexagonal plate when organic amino groups are added. Zinc oxide nanoplates, for instance, have been created while hexamethylenetetramine and diethylenetriamine are present [53–55].

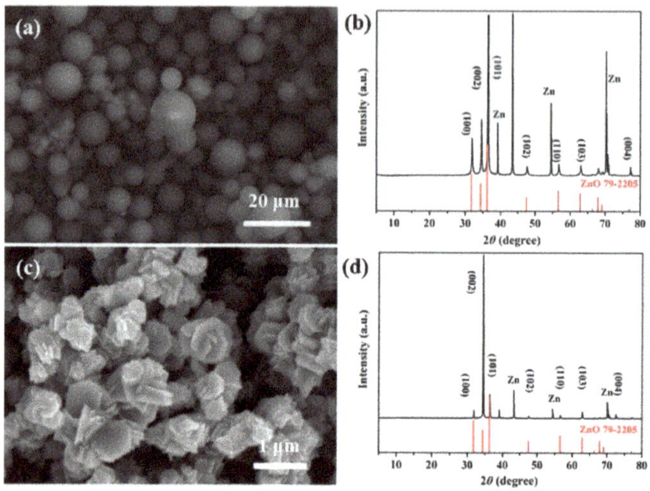

Figure 7. (**a**) SEM image and (**b**) XRD patterns of zinc foil treated with the methanol-based protocol; (**c**) SEM image and (**d**) XRD patterns of zinc foil treated with DMF-based synthesis without mim.

3.4. Formation Mechanisms of ZIF-8 Film

The direct manufacturing of MOF film on metal substrates at large geometrical scales has been tried in a few prior investigations. But only three methods—(i) forming metal oxide overlayer on metal substrates via an earlier oxidation step [27], (ii) adding oxidation reagents in the synthesis solution to oxidize the metal surface [56], or (iii) reducing solvents or introducing reducing gases [21]—can successfully construct ZIF-8 film. Note that no extra chemicals or preparation procedures were used in the construction of the ZIF-8 films given in this study; instead, they were created utilizing a one-step in-situ method. The solvothermal technique is environmentally friendly since it only uses organic ligands and metal substrates. The ZIF-8 film production techniques in this work differ significantly from the earlier methods. ZnO formation and ZIF-8 synthesis processes can occur on the surface of zinc substrates in an aqueous precursor solution. The feasible creation methods of the ZIF-8 film might be postulated as represented in Figure 8.

In the initial stage, the zinc substrate, amphoteric metal, was etched by amine derived from DMF decomposition to release zinc ions. A rich hydroxyl species solution environment is created when the amine is present. Therefore, it is expected that the OH- will react to create $Zn(OH)_2$ with the Zn^{2+} ions from the zinc reagent and etched zinc substrate. The coordination compounds can be adsorbed on the zinc substrate surface due to electrostatic effects [57]. ZnO is then formed when these hydroxyl species are dehydrated. Additionally, the ligand (mim) also enhances the production of ZnO.

In general, the nucleation of ZIF-8 crystals is affected by deprotonation equilibrium and coordination [58]. According to Cravillon et al. [58], the ZIF-8′s deprotonation equilibrium can be controlled by a modulator. More basic modulating ligands (pKa > 10.3) can deprotonate ligands effectively, resulting in a high nucleation rate, but less basic modulating ligands (pKa < 10.3) cannot. The organic amines from decomposed DMF deprotonate the organic ligands, which act as protonation agents in the synthesis solution. In our case, deprotonation of mim is inhibited by consumed amines for ZnO. Moreover, the amine may hinder the synthesis of ZIF-8 by competing with the mim ligand to coordinate with the Zn^{2+} on the substrate to generate coordination compounds. Therefore, the ZnO layer's growth supersedes the ZIF-8 film. The growth of ZnO continues until it is completely covered with ZIF-8.

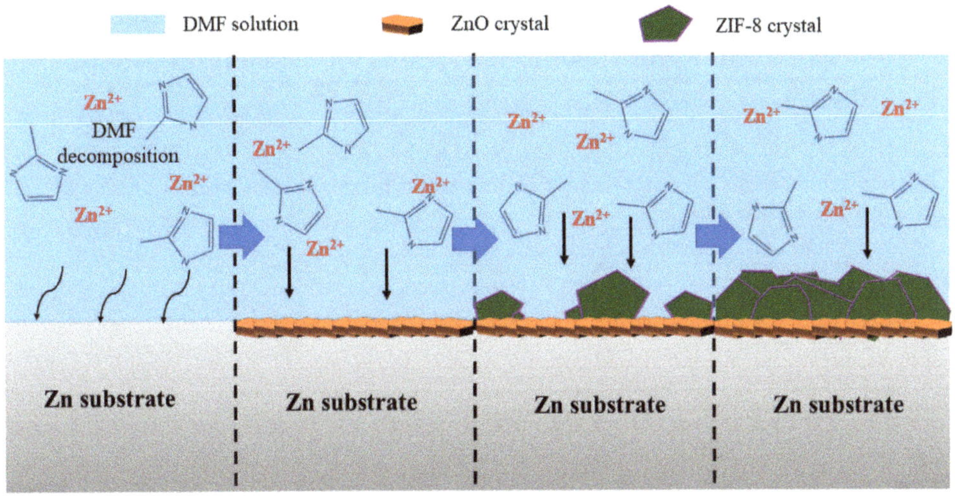

Figure 8. Schematic diagram of the formation mechanism of ZIF-8 film.

During this process, nucleation of ZIF-8 may take place once the deprotonated ligand binds to Zn^{2+} in the solution. Then the heterogeneous nucleation and crystallization of ZIF-8 on the ZnO surface are triggered. The crystallization of ZIF-8 immediately resulted in the formation of nanocrystals on the ZnO surface due to the tendency of ZIFs to preferentially nucleate on ZnO crystals [59] and supersaturated interface near the ZnO surface. The nanocrystals acting as seed crystals are further crystallized, and then microcrystals are gradually formed on the surface of ZnO. According to the SEM and XRD results, ZIF-8 crystals grow gradually after 3 h treatment. The dense and continuous ZIF-8 layer is formed by the fusing of nanoparticles between the big particles, which seals the grain boundaries and imperfections and ensures effective ZIF-8 surface coverage. In addition, ZnO crystals may be etched by mim ligand in this process, resulting in more mobile Zn^{2+} for ZIF-8 crystal growth.

3.5. Adhesion Test of the ZIF-8 Film on Metal Substrate

For prospective uses, ZIF film must have strong adhesion between the film and its substrate. Therefore, the adhesion strength of ZIF-8 film to the substrate was evaluated by a sonication method [27]. Additionally, during the ultrasonic treatment, we employed an SB-5200 ultrasonic cleaner from Ningbo Xinzhi Bio-Technology Co., Ltd., Ningbo, China. The ultrasonic procedure involved placing the substrate with the film into a beaker containing ethanol and then placing the beaker in the ultrasonic cleaner for treatment. Figure 9 shows the morphology of the ZIF-8 film before and after 30 min of sonication treatment in ethanol. As compared with that of a sample without sonication treatment (Figure 9a–c), no obvious shedding and damage is observed on the ZIF-8 film surface (Figure 9d–f), indicating that the prepared ZIF-8 film remained intact and firmly attached to the zinc substrate. The strong adhesion of the ZIF-8 film to the substrate can be attributed to the extra binding strength stemming from the in-situ grown ZnO layer. The ZnO layer served as an anchoring site between the ZIF-8 film and substrate for reinforcing the film's stability.

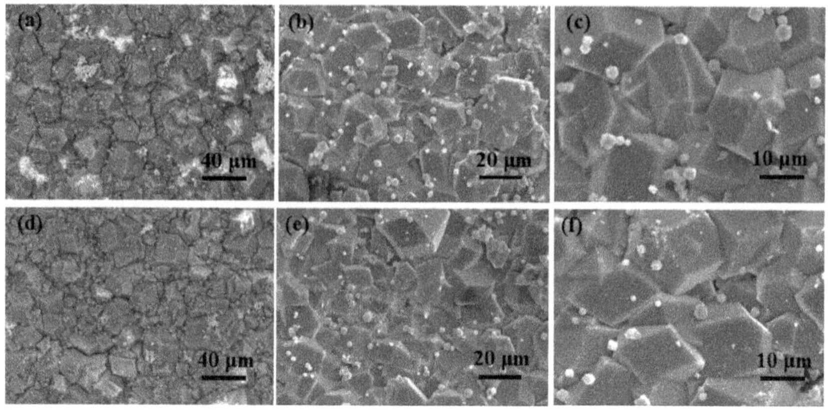

Figure 9. Micrographs of ZIF-8 coatings before and after ultrasonic testing: (**a–c**) before testing and (**d–f**) after testing.

4. Conclusions

In summary, a strong bonding ZIF-8 film is successfully constructed on a zinc substrate by a direct in-situ solvothermal process. The strong bonding ZIF-8 film can be attributed to the spontaneous in-situ grown ZnO interlayer between the ZIF-8 and substrate. The growth process demonstrated that the ZnO interlayer is preferentially formed on the zinc substrate and subsequently covered by ZIF-8 crystals. The zinc ions and the solvent's breakdown products combine to form the ZnO interlayer. The ZnO interlayer not only acts as an active seed for the nucleation and continuous growth of ZIF-8 film but also acts as an anchoring site between ZIF-8 film and substrate for reinforcing the film adhesion to the substrate. The zinc reagent and solvent's breakdown products encourage the formation of the ZnO interlayer. We firmly believe that the strategy opens up a simple direct growth route for achieving a strong bonding ZIF-8 film on a metal substrate, and a basic understanding of the formation mechanisms of ZIF film is beneficial for the efficient fabrication of advanced MOF film on a metal substrate. Future research could build upon this method to prepare more efficient and strongly bonded thin films on metal substrates. Additionally, this approach can be experimentally applied to other metal substrates to further investigate the complete thin film formation mechanism. Such studies would provide a more comprehensive theoretical foundation for the preparation of thin films on metal substrates.

Supplementary Materials: The following are available online at https://www.mdpi.com/article/10.3390/met14121403/s1, Figure S1: Illustration of the crystal morphology evolution with time, Figure S2: (a) SEM images of Zn foil after 3 h and (b) elemental mapping by energy-dispersive X-ray spectroscopy (EDS), Figure S3 XRD patterns of Zn foil after 1 h, and Figure S4. (a) SEM image of zinc foil treated with the methanol-based protocol and (b) SEM image of zinc foil treated with DMF-based synthesis without mim.

Author Contributions: Conceptualization, H.W. methodology, B.L. and Y.Z. validation, P.W., Z.Z. and X.W. data curation, H.W., Y.Z. and X.W. visualization, Z.Z., X.R. and J.L.; formal analysis, H.W. resources, H.W. writing—original draft preparation, H.W.; writing—review and editing, J.L. supervision; Y.Z. funding acquisition, B.L., Y.Z., P.W. and X.R. All authors have read and agreed to the published version of the manuscript.

Funding: The authors gratefully acknowledge the Central Guiding Science and Technology Development of Local Fund (Grant No. YDZJSK20231A046), National Natural Science Foundation of China (Grant No. 52071227), Shanxi Province Patent Transformation Project (Grant Nos. 202402004, 202402019), and Fundamental Research Program of Shanxi Province (Grant Nos. 202203021222188 and 202303021211166).

Data Availability Statement: The original contributions presented in the study are included in the article/Supplementary Material, further inquiries can be directed to the corresponding authors.

Conflicts of Interest: The authors declare no conflict of interest.

References

1. Mohammadkhani, R.; Sharifi, K.; Fedel, M.; Ramezanzadeh, B. Fabricating epoxy composite coating having self-healing/barrier anti-corrosion functions utilizing ion-exchange/pH-sensitive phosphate-doped ZIF8 MOF decorated Zn-Al-LDH nano-layers. *Surf. Coat. Technol.* **2024**, *477*, 130284. [CrossRef]
2. Yu, Y.; Wang, D.; Luo, J.; Xiang, Y. First-principles study of ZIF-8 as anode for Na and K ion batteries. *Colloids Surf. A* **2023**, *659*, 130802. [CrossRef]
3. Tian, Y.; Li, J.; Li, X.; Wang, R.; Zhang, Y.; Wang, G.; Shi, G. ZIF-8/ZIF-67 solid electrolyte ozone sensor at room temperature. *Sens. Actuators A Phys.* **2023**, *354*, 114281. [CrossRef]
4. Lai, N.; Chang, G.; Yang, Y.; He, M.; Tang, W.; Huang, Q.; Liao, J.; Yang, Y.; Wang, C.; Wang, R. $CsPbX_3$ quantum dots@ZIF-8 composites with enhanced luminescence emission and stability. *J. Luminesc.* **2024**, *266*, 120280. [CrossRef]
5. Liu, L.; Ji, T.; Hu, W.; Sun, Y.; He, Y.; Yan, J.; He, G.; Liu, Y. Epitaxial supercritical fluid processing of ZIF-8 membranes towards efficient C_3H_6/C_3H_8 separation. *J. Membr. Sci.* **2023**, *669*, 121300. [CrossRef]
6. Fatima, H.; Azharb, M.R.; Zhong, Y.; Arafat, Y.; Khiadani, M.; Shao, Z. Rational design of ZnO-zeolite imidazole hybrid nanoparticles with reduced charge recombination for enhanced photocatalysis. *J. Colloid Interface Sci.* **2022**, *614*, 538–546. [CrossRef]
7. Luo, S.; Chen, R.; Wang, J.; Xiang, L. Conductometric methane gas sensors based on ZnO/Pd@ZIF-8: Effect of dual filtering of ZIF-8 to increase the selectivity. *Sens. Actuators B Chem.* **2023**, *383*, 133600. [CrossRef]
8. Huang, W.; Li, Y.; Chang, N.; Hong, W.; Wang, C.; Huang, C. Highly stable and selective H_2 gas sensors based on light-activated a-IGZO thin films with ZIF-8 selective membranes. *Sens. Actuators B Chem.* **2024**, *417*, 136175. [CrossRef]
9. Yang, Y.; Yue, J.; Zhang, X.; Ren, B.; Fu, S.; Sun, Y.; Luo, Z. Accordion-like ZIF-8/MoO_3 composite gas sensor for highly selective and sensitive H_2S detection. *Ceram. Int.* **2024**, *50*, 38253–38262. [CrossRef]
10. Peng, C.; Wang, C.; Li, Z.; Wang, Z. A novel approach to detecting doping agents in food using electrochemical sensor based on zinc oxide/graphene oxide nanocomposites. *J. Food Meas. Charact.* **2024**, *18*, 6770–6781. [CrossRef]
11. Nguyen, T.; Chen, J.; Pham, M.; Bui, H.; Hu, C.; You, S.; Wang, Y. A high-performance ZIF-8 membrane for gas separation applications: Synthesis and characterization. *Environ. Technol. Innov.* **2023**, *31*, 103169. [CrossRef]
12. Shan, Y.; He, M.; Zhang, F.; Wang, Y.; Liu, Y.; Yang, Y.; Wang, Z.; Chen, X. Plasma-assisted synthesis of ZIF-8 membrane for hydrogen separation. *Sep. Purif. Technol.* **2023**, *317*, 123871. [CrossRef]
13. Zhu, R.; Wang, L.; Zhang, H.; Liu, C.; Wang, Z. ZIF-8 membranes on ZIF-8-PVDF/PVDF dual-layer polymeric hollow fiber supports for gas separation. *Sep. Purif. Technol.* **2024**, *335*, 126209. [CrossRef]
14. Liu, Y.; Matsuda, R.; Kusaka, S.; Hori, A.; Ma, Y.; Kitagawa, S. Insights into inorganic buffer layer-assisted in situ fabrication of MOF films with controlled microstructures. *CrystEngComm* **2018**, *20*, 6995–7000. [CrossRef]
15. Shi, P.; Wang, C.; Wang, H.; Lei, X.; Wang, B.; Liu, X.; You, J.; Guo, R. Characteristics of zeolitic imidazolate framework-L and application of its derivatives in oxygen evolution reaction: Recent trends. *J. Alloys Compd.* **2024**, *1006*, 176293. [CrossRef]
16. Sadughi, M.M.; Mazani, A.; Varnaseri, M. Synthesis of magnetic nanocomposites based on imidazole zeolite-8 framework doped with silver nanoparticles for effective removal of norfloxacin from effluents. *J Clust. Sci.* **2024**, *35*, 2991–3009. [CrossRef]
17. Su, Z.; Shaw, W.L.; Miao, Y.R.; You, S.; Dlott, D.D.; Suslick, K.S. Shock wave chemistry in a metal–organic framework. *J. Am. Chem. Soc.* **2017**, *139*, 4619. [CrossRef] [PubMed]
18. Hou, J.; Sutrisna, P.D.; Zhang, Y.; Chen, V. Formation of ultrathin, continuous metal-organic framework membranes on flexible polymer substrates. *Angew. Chem. Int. Ed.* **2016**, *55*, 3947. [CrossRef] [PubMed]
19. Li, Q.; Li, J.; Fang, X.; Liao, Z.; Wang, D.; Sun, X.; Shen, J.; Han, W.; Wang, L. Interfacial growth of metal–organic framework membranes on porous polymers via phase transformation. *Chem. Commun.* **2018**, *54*, 3590–3593. [CrossRef]
20. Hamid, M.R.A.; Park, S.; Kim, J.S.; Lee, Y.M.; Jeong, H.-K. In situ formation of zeolitic-imidazolate framework thin films and composites using modified polymer substrates. *J. Mater. Chem. A* **2019**, *7*, 9680. [CrossRef]
21. Li, B.; Xu, C.; Yu, D.; Qi, Z.; Wang, Y.; Peng, Y. Enhanced phosphate remediation of contaminated natural water by magnetic zeolitic imidazolate framework-8@engineering nanomaterials (ZIF8@ENMs). *J. Colloid Interface Sci.* **2022**, *613*, 71–83. [CrossRef] [PubMed]
22. Papporello, R.L.; Miro, E.E.; Zamaro, J.M. Secondary growth of ZIF-8 films onto copper-based foils. Insight into surface interactions. *Microporous Mesoporous Mater.* **2015**, *211*, 64–72. [CrossRef]
23. Bradshaw, D.; Garai, A.; Huo, J. Metal–organic framework growth at functional interfaces: Thin films and composites for diverse applications. *Chem. Soc. Rev.* **2012**, *41*, 2344–2381. [CrossRef]
24. Li, Y.; Liang, F.; Bux, H.; Feldhoff, A.; Yang, W.; Caro, J. Molecular sieve membrane: Supported metal–organic framework with high hydrogen selectivity. *Angew. Chem. Int. Ed.* **2010**, *49*, 548–551. [CrossRef]

25. Sun, Y.; Yang, F.; Wei, Q.; Wang, N.; Qin, X.; Zhang, S.; Wang, B.; Nie, Z.; Ji, S.; Yan, H.; et al. Oriented nano-microstructure-assisted controllable fabrication of metal-organic framework membranes on nickel foam. *Adv. Mater.* **2016**, *28*, 2374–2381. [CrossRef] [PubMed]
26. Okada, K.; Ricco, R.; Tokudome, Y.; Styles, M.J.; Hill, A.J.; Takahashi, M.; Falcaro, P. Copper conversion into $Cu(OH)_2$ nanotubes for positioning $Cu_3(BTC)_2$ MOF crystals: Controlling the growth on flat plates, 3D architectures, and as patterns. *Adv. Funct. Mater.* **2014**, *24*, 1969–1977. [CrossRef]
27. Abuzalat, O.; Wong, D.; Elsayed, M.; Park, S.; Kim, S. Sonochemical fabrication of Cu(II) and Zn(II) metal-organic framework films on metal substrates. *Ultrason. Sonochem.* **2018**, *45*, 180–188. [CrossRef]
28. Zhu, C.; Hou, J.; Wang, X.; Wang, S.; Xu, H.; Hu, J.; Jing, L.; Wang, S. Optimizing ligand-to-metal charge transfer in metal–organic frameworks to enhance photocatalytic performance. *Chem. Eng. J.* **2024**, *499*, 156527. [CrossRef]
29. Chen, M.; Cheng, X.; Li, J.; Wang, N. Bifunctional polyimide/ZIF8 composite nanofibrous membranes with controllable bilayer structure for bioprotective application and high-efficiency oil/water separation. *J. Environ. Chem. Eng.* **2023**, *11*, 110913. [CrossRef]
30. Zhang, X.; Liu, Y.; Li, S.; Kong, L.; Liu, H.; Li, Y.; Han, W.; Yeung, K.L.; Zhu, W.; Yang, W.; et al. New membrane architecture with high performance: ZIF-8 membrane supported on vertically aligned ZnO nanorods for gas permeation and separation. *Chem. Mater.* **2014**, *26*, 1975–1981. [CrossRef]
31. Tian, L.; Sun, Y.; Huang, H.; Guo, X.; Qiao, Z.; Meng, J.; Zhong, C. Porous ZIF-8 thin layer coating on ZnO hollow nanofibers for enhanced acetone sensing. *Chemistryselect.* **2020**, *5*, 2401–2407. [CrossRef]
32. Dong, X.L.; Lin, Y.S. Synthesis of an organophilic ZIF-71 membrane for pervaporationsolvent separation. *Chem. Commun.* **2013**, *49*, 1196–1198. [CrossRef] [PubMed]
33. Dong, X.; Huang, K.; Liu, S.; Ren, R.; Jin, W.; Lin, Y.S. Synthesis of zeolitic imidazolate framework-78 molecular-sieve membrane: Defect formation and elimination. *J. Mater. Chem.* **2012**, *22*, 19222–19227. [CrossRef]
34. Kasik, A.; James, J.; Lin, Y.S. Synthesis of ZIF-68 membrane on a ZnO modified α-Alumina support by a modified reactive seeding method. *Ind. Eng. Chem. Res.* **2016**, *55*, 2831–2839. [CrossRef]
35. Wang, X.; Sun, M.; Meng, B.; Tan, X.; Liu, J.; Wang, S.; Liu, S. Formation of continuous and highly permeable ZIF-8 membranes on porous alumina and zinc oxide hollow fibers. *Chem. Commun.* **2016**, *52*, 13448–13451. [CrossRef]
36. Meckler, S.M.; Li, C.; Queen, W.L.; Williams, T.E.; Long, J.R.; Buonsanti, R.; Milliron, D.J.; Helms, B.A. Sub-micron polymer–zeolitic imidazolate framework layered hybrids via controlled chemical transformation of naked ZnO nanocrystal films. *Chem. Mater.* **2015**, *27*, 7673–7679. [CrossRef]
37. Wang, R.; Chang, Y.; Li, J.; Yang, S.; Zhu, T.; Bi, Y.; Cui, J. Carbonic anhydrase-embedded ZIF-8 membrane reactor with improved the recycling and stability for efficient CO_2 capture. *Int. J. Biol. Macromol.* **2024**, *280*, 136083. [CrossRef]
38. Kong, L.; Zhang, X.; Liu, H.; Qiu, J. Synthesis of a highly stable ZIF-8 membrane on a macroporous ceramic tube by manual-rubbing ZnO deposition as a multifunctional layer. *J. Membr. Sci.* **2015**, *490*, 354–363. [CrossRef]
39. Tian, F.; Cerro, A.M.; Mosier, A.M.; Wayment-Steele, H.K.; Shine, R.S.; Park, A.; Webster, E.R.; Johnson, L.E.; Johal, M.S.; Benz, L. Surface and stability characterization of a nanoporous ZIF-8 thin Film. *J. Phys. Chem. C* **2014**, *118*, 14449–14456. [CrossRef]
40. Ai, S.; Guo, X.; Zhao, L.; Yang, D.; Ding, H. Zeolitic imidazolate framework-supported Prussian blue analogues as an efficient Fenton-like catalyst for activation of peroxymonosulfate. *Colloids Surf. A* **2019**, *581*, 123796. [CrossRef]
41. Kwon, H.T.; Jeong, H.-K.; Lee, A.S.; An, H.S.; Lee, J.S. Heteroepitaxially grown zeolitic imidazolate framework membranes with unprecedented propylene/propane separation performances. *J. Am. Chem. Soc.* **2015**, *137*, 12304–12311. [CrossRef]
42. Hillman, F.; Brito, J.; Jeong, H.-K. Rapid one-pot microwave synthesis of mixed-linker hybrid zeolitic-imidazolate framework membranes for tunable gas separations. *ACS Appl. Mater. Interfaces* **2018**, *10*, 5586–5593. [CrossRef] [PubMed]
43. Xu, B.; Mei, Y.; Xiao, Z.; Kang, Z.; Wang, R.; Sun, D. Monitoring thermally induced structural deformation and framework decomposition of ZIF-8 through in situ temperature dependent measurements. *Phys. Chem. Chem. Phys.* **2017**, *19*, 27178–27183. [CrossRef] [PubMed]
44. Schejn, A.; Balan, L.; Falk, V.; Aranda, L.; Medjahdi, G.; Schneider, R. Controlling ZIF-8 nano- and microcrystal formation and reactivity through zinc salt variations. *CrystEngComm* **2014**, *16*, 4493–4500. [CrossRef]
45. Cravillon, J.; Schröder, C.A.; Bux, H.; Rothkirch, A.; Caro, J.; Wiebcke, M. Formate modulated solvothermal synthesis of ZIF-8 investigated using time-resolved in situ X-ray diffraction and scanning electron microscopy. *CrystEngComm* **2012**, *14*, 492–498. [CrossRef]
46. Hillman, F.; Zimmerman, J.M.; Paek, S.-M.; Hamid, M.R.A.; Lim, W.T.; Jeong, H.-K. Rapid microwave-assisted synthesis of hybrid zeolitic–imidazolate frameworks with mixed metals and mixed linkers. *J. Mater. Chem. A* **2017**, *5*, 6090–6099. [CrossRef]
47. Lee, M.J.; Hamid, M.R.A.; Lee, J.; Kim, J.S.; Lee, Y.M.; Jeong, H.-K. Ultrathin zeolitic-imidazolate framework ZIF-8 membranes on polymeric hollow fibers for propylene/propane separation. *J. Membr. Sci.* **2018**, *559*, 28–34. [CrossRef]
48. Enomoto, T.; Ueno, S.; Hosono, E.; Hagiwara, M.; Fujihara, S. Size-controlled synthesis of ZIF-8 particles and their pyrolytic conversion into ZnO aggregates as photoanode materials of dye-sensitized solar cells. *CrystEngComm* **2017**, *19*, 2844–2851. [CrossRef]
49. Zhang, C.; Wang, X.; Hou, M.; Li, X.; Wu, X.; Ge, J. Immobilization on metal–organic framework engenders high sensitivity for enzymatic electrochemical detection. *ACS Appl. Mater. Interfaces* **2017**, *9*, 13831–13836. [CrossRef]
50. Muzart, J. N,N-Dimethylformamide: Much more than a solvent. *Tetrahedron* **2009**, *65*, 8313–8323. [CrossRef]

51. Zhu, M.; Venna, S.R.; Jasinski, J.B.; Carreon, M.A. Room-temperature synthesis of ZIF-8: The coexistence of ZnO nanoneedles. *Chem. Mater.* **2011**, *23*, 3590–3592. [CrossRef]
52. Nian, P.; Liu, H.; Zhang, X. Bottom-up synthesis of 2D Co-based metal–organic framework nanosheets by an ammonia-assisted strategy for tuning the crystal morphology. *CrystEngComm* **2019**, *21*, 3199–3208. [CrossRef]
53. Khan, M.A.; Magnone, E.; Kang, Y.M. Elucidation of optoelectronic properties of the sol-gel-grown Al-doped ZnO nanostructures. *J. Sol-Gel Sci. Technol.* **2015**, *77*, 642–649. [CrossRef]
54. Gobo, F.; Goto, T.; Long, T.; Yin, S.; Sato, T. Mild solution synthesis of plate-like and rod-like ZnO crystals. *Res. Chem. Intermed.* **2011**, *37*, 211–217. [CrossRef]
55. Xu, F.; Lu, Y.; Xie, Y.; Liu, Y. Controllable morphology evolution of electrodeposited ZnO nano/micro-scale structures in aqueous solution. *Mater. Des.* **2009**, *30*, 1704–1711. [CrossRef]
56. Ji, H.; Hwang, S.; Kim, K.; Kim, C.; Jeong, N.C. Direct in situ conversion of metals into metal–organic frameworks: A strategy for the rapid growth of MOF films on metal substrates. *ACS Appl. Mater. Interfaces* **2016**, *8*, 32414–32420. [CrossRef]
57. Chen, H.; Wang, L.; Yang, J.; Yang, R. Investigation on hydrogenation of metal–organic frameworks HKUST-1, MIL-53, and ZIF-8 by hydrogen spillover. *J. Phys. Chem. C* **2013**, *117*, 7565–7576. [CrossRef]
58. Cravillon, J.; Nayuk, R.; Springer, S.; Feldhoff, A.; Huber, K.; Wiebcke, M. Controlling zeolitic imidazolate framework nano-and microcrystal formation: Insight into crystal growth by time-sesolved in situ static light scattering. *Chem. Mater.* **2011**, *23*, 2130–2141. [CrossRef]
59. Hu, H.; Zhang, X.; Liu, W.; Hou, Q.; Wang, Y. Advances in bioinspired and multifunctional biomaterials made from chiral cellulose nanocrystals. *Chem. Eng. J.* **2023**, *474*, 145980. [CrossRef]

metals

MDPI

Article

Microstructure and Properties of Mooring Chain Steel Prepared by Selective Laser Melting

Xiaojie Cui †, Xiaoxin Li †, Changqing Hu *, Dingguo Zhao, Yan Liu and Shuhuan Wang

School of Metallurgy and Energy, Metallurgical Engineering, North China University of Science and Technology, Tangshan 063210, China; cuixiaojie99@163.com (X.C.); li15176555391@163.com (X.L.); zhaodingguo@ncst.edu.cn (D.Z.); liuyan1342991123@163.com (Y.L.); wshh88@ncst.edu.cn (S.W.)

* Correspondence: hiq-73@163.com

† These authors contributed equally to this work.

Abstract: 22MnCrNiMo steel, a high-strength low-alloy material, is primarily used in the production of mooring chains for offshore oil platforms, offshore wind turbines, and ships. The application of additive manufacturing technology allows for the direct fabrication of seamless mooring chains. This paper investigates the selective laser melting (SLM) process parameters for 22MnCrNiMo mooring chain steel, analyzing the effects of different process parameters on the microstructure, phase composition, and mechanical properties of the steel. The experimental results demonstrate that under the laser parameters of 200 W laser power, 800 mm/s scanning speed, 30 μm layer thickness, and 110 μm scanning spacing, the SLM-formed parts exhibit the best comprehensive mechanical properties, with a microhardness of 513.2 HV0.5, a tensile strength of 1223 MPa, a yield strength of 1114 MPa, an elongation of 8.5%, and an impact energy of 127 J. This study reveals the microstructure evolution and the mechanism of enhanced mechanical properties in SLM-fabricated 22MnCrNiMo steel, providing a new approach for the preparation of high-performance mooring chains using 22MnCrNiMo steel.

Keywords: mooring chain steel; selective laser melting; microstructure; machine property

1. Introduction

Mooring chain steel belongs to the category of high-strength low-alloy steel (HSLA) and is traditionally produced in steelworks through melting, casting, and rolling processes, resulting in steel billets. These billets are then heated at anchor chain processing plants, formed into links through bending and welding, and finally assembled into high-strength, corrosion-resistant mooring chains. Mooring chains are a critical component in the anchoring systems of offshore floating structures used in oil and gas extraction and wind energy generation [1–4].

Mooring chains have extensive application in offshore platforms, ship anchoring, and other marine engineering projects [5–7], particularly in R3- to R5-grade chains, whose mechanical and corrosion resistance properties have been extensively studied. R6-grade mooring chain steel has also been developed and is now in use. Continuously improving the mechanical properties of mooring chain steel can effectively enhance the performance of mooring chains [8–10]. However, the conventional manufacturing process of mooring chain steel involves bending, forming links, and welding, resulting in welds at each link, which typically exhibit lower impact toughness and corrosion resistance.

In the field of additive manufacturing of metal materials, selective laser melting (SLM) technology has emerged as an efficient, high-precision, and integrated method for the fast production of metal components. SLM can produce highly complex geometric parts with minimal fabrication steps and shorter production cycles, without the need for specific molds or pre-production costs [11–13]. Research by Chen et al. [14], Liverani et al. [15], and Karlsson et al. [16] highlights that parameters such as laser power, scanning speed, scanning strategies, and sample orientation significantly affect the microstructure, thereby influencing the mechanical properties of the fabricated components. Zhang et al. [17], Qiu et al. [18], and Montero-Sistiaga et al. [19] have studied the impact of laser power on porosity and grain structure development during manufacturing, which in turn affects the mechanical properties of the samples. Additionally, the composition of microstructures significantly influences mechanical properties; for instance, Wan et al. [20] proposed that both martensite and bainite structures contain high densities of dislocations, which, when moving, interact and increase stress, thereby enhancing the hardness and strength of steel samples. Sun et al. [21] demonstrated that the stress required for dislocation slip in crystal structures with twinning is 1 to 1.3 times higher than that without twinning.

Inspired by additive manufacturing technologies, the application of SLM in mooring chain fabrication enables the production of structurally integrated components with optimized performance. By leveraging new design strategies to align the relationship between the structure and properties of mooring chains, this paper employs SLM to fabricate high-performance 22MnCrNiMo mooring chain steel. The study focuses on altering the laser energy density parameters utilized for printing the steel and investigates the resulting phase organization and mechanical properties.

2. Experimental Materials and Methods

2.1. Mooring Chain Steel Experimental Materials

The experimental material was prepared using a vacuum induction melting gas atomization (VIGA) method to produce 22MnCrNiMo steel powder. This powder is produced by Foshan Chengfeng Material Technology Co., Ltd. (Foshan, China).

The primary chemical composition was analyzed using ICP-OES (Inductively Coupled Plasma Optical Emission Spectrometry) and a carbon–sulfur analyzer, as shown in Table 1. The steel belongs to the R4-grade mooring chain steel category.

Table 1. Chemical composition of 22MnCrNiMo steel powder (wt%).

Element	C	Si	Mn	S	P	Cr	Mo	Ni	Nb	Cu	Al	Fe
Actual composition	0.229	0.236	1.396	0.001	0.003	0.946	0.390	0.815	0.001	0.026	0.028	bal. *
R4 standard ingredient	0.18~0.28	0.15~0.30	1.20~1.75	≤0.025	≤0.025	0.40~1.30	0.20~0.60	0.40~1.40	≤0.06	≤0.20	0.020~0.05	bal. *

* bal. refers to the balance amount, meaning the remaining composition is primarily Fe (iron).

The gas-atomized raw powder has a relatively fine particle size. The powder particle size distribution was measured using a Malvern laser particle size analyzer, and the resulting particle size distribution graph is shown in Figure 1. It can be observed that the proportion of powder with particle sizes suitable for 3D printing (15~53 μm) is the highest. The powder flowability was measured using a Hall flow meter, yielding a flow time of 18.6 s per 50 g. The apparent density of the powder was measured using a Topsizer particle size analyzer, resulting in a bulk density of 4.16 g/cm^3.

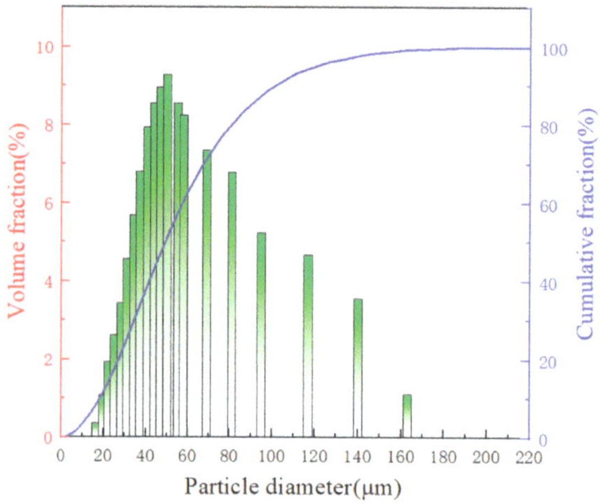

Figure 1. Particle size distribution of powder. The red line represents the percentage of each size, while the blue line shows the cumulative percentage, which totals 100%.

The microstructure of the powder is shown in Figure 2. It can be observed that the powder exhibits a high sphericity with no significant agglomeration, making it suitable for SLM (selective laser melting) for the production of printed parts. The atomized powder was dried in a vacuum drying oven with drying parameters set at 80 °C for 6 h. After screening the powder through a specialized sieve, powder with a particle size range of 15–53 μm and an average particle size of 30.3 μm was selected as the raw material for additive manufacturing of mooring chain steel.

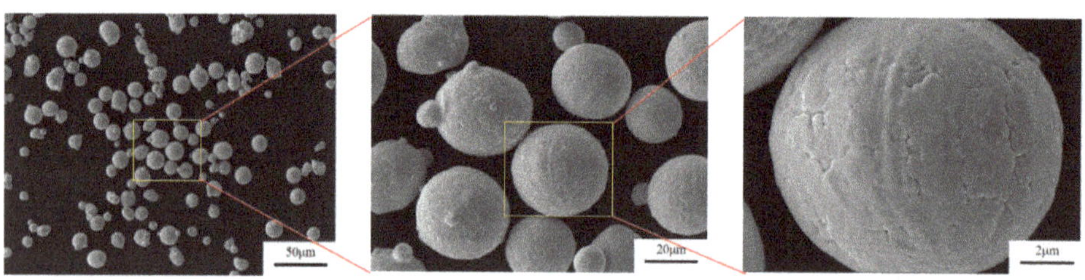

Figure 2. Microscopic morphology of 22MnCrNiMo steel powder prepared by aerosolization.

2.2. Method for Preparing Mooring Chain Steel by Selective Laser Melting

This study employed the EP-M150 small-scale metal laser additive manufacturing machine by Eplus3D (Beijing, China) to fabricate mooring chain steel components, as shown in Figure 3. The base plate of the equipment is circular with a diameter of 150 mm. The selective laser melting (SLM) equipment was equipped with a continuous single-mode ytterbium fiber laser (maximum power of 500 W, wavelength of 1080 ± 5 nm). The focusing optical system utilized an F-θ lens with a focal length of 254 mm, which produces a focused laser beam spot diameter of approximately 90 μm. The experimental equipment was also equipped with a base plate preheating device capable of reaching a preheating temperature of 150 °C. During the SLM process, argon gas was used as the protective gas to prevent oxidation of the samples, ensuring that the oxygen volume fraction inside the printer's build chamber is less than 0.01%.

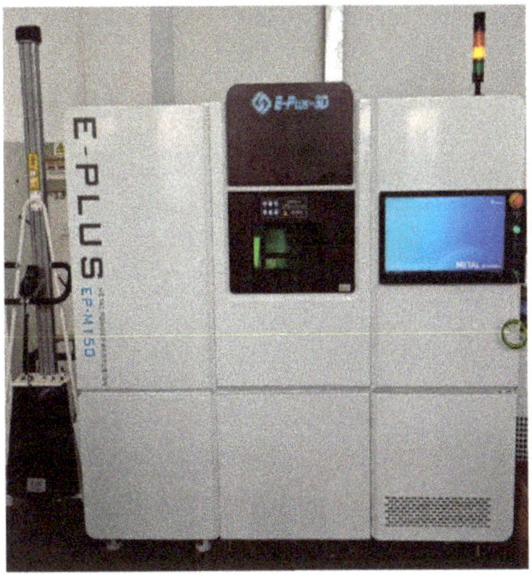

Figure 3. Additive manufacturing equipment for SLM experiment.

The schematic diagram of the process for preparing mooring chain steel by SLM is shown in Figure 4. After each layer of powder was melted, the forming cylinder descended by one layer. Then, the powder was re-spread on the powder bed, and the selective scanning continued for layer-by-layer printing. To minimize porosity and further melt the unmelted powder, a cross-scanning strategy was adopted, in which the laser scanning direction rotated periodically by 67° between adjacent layers.

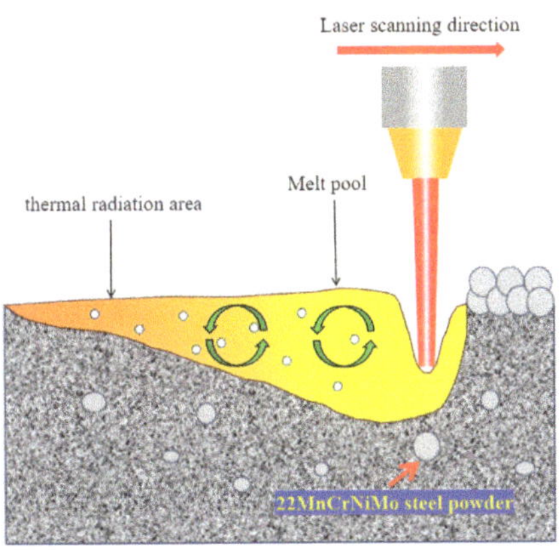

Figure 4. Schematic diagram of SLM technology preparation.

In the experimental plan, the powder bed was set with a layer thickness (t) of 30 μm and a scan spacing (h) of 110 μm. The laser processing parameters include laser power (P) and scan speed (v). The laser power was optimized within a range of 175–325 W, while three scan speeds—800, 1000, and 1200 mm/s—were chosen. As shown in Table 2, the parameter combinations studied and their corresponding energy densities are listed. Typically, in SLM, the laser energy density E (J/mm^3) is used as an evaluation metric for printing parameters, and the formula is shown in Equation (1) [22]:

$$E = P/(vhd) \tag{1}$$

where P is the laser power, v is the laser scanning speed, h is the thickness of the powder layer, and d is the laser scanning spacing.

Table 2. Experimental scheme and corresponding SLM process parameters.

Serial Number	Laser Power/W	Scanning Speed/mm·s^{-1}	Scanning Spacing/μm	Layer Thickness/μm	Energy Density/J·mm^{-3}
T1	175				66.3
T2	200				75.8
T3	225				85.2
T4	250	800			94.7
T5	275				104.2
T6	300				113.6
T7	325				123.1
T8	175				53.0
T9	200				60.6
T10	225				68.2
T11	250	1000	110	30	75.8
T12	275				83.3
T13	300				90.9
T14	325				98.5
T15	175				44.2
T16	200				50.5
T17	225				56.8
T18	250	1200			63.1
T19	275				69.4
T20	300				75.8
T21	325				82.1

A portion of the cube specimens, tensile specimens, and impact specimens obtained from the experiments is shown in Figure 5. As can be observed, the printed components exhibit good forming quality, with no macroscopic cracks or pores. The specimens have a high degree of density.

Figure 5. Cubic samples, tensile samples, and impact samples.

2.3. Testing Methods and Equipment for Microstructure and Properties of 22MnCrNiMo Steel

This study evaluates the mechanical properties and microstructural observations of three types of fabricated samples. These include 10 mm × 10 mm × 10 mm cubic specimens for microstructural observation and hardness testing, 70 mm × 31 mm × 5 mm tensile specimens, and 10 mm × 10 mm × 55 mm impact specimens, which have a 2 mm V-notch machined into them, as shown in Figure 6a–c.

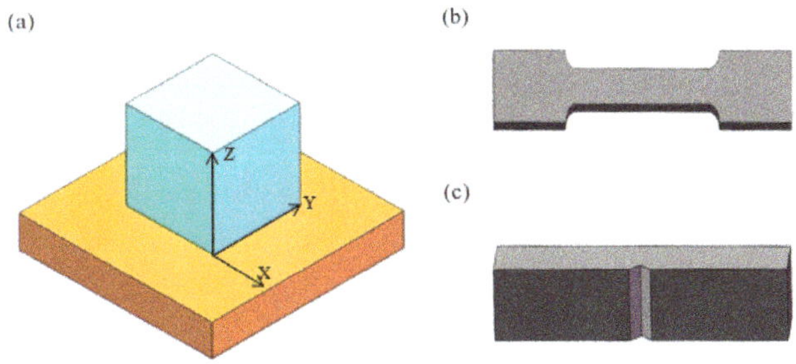

Figure 6. Schematic diagram of SLM-formed parts: (**a**) schematic diagram of cubic samples and observation direction, (**b**) schematic diagram of tensile samples, (**c**) schematic diagram of impact samples.

Before conducting optical and scanning electron microscopy (SEM) observations, the samples were polished using sandpaper with grits ranging from 240 to 2000, followed by 20 min of polishing with 1 μm diamond paste. The polished samples were then observed under a LEICA DM2500M (Suzhou Nan Guang Electronics Technology Co., Ltd., Suzhou, China) wide-field metallographic optical microscope to examine the XY and XZ planes, with the orientation of the observation surfaces shown in Figure 6a.

The samples were analyzed using a ZEISS GEMINI300 (Guoyi Quantum Technology Co., Ltd., Hefei, China) field emission scanning electron microscope (SEM) to observe their metallographic and micro- structures. The SEM was equipped with an Oxford Nordlys3 fast electron backscatter diffraction (EBSD) system (Guoyi Quantum Technology Co., Ltd., Hefei, China). For metallographic etching, a 4% nitric acid alcohol solution was used. The etching solution was applied to the sample surface for 5 s, followed by rinsing with anhydrous ethanol. Prior to EBSD experiments, the samples were ion-etched for 60 min using a PECSII 685 (Gatan Company, Pleasanton, CA, USA) device.

Tensile specimens were designed according to the ISO 6892-1 standard and tested using an AG-X 100 kN electronic universal testing machine at room temperature (Shimadzu Corporation, Kyoto, Japan) with a tensile speed of 0.5 mm/min. The schematic of the tensile specimen is shown in Figure 6b. Impact specimens were tested using a PIT-450 pendulum impact testing machine at −20 °C, with the schematic shown in Figure 6c. The fracture morphologies of the tensile and impact specimens were observed using an SEM.

An FE-800 microhardness tester (Suzhou Nan Guang Electronics Technology Co., Ltd., Suzhou, China) was used to measure the Vickers hardness (HV) of the polished samples at room temperature. The tester is suitable for microscopic analysis, with a fixed loading force of 1 kg and a loading time of 15 s. Five random points were selected on each sample surface, and the average value of the measurements was taken as the hardness of the sample.

3. Experimental Results and Analysis

3.1. Parameter Optimization and Micro-Molten Pool Morphology of 22MnCrNiMo Steel Prepared by SLM

The P–V diagram of the formed 22MnCrNiMo steel samples prepared under different laser power and scanning speed conditions is shown in Figure 7a. Improper laser forming process parameters can easily lead to defects in the samples, such as pores, unmelted powder, and microcracks [23]. As can be seen in the figure, the number of defects in the samples increased with decreasing energy density.

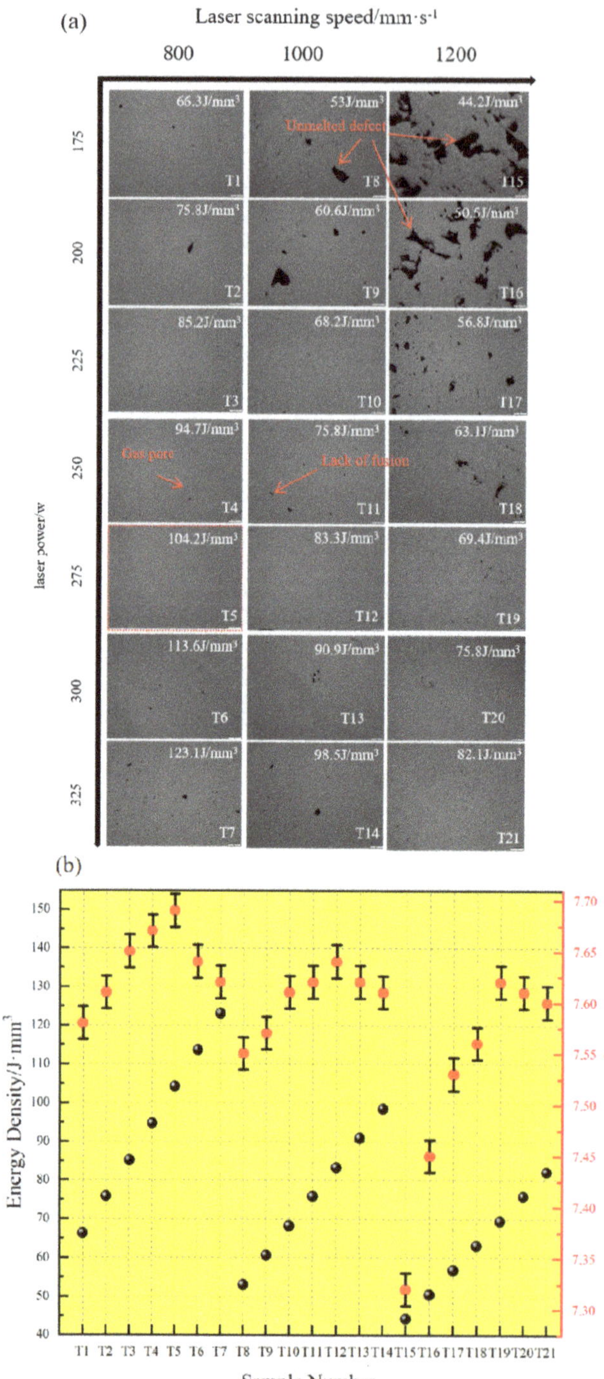

Figure 7. Optimization of parameters for SLM: (**a**) microstructure of samples, (**b**) statistics of density and energy density. The red dot represents density, and the black dot represents energy density.

Samples with excessively low laser energy density, such as sample T15, exhibit many defects. This is because the insufficient energy density prevented the powder from fully melting, resulting in unmelted material defects [24]. During the SLM (selective laser melting) process, where the laser beam cannot adequately penetrate previously deposited powder layers, defects can form between the deposition layers, as seen in sample T11. In samples T3 and T4, a few spherical micro-pores are present, primarily due to rapid solidification during the preparation process, which traps gases and often serves as a source of microcracks.

Controlling the energy density is crucial for reducing defects, and selecting an appropriate energy density can produce high-density components. Increasing the energy density can mitigate material defect formation and reduce porosity in the material. However, excessively high energy density can promote grain growth and may lead to overheating, which in turn affects material strength [25] and results in additional energy consumption.

Testing the sample density can also reflect the internal defect situation. The sample density ρ_s is calculated using Equation (2) [26]. The variation in sample density ρ_s is shown in Figure 7b, revealing that as the energy density increases, the density first rises and then decreases.

$$\rho_s = \frac{m_1}{m_1 - m_2} \times \rho_0 \quad (2)$$

In the formula, m_1 and m_2 are the mass of the sample in air and ethanol liquid (concentration of 99.7%), g; ρ_0 is the density of ethanol liquid, g/cm^3.

Taking into account all factors, we conducted in-depth experiments using the preparation process parameters for samples T1 to T7.

Formed parts of 22MnCrNiMo steel with a laser power of 275W were selected for observation using optical microscopy (OM) to analyze the morphology of the SLM (selective laser melting) micro-melt pools. Figure 8a shows the XY cross-section, illustrating the scanning traces of different orthogonal deposition layers. SLM is a layer-by-layer construction process, and, due to the scanning strategy involving 67° rotation for each layer, the micro-melt pools overlap and interconnect, resulting in a dense formation.

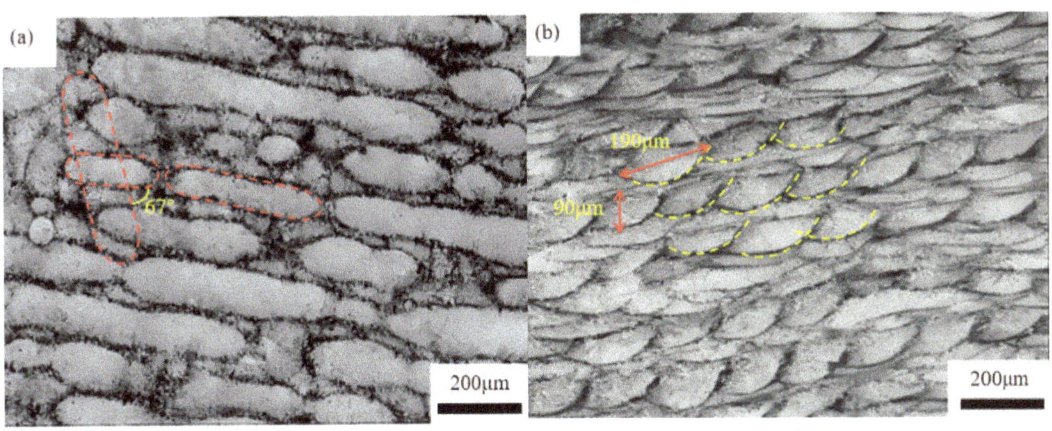

Figure 8. OM photo of 22MnCrNiMo steel sample melted by selective laser melting: (**a**) XY section, (**b**) XZ section.

As seen in Figure 8b, the XZ plane features numerous "fish scale"-like micro-melt pools with diameters ranging from 150 to 200 μm and depths of 80 to 120 μm. This is typical of the lateral micro-melt pool morphology in selective laser melting.

3.2. Microstructure of 22MnCrNiMo Steel Prepared by SLM

Figure 9 presents the XRD (X-ray Diffraction) patterns of 22MnCrNiMo steel samples prepared under different laser powers. The results indicate that the primary phase in the samples is α-FeM (where M represents elements such as Mo, Ni, and Cr), and the reinforcing phase includes Fe_3C, among others. Compared with the standard diffraction peaks, the main peaks of the samples are shifted to the left by approximately 4°. This shift is attributed to lattice distortion caused by alloy atoms in the solid solution and thermal stresses induced by the laser forming process.

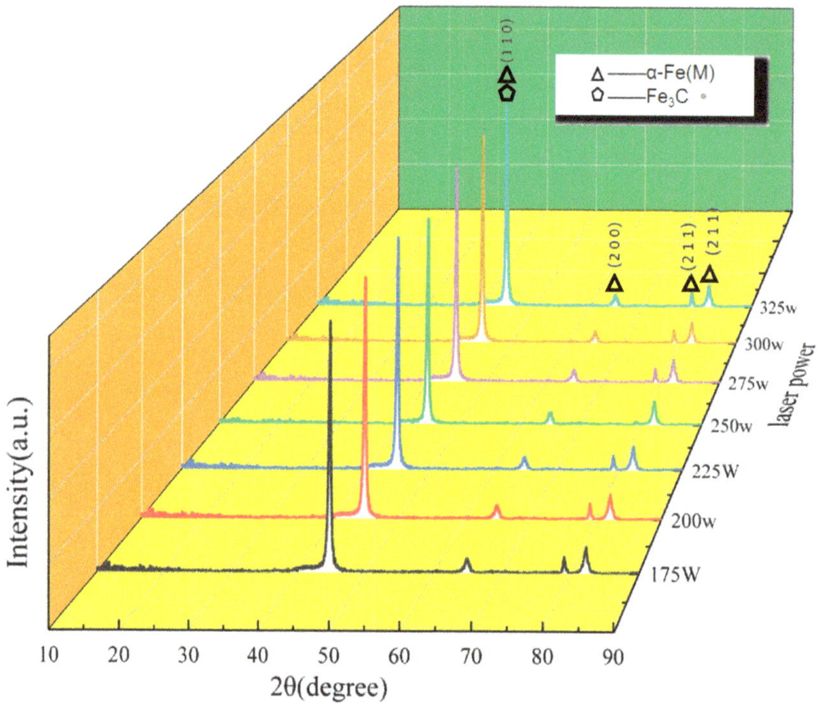

Figure 9. XRD patterns of 22MnCrNiMo steel samples under different laser powers.

Since the experimental material composition was identical, the density increased from T1 to T5. Figure 10 shows the microstructures of typical samples T2 and T5, highlighting the microstructural characteristics of 22MnCrNiMo steel produced by SLM (selective laser melting). The predominant microstructures include lath martensite (M) and bamboo-leaf-shaped lower bainite (LB). The martensite exhibits a fine lath structure, while the lower bainite appears in the form of a layered or needle-like morphology.

Figure 10a,c display the XY and XZ cross-section microstructures of sample T2, which are predominantly composed of lath martensite and needle-like lower bainite. The average grain size is approximately 3–4 μm. No significant precipitated phases were observed at the grain boundaries. Due to the more pronounced grain organization in the XZ plane, ImageJ (1.52P) software was used to calculate the area ratios of martensite and lower bainite in the XZ plane, which are 47% and 53%, respectively, as shown in Figure 11a. In the figure, the red areas represent lower bainite, while the gray areas represent martensite. The alternating arrangement of martensite and lower bainite is evident.

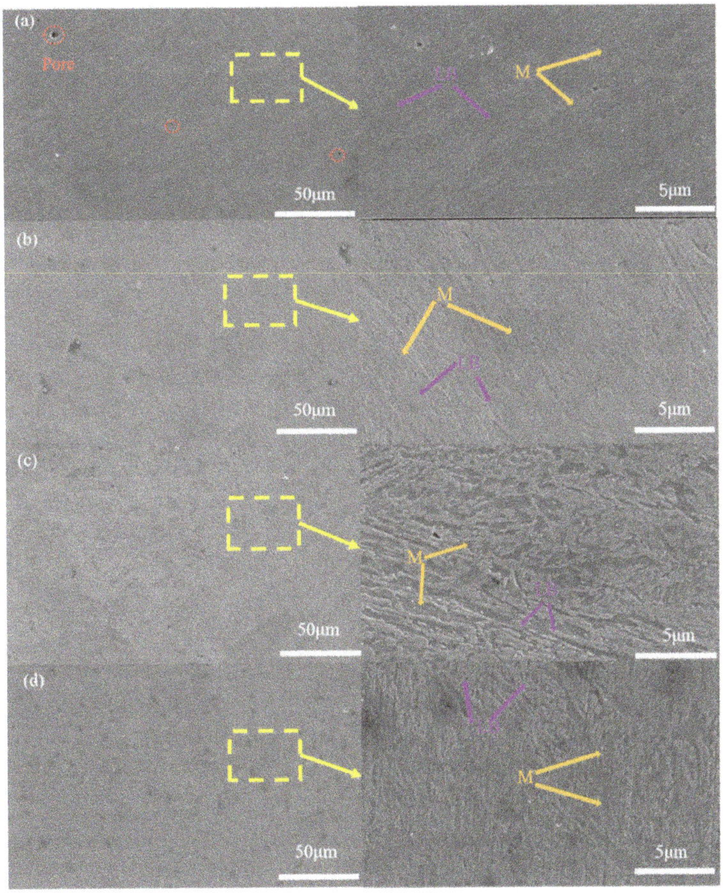

Figure 10. Scanning electron microscope images of 22MnCrNiMo steel; (**a**) T2 XY plot (**b**) T5 XY plot (**c**) T2 XZ plot (**d**) T5 XZ plot.

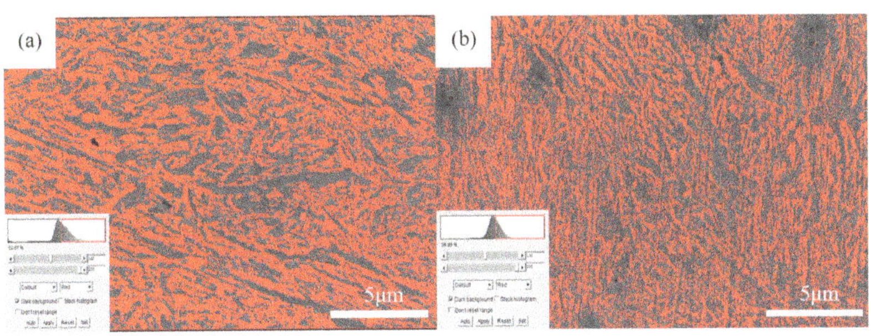

Figure 11. Statistical chart of the proportion of lower bainite area: (**a**) T2, (**b**) T5.

Figure 10b,d show the XY and XZ cross-section microstructures of sample T5, which are also primarily composed of lath martensite and needle-like lower bainite. Compared to sample T2, the microstructural composition of T5 shows little variation. However, the area ratios of martensite and lower bainite calculated for the XZ plane using ImageJ are 60% and 40%, respectively, as illustrated in Figure 11b.

Martensite enhances the strength and hardness of mooring chain steel, while the lower bainite structure, in addition to providing high strength, reduces the notch sensitivity of the steel, thereby enabling the mooring chain steel to better withstand impact loads. Therefore, theoretically, the microstructure at 200 W should exhibit higher strength and impact toughness. As observed in the XZ plane, the two types of structures are interspersed, which is similar to the microstructural characteristics of 24CrNiMo steel fabricated using a higher laser energy density [27].

In this study, the lower laser energy density employed resulted in shorter solidification times, which may have contributed to the formation of lower bainite and martensite in the samples, thereby improving the strength of the 22MnCrNiMo steel. During the SLM process, the rapid cooling rate of the melt pools, as high as 105~106 K/s, easily lead to the formation of martensitic structures [28].

Figure 12 shows the EBSD (Electron Backscatter Diffraction) maps of 22MnCrNiMo steel on the XY and XZ planes at different laser powers (200 W and 275 W). The images reveal that SLM-fabricated 22MnCrNiMo steel exhibits a microstructure of columnar and equiaxed grains. The XY plane is dominated by lath martensite and needle-like bainite, with an average grain size of approximately 3–4 μm. No significant precipitated phases were observed at the grain boundaries, and no preferred orientations were present in the crystal structure. The grain size of the XZ plane is smaller compared to the XY plane, and the structure is predominantly single-phase BCC, with almost no FCC structure, which is a typical characteristic of SLM-fabricated materials. For cubic crystal structures, the EBSD results indicate the absence of significant preferred orientations.

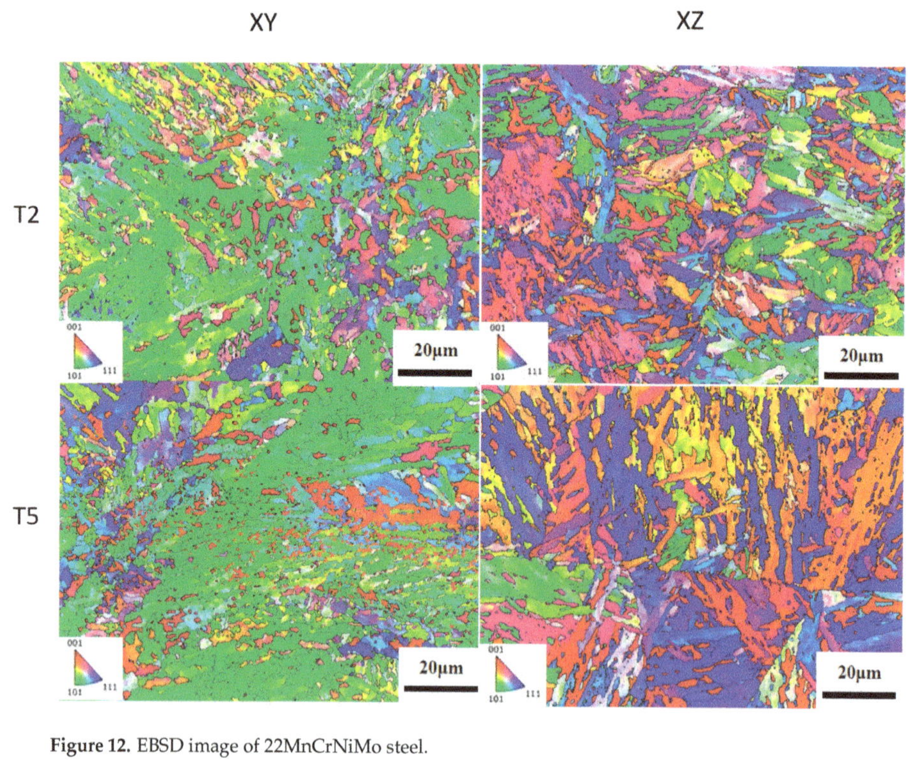

Figure 12. EBSD image of 22MnCrNiMo steel.

The grain size of the EBSD maps was specifically analyzed using the Aztec crystal 2.1 software, which yielded an average grain size of 3.2 μm for sample T2 and 3.8 μm for sample T5.

Figure 13 illustrates the EBSD grain boundary analysis results of 22MnCrNiMo steel under different laser powers. Grain boundaries with orientation angles between 2° and 15° are classified as small-angle grain boundaries (LAGBS), represented by black lines in the figure. Grain boundaries with orientation angles greater than 15° are classified as large-angle grain boundaries (HAGBS), represented by red lines in the figure.

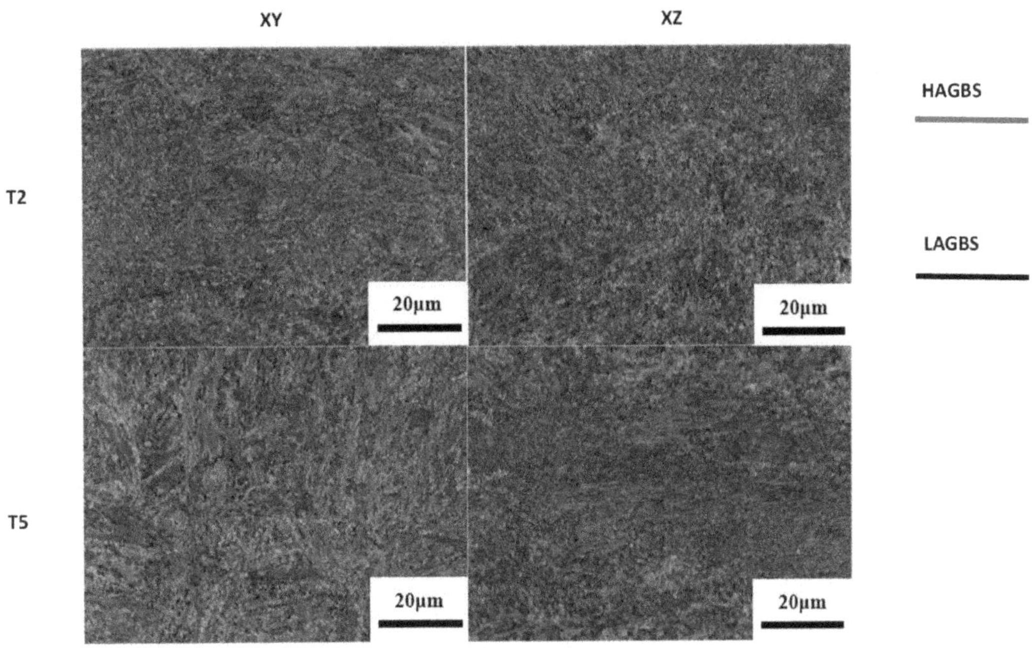

Figure 13. Distribution of grain boundaries of 22MnCrNiMo steel.

The microstructures under different processing parameters exhibit varying proportions of large-angle and small-angle grain boundaries. Specifically, for sample T2, the average proportion of HAGBs on the XY and XZ planes is 56%, while the average proportion of LAGBS is 44%. For sample T5, the average proportion of HAGBS on the XY and XZ planes is 44%, while the average proportion of LAGBS is 56%. Consequently, LAGBs dominates in sample T2, corresponding to the dark gray regions in the figure, while HAGBs dominates in sample T5, corresponding to the red regions.

LAGBS serve as an indicator of dislocations and residual stresses, whereas the presence of HAGBs can significantly enhance the toughness of the steel by effectively inhibiting the propagation of linear cracks [29].

Figure 14 shows a comparison of the average grain size of the XY and XZ planes of the samples under different laser powers. It can be seen from the figure that the average grain size of both the XY and XZ planes in the T2 sample is smaller than T5. Therefore, it can be concluded that the average grain size of the samples increases with the increase in laser energy density during SLM additive manufacturing.

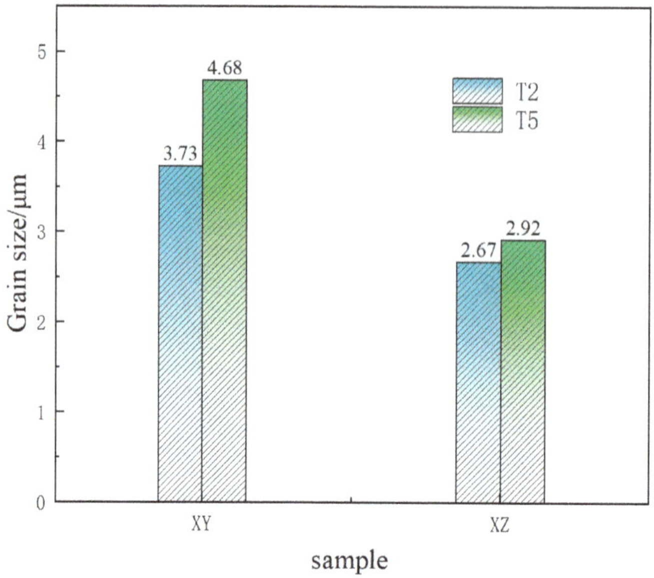

Figure 14. Grain distribution of 22MnCrNiMo steel.

The variation in the grain size of the samples is closely related to the nucleation density, m, and undercooling degree, ΔT, during the solidification process of the micro-melt pools, as expressed by Equation (3) shown below [30]:

$$m(\Delta T) = \int_0^{\Delta T} \frac{dn}{d(\Delta T)} d(\Delta T) = \frac{m_{max}}{\Delta T_\sigma} \int_0^{\Delta T} exp[-\frac{(\Delta T - \Delta T_N)^2}{2(\Delta T_\sigma)^2}] d(\Delta T) \tag{3}$$

where m_{max} is the maximum nucleation density, ΔT_σ is the standard deviation, and $\Delta T_{\rm N}$ is the average undercooling.

When the laser energy density is low, the temperature gradient is small, but the cooling rate is high, leading to a greater degree of undercooling. This results in a higher nucleation density and smaller grain size. Conversely, as the laser energy density increases, the cooling rate decreases, and the undercooling degree reduces, leading to a lower nucleation density and larger grain size.

Compared to low-alloy steels produced through forging or rolling, low-alloy steels prepared using the SLM process exhibit smaller grain sizes and a higher proportion of high-angle grain boundaries. The smaller grain size improves the strength and hardness of the samples.

3.3. *Mechanical Properties of 22MnCrNiMo Steel*

3.3.1. Tensile Properties of Mooring Chain Steel

Table 3 presents the mechanical properties data of 22MnCrNiMo steel fabricated under different laser parameters. As seen in the table, the tensile strength of 22MnCrNiMo steel produced by selective laser melting far exceeds the standard value of 860 MPa, averaging over 38% higher. Additionally, the variations in the strength and elongation of samples T2 and T5 align with the influence of the proportions of high-angle and low-angle grain boundaries on performance.

Table 3. Mechanical properties of 22MnCrNiMo steel with different laser parameters.

Sample	Laser Power/W	Tensile Strength/MPa	Yield Strength/MPa	Elongation/%
T1	175	1281	1121	6.7
T2	200	1223	1114	8.5
T3	225	1195	1089	8.9
T4	250	1184	1035	9.2
T5	275	1175	1010	10.2
T6	300	1190	1046	7.2
T7	325	1194	1051	6.9
S1	standard	>860 MPa	>580 MPa	>12
O1	Liang et al. [31]	1009	929.5	18
O2	Mainier et al. [32]	857	937	16.1

Experimental results also show that the elongation of SLM-formed steel is slightly lower than the standard value. Elongation mainly affects the plasticity of formed parts and their ductility during reprocessing. Since selective laser melting (SLM) can directly fabricate the desired shape without requiring additional reprocessing or shaping of specific forms, the influence of elongation in the SLM process is relatively insignificant.

As shown in Figure 15, a comparison of the schematic diagrams for the traditional manufacturing process and the new additive manufacturing process of mooring chains reveals that the traditional process for mooring chains involves smelting–casting–rolling–bending–welding, while the new additive manufacturing process eliminates the rolling, bending, and welding steps. This makes the requirement for elongation less critical. Therefore, the elongation requirements for 22MnCrNiMo mooring chain steel prepared by the new additive manufacturing process merit further evaluation, and the use of selective laser melting (SLM) technology for manufacturing mooring chains has pioneering and great significance.

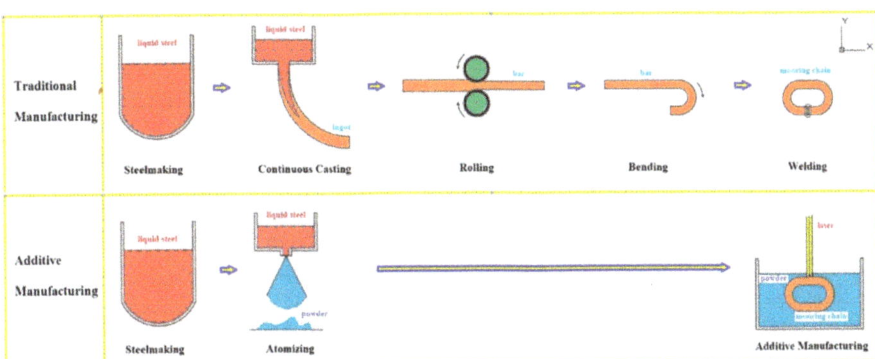

Figure 15. Schematic diagrams of Traditional Manufacturing (TM) process and Additive Manufacturing (AM) process for mooring chains.

Figure 16 shows the engineering stress–strain curves of 22MnCrNiMo steel specimens prepared under three typical laser powers: low, medium, and high (200 W, 250 W, 300 W, respectively). As can be seen in Figure 16a, the elongation of T5 is significantly higher than that of T2 and T6, while the maximum engineering stress of T2 is significantly higher than that of T5 and T6.

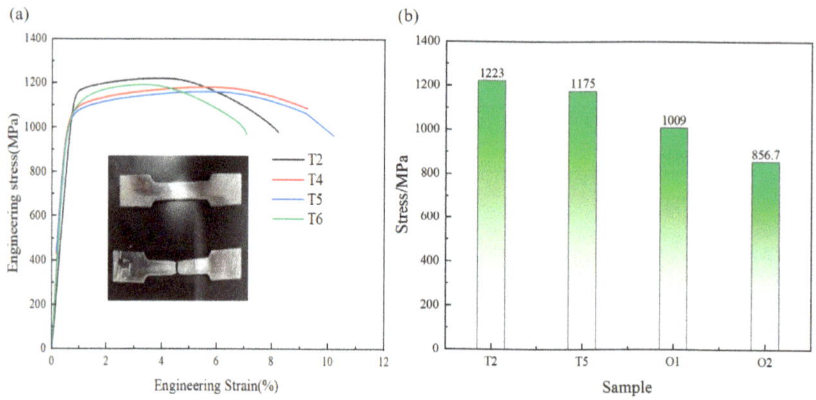

Figure 16. Mechanical properties of 22MnCrNiMo steel under different manufacturing processes: (**a**) mechanical properties of 22MnCrNiMo steel via novel additive manufacturing processes; (**b**) tensile strength of mooring chain steel under different conventional manufacturing processes.

The T1 specimen with a laser power of 175W has the highest strength, with a tensile strength of 1227 MPa, a yield strength of 1116 MPa, and an elongation of 6.7%. The T4 specimen with a laser power of 275 W has the best plasticity, with an elongation of 10.2%, a tensile strength of 1175 MPa, and a yield strength of 1010 MPa. The elongation of the 22MnCrNiMo steel specimens first increases and then decreases with the increase in the laser power, while the tensile strength and yield strength show an opposite trend to the change in elongation.

The new additive manufacturing process produces mooring chain steel with superior performance compared with traditional manufacturing processes, as illustrated in Figure 16b. Qiu et al. [31] used 22MnCrNiMo steel sourced from 130 mm diameter annealed rolled bars for their experiments. The manufacturing process included electric furnace smelting, LF + VD, continuous casting, red material cold drawing and annealing, material production, pit cooling and holding, and final annealing. The final annealing temperature was 670 ± 10 °C, followed by air cooling to 500 °C at a cooling rate less than 30 °C/h. The samples were then held at 910 °C for 2.5 h before water quenching and tempered at 610 °C for 3 h followed by another water quench. Under these conditions, the experimental steel achieved a maximum tensile strength of 1009 MPa.

Mainier et al. [32] used 22MnCrNiMo steel for their experiments. The steel ingots were homogenized in a heat treatment furnace at 1200 °C for 15 h, then hot rolled into cylinders with a nominal diameter of 123 mm and a reduction ratio of 28:1. The samples were held at 900 °C for 1 h, water quenched, reheated to 650 °C, held for 2 h, and water quenched again. Under these conditions, the experimental steel achieved a maximum tensile strength of 857 MPa.

In comparison to 22MnCrNiMo steel produced by traditional manufacturing processes, the strength of steel produced by the new additive manufacturing process increased by over 18%. The SLM-fabricated 22MnCrNiMo steel samples had a high density, fewer internal defects, closely packed martensitic structures, a higher percentage of lower bainite, good impact toughness, and maintained high strength.

Figure 17 shows the room temperature tensile fracture morphologies of 22MnCrNiMo steel samples formed under different laser energy densities. With increasing laser power, the tensile specimens exhibited ductile fracture characteristics, with dimples evenly distributed across the fracture surfaces.

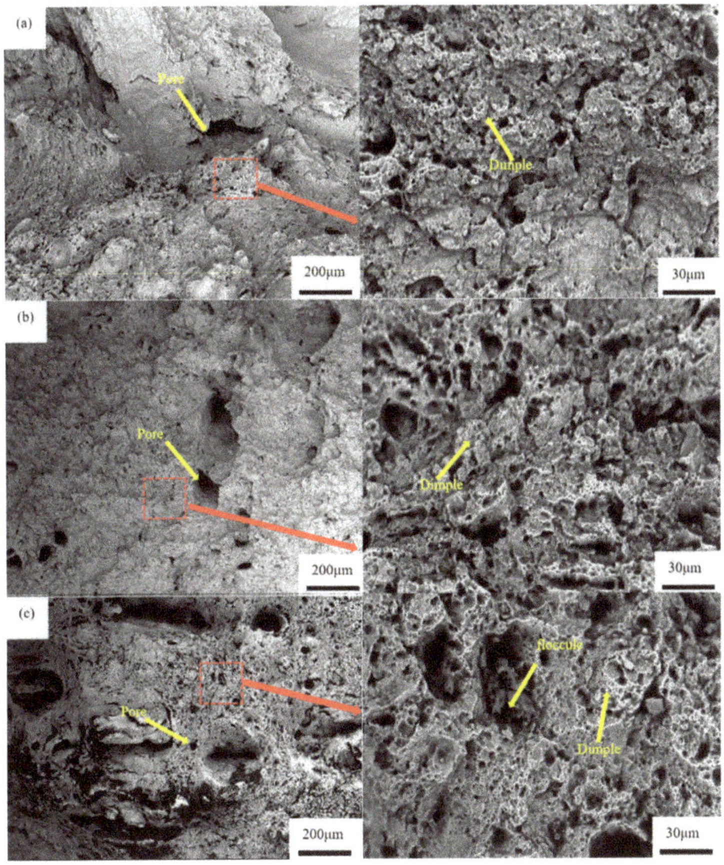

Figure 17. Tensile fracture morphology of 22MnCrNiMo steel: (**a**) T2, (**b**) T4, (**c**) T6.

From Figure 17a, it can be seen that the sample presents a smooth layered section with small pores. At this time, the fracture surface has both ductile dimples and a smooth section, belonging to a mixed ductile and brittle section. Therefore, the elongation of the T2 sample is relatively low.

From Figure 17b, when the laser power reaches 250 W, the fracture surface of the sample shows no obvious unmelted or over-burned materials, but rather a dense and fine pore defect structure. Additionally, there are densely distributed dimples, indicating ductile fracture. Therefore, theoretically, the elongation of sample T4 should be higher than that of T2 and T6, which aligns with the experimental results.

From Figure 17c, sample T6 exhibits a tensile fracture surface with numerous large internal pore defects and some flocculent material. This is due to excessive laser power causing over-burning, which not only induces stress concentration during the tensile process, triggering microcracks and reducing the tensile strength of the sample, but also diminishes the sample's ductility, adversely affecting its overall performance.

The grain size of 22MnCrNiMo steel manufactured by the traditional process is relatively large. For example, Liang Qiu et al. [31] measured that the grain size of the specimen processed by the annealing and rolling process was approximately 9.1 μm, with a tensile strength of 1006 MPa. Mainier et al. [32] measured that the grain size of the sample processed by the hot rolling process was approximately 10.5 μm, with a tensile strength of 856 MPa. Compared with the grain size of 9–11 μm of the samples processed by the traditional process, the grain size of the SLM process was 3–4 μm, a reduction of 70%. According to the principle of fine grain strengthening, the reduction in the metal grain size has a significant impact on the increase in strength. Compared with the traditional manufacturing process, the strength of the samples prepared by the new additive manufacturing process was increased by 20–50%. Figure 18 shows the relationship between the grain size and strength of samples under different processes.

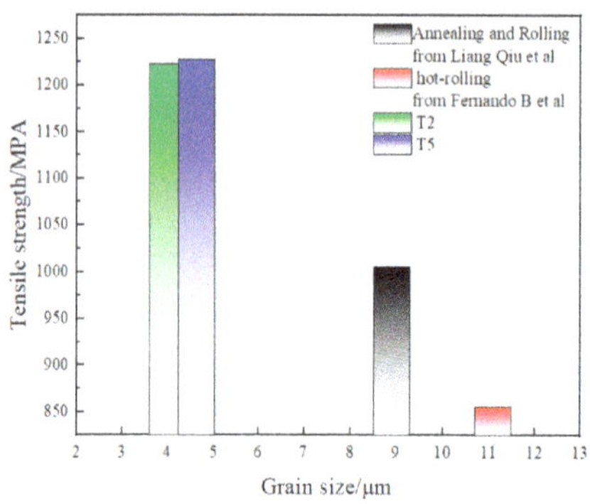

Figure 18. Relationship between grain size and strength of samples under different processes [31,32].

3.3.2. Impact Toughness of Mooring Chain Steel

The impact toughness requirement for grade R4 22MnCrNiMo mooring chain steel is greater than 50 J at −20 °C. The impact performance of mooring chain samples produced by the SLM process under different laser powers, as shown in Figure 19, reveals the following: at laser powers of 175 W and 200 W, the absorbed impact energies of the samples are 121 J and 127 J, respectively, both exceeding the standard requirement. From 200 W to 325 W, the absorbed impact energy decreases with increasing laser power, but even at 325 W, the impact energy still meets the standard requirement. Notably, the impact energy at 200 W is significantly higher than at 275 W, which aligns with the influence of the area fraction of lower bainite and martensite. Therefore, the impact toughness of 22MnCrNiMo steel produced by the selective laser melting (SLM) process meets the requirements for grade R4 mooring chains.

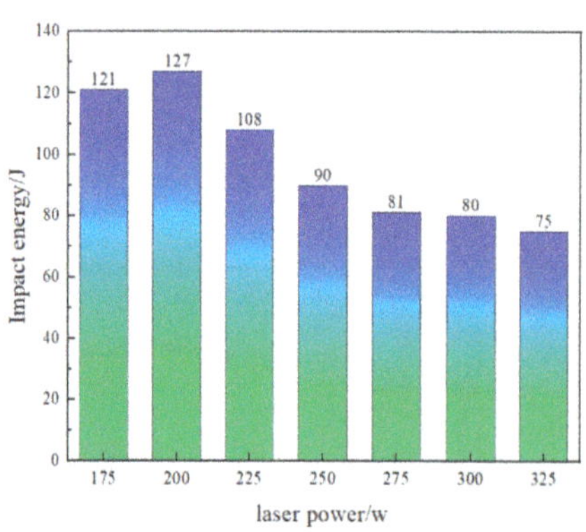

Figure 19. Impact absorption energy of 22MnCrNiMo mooring chain samples under different laser powers.

Figure 20 shows the impact fracture morphology of the 22MnCrNiMo specimen. It can be seen from the figure that there are a large number of dimples on the impact fracture surface. During the impact process, a specimen undergoes local plastic deformation. Microscopically, it manifests as the continuous expansion and connection of micro-cracks inside the material under the action of shear stress, ultimately forming small pits. The size and depth of the dimples reflect the plastic deformation ability of the material during the impact process.

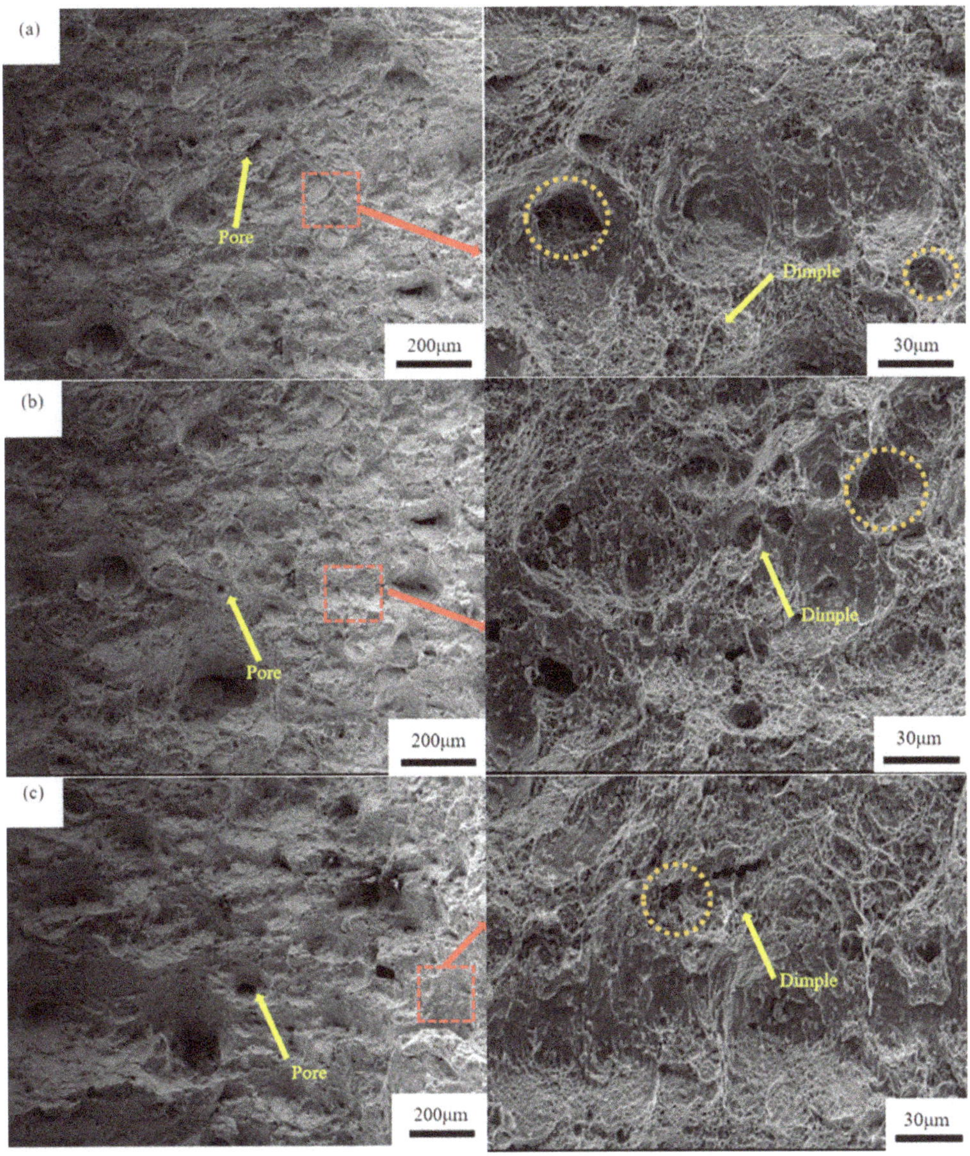

Figure 20. Impact fracture morphology of 22MnCrNiMo sample: (**a**) T2, (**b**) T4, (**c**) T6. The red squares indicate the position of the enlarged image.

In the area where the dimples are deep and large, the material has experienced a large amount of plastic deformation before fracture and has good impact toughness. As shown in the area marked by the orange circle in Figure 20, deep and large dimples can be clearly seen. The number and size of pores on the fracture surface of the specimen under the laser power of 200 W were significantly smaller than those of the specimens under the laser powers of 250 W and 300 W. Under the action of impact load, stress concentration is likely to occur around the pore defects, leading to the initiation and propagation of micro-cracks.

3.3.3. Hardness Properties of Mooring Chain Steel

Figure 21 illustrates the distribution of microhardness across the XZ section of 22MnCrNiMo steel samples prepared under different laser powers. The average hardness values for the samples prepared at laser powers of 175 W, 200 W, 225 W, 250 W, 275 W, 300 W, and 325 W are 514.4 HV0.5, 513.2 HV0.5, 436 HV0.5, 427 HV0.5, 416.4 HV0.5, 409.4 HV0.5, and 400.8 HV0.5, respectively. It can be observed that as the laser power increases, the microhardness of the samples exhibits a decreasing trend. Notably, the hardness of samples T1 and T2 is significantly higher than that of the other samples.

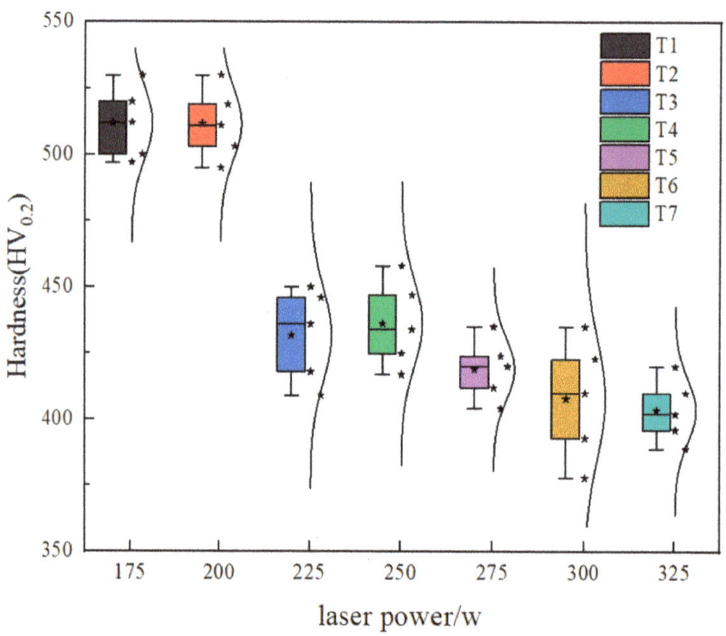

Figure 21. Hardness properties of 22MnCrNiMo steel under different laser powers. *—Represents the hardness values of the same item sampled at random points, with each sample taking 5 data points, and the average value being the final result.

Usually, the grain size and density of a sample are the key factors determining its microhardness, and the relationship between microhardness (HV) and grain size (d) can be expressed by the Hall–Petch formula [33]:

$$HV = H_0 + Kd^{-1/2} \tag{4}$$

where H_0 is the pure iron hardness determined by the crystal structure and dislocation density, and K represents the slope of the curve.

As can be seen from Equation (4), the finer the metal grains, the higher the micro-hardness. Therefore, among the parameters in this study, the specimen with an energy density of 66.29 J/mm^3 should theoretically have the highest micro-hardness, and in fact, it is the parameter with the highest hardness. On the other hand, internal defects of the material, such as pores, cracks, and lack of fusion, are weak points under load. Thus, when the energy density is 123.11 J/mm^3, it should also theoretically have a relatively high micro-hardness.

However, based on the experimental results, it is evident that grain refinement plays the most significant role in determining the microhardness of 22MnCrNiMo steel fabricated via SLM. Although samples prepared at higher energy densities exhibit relatively dense microstructures, the coarsening of their grains leads to a decrease in microhardness. This observation aligns with the findings of AlMangour et al. [34], who reported similar results for SLM-prepared SS316.

In summary, to improve the overall performance of 22MnCrNiMo steel produced via SLM, the optimal additive manufacturing parameters should be based on those used for sample T2. By selecting these parameters, the desired microstructure can be achieved, ensuring grain refinement and high densification, which are essential for the performance of mooring chain steel.

3.4. Grain Refinement Enhancement Mechanism of SLM Forming for Mooring Chain Steel

In the SLM (selective laser melting) process for fabricating mooring chain steel, the size of the molten pool formed is extremely small, often in the micrometer range. Due to the limited volume of liquid metal within the micro-melt pool, the space available for grain growth and the supply of liquid metal during the solidification process are both constrained. Consequently, the grain sizes formed are relatively fine. In contrast, traditional manufacturing processes involve larger melt pools, which provide sufficient space and liquid metal for continuous grain growth, resulting in relatively coarse grains.

Within the micro-melt pool, there exists a significant temperature gradient. The center of the melt pool is at a higher temperature, while the edges are relatively cooler. As the energy density increases, the depth of the melt pool increases, and the range of the temperature gradient expands. This temperature gradient causes the growth rate of grains near the edges to be relatively slow, while the growth of grains at the center is restricted by the edges. Under this non-uniform thermal field, grain growth occurs within a microscopic range.

During the SLM additive manufacturing process, the cooling rate of the micro-melt pool is extremely high, typically reaching 10^5 K/s or even higher. In contrast, the cooling rate during traditional water quenching after heat treatment is generally around 10^3 K/s [35]. Figure 22 illustrates the schematic variation in grain size under different manufacturing processes. The cooling rate in traditional manufacturing processes is much slower than that in SLM, which provides ample space for recrystallization, leading to coarse grains. Tan et al. [36] highlighted the unique cyclic heat treatment history of laser additive manufacturing imposes an intrinsic heat treatment (IHT) on the material during deposition. IHT involves a cyclic process of rapid cooling and heating, with high solidification rates. The rapid solidification characteristic of additive manufacturing prevents the atoms in the liquid metal from undergoing sufficient diffusion and arrangement, leaving insufficient time for grain growth. Within an extremely short timeframe, the liquid metal transforms into a solid state, and atoms can only form nuclei and grow within a small range, resulting in fine grain sizes.

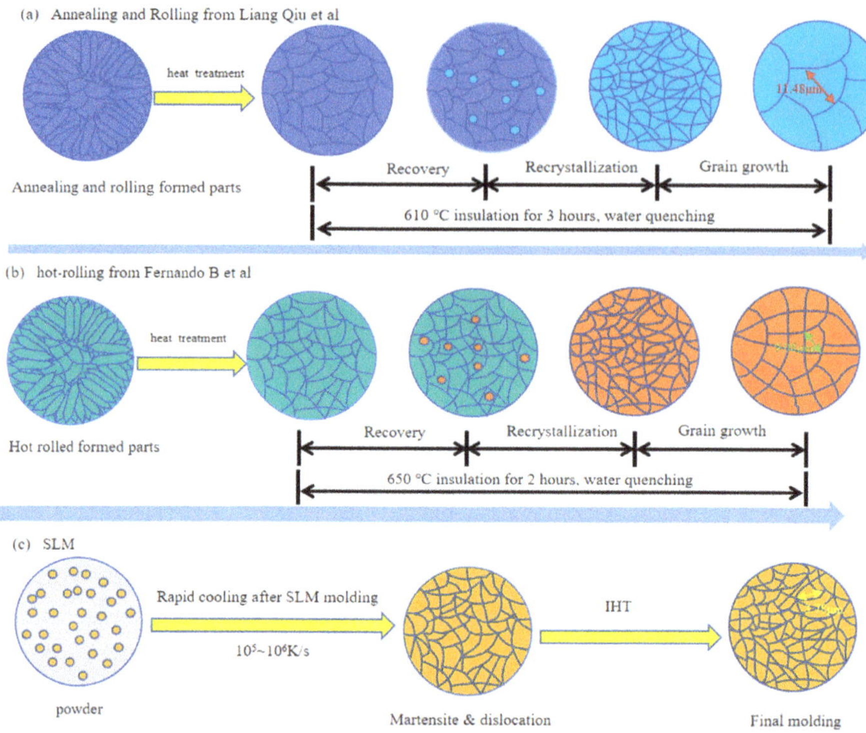

Figure 22. Schematic diagram of grain size changes in different manufacturing processes [31,32]: (**a**) annealing and rolling, (**b**) hot rolling, (**c**) SLM.

The rapid solidification inherent in additive manufacturing processes significantly increases the nucleation rate of liquid metal. Under rapid solidification conditions, the undercooling of the liquid metal increases. Greater undercooling reduces the energy barrier for nucleation in the liquid phase, making it easier to form nuclei. The formation of a large number of nuclei leads to an increase in the number of grains. However, with each grain having a reduced growth space and limited supply of liquid metal, the overall grain size becomes finer.

This phenomenon is a direct consequence of the high cooling rates associated with additive manufacturing, which not only restrict the growth of individual grains but also enhance the nucleation process, ultimately leading to a finer microstructure. These fine grains contribute to improved mechanical properties, which are critical for the performance of materials like mooring chain steel fabricated via SLM.

4. Conclusions

When manufacturing mooring chains using traditional welding processes, the toughness at the weld seams is often inferior, leading to issues such as fracture or corrosion at these locations. In contrast, the novel additive manufacturing process utilizing SLM (selective laser melting) technology produces mooring chain steel with uniform material properties throughout, eliminating problems associated with weld seams. Additionally, the traditional manufacturing of complex-structure mooring chains typically requires multiple steps and intricate molds, which are costly and challenging. SLM technology, however, can directly manufacture mooring chains based on design models without the need for molds, enabling seamless and integrated production. Furthermore, SLM can produce mooring chains with complex internal cavities and unique shapes that are difficult to achieve with traditional processes, providing greater design flexibility for mooring chain applications. The main conclusions of this study are as follows:

1. The microstructure of 22MnCrNiMo steel primarily consists of martensite and lower bainite. As the laser power increases, the area of the heat-affected zone (HAZ) expands, leading to a higher proportion of bainite structures. Optimal results are achieved at a laser power of 200 W, where the grain size becomes fine and the grain orientation is irregular, resulting in a uniform sample microstructure. This uniformity is beneficial for enhancing the mechanical properties of the steel.
2. Under the laser parameters of 200 W laser power, a scanning speed of 800 mm/s, a layer thickness of 30 μm, and a scan spacing of 110 μm, the 22MnCrNiMo steel samples exhibit the best comprehensive mechanical properties. Specifically, the microhardness reaches 513.2 HV0.5, the tensile strength is 1223 MPa, the yield strength is 1114 MPa, the elongation is 8.5%, and the impact energy is 127 J.
3. The selective laser melting (SLM) process creates micro-melt pools with a high solidification rate, which refines the grains of the mooring chain steel, significantly enhancing the sample's strength. Compared with traditional manufacturing processes, the strength of 22MnCrNiMo steel produced using the SLM additive manufacturing process is increased by over 20%. This enhanced strength lays the foundation for the production of high-strength and high-toughness mooring chains.

Author Contributions: Methodology, C.H.; software, D.Z.; data curation, Y.L. and S.W.; writing—original draft, X.C.; writing—review and editing, X.L. All authors have read and agreed to the published version of the manuscript.

Funding: The work was supported by the National Natural Science Foundation of China (No. 52274336), Hebei Province Innovation Ability Improvement Plan (No. 24461002D).

Data Availability Statement: The datasets presented in this article are readily available.

Conflicts of Interest: The authors declare no conflicts of interest.

References

1. Qvale, P.; Nordhagen, H.O.; Ås, S.K.; Skallerud, B.H. Effect of long periods of corrosion on the fatigue lifetime of offshore mooring chain steel. *Mar. Struct.* **2022**, *85*, 103236. [CrossRef]
2. Fernandez, J.; Storesund, W.; Navas, J. Fatigue performance of grade R4 and R5 mooring chains in seawater. In Proceedings of the International Conference on Offshore Mechanics and Arctic Engineering, Rio de Janeiro, Brazil, 1–6 July 2012; American Society of Mechanical Engineers: New York, NY, USA, 2014; pp. 1–6.
3. Ma, K.T.; Luo, Y.; Kwan, T.; Wu, Y. Mooring for floating wind turbines. *Mooring Syst. Offshore Struct.* **2019**, *392*, 299–315.
4. Kang, J.; Sun, L.; Sun, H.; Wu, C. Risk assessment of floating offshore wind turbine based on correlation-FMEA. *Ocean. Eng.* **2017**, *129*, 382–388. [CrossRef]
5. Pham, H.D.; Schoefs, F.; Cartraud, P.; Soulard, T.; Pham, H.H.; Berhault, C. Methodology for modeling and service life monitoring of mooring lines of floating wind turbines. *Ocean. Eng.* **2019**, *193*, 106603. [CrossRef]
6. Gotoh, K.; Murakami, K.; Nakagawa, M.; Utsunomiya, T. Wear performance of the mooring chain used in floating wind turbines. In Proceedings of the International Conference on Offshore Mechanics and Arctic Engineering, Rio de Janeiro, Brazil, 1–6 July 2012; American Society of Mechanical Engineers: New York, NY, USA, 2017; pp. 1–6.
7. Barrera, C.; Battistella, T.; Guanche, R.; Losada, I.J. Mooring system fatigue analysis of a floating offshore wind turbine. *Ocean Eng.* **2020**, *195*, 106670. [CrossRef]
8. Sidhar, H.; Fenske, J.A.; Xu, W.; Baker, D.A.; Jin, H.W.; Neerav, V.; Xin, Y. High Manganese Steel for Mooring Chain Applications. *Berg. Huettenmaenn. Monatsh.* **2022**, *167*, 530–533. [CrossRef]
9. Ankit, S.; Yoo Sang, C. Mooring chain strength tests and ductile failure modeling using micromechanics and phenomenology based failure models. *Ocean. Eng.* **2020**, *195*, 106663.
10. Bergara, A.; Arredondo, A.; Altuzarra, J.; Martínez-Espanola, J.M. Calculation of stress intensity factors in offshore mooring chains. *Ocean. Eng.* **2020**, *214*, 107762. [CrossRef]

11. Wei, M.W.; Chen, S.Y.; Xi, L.Y.; Liang, J.; Liu, C.S. Selective laser melting of 24CrNiMo steel for brake disc: Fabrication efficiency, microstructure evolution, and properties. *Laser Technol.* **2018**, *107*, 99–109. [CrossRef]
12. Guenther, D.; Heymel, B.; Guenther, J.F.; Ederer, I. Continuous 3D-printing for additive manufacturing. *Rapid Prototyp.* **2014**, *20*, 320. [CrossRef]
13. Kong, Z.; Wang, X.; Hu, N.; Jin, Y.; Tao, Q.; Xia, W.; Lin, X.-M.; Vasdravellis, G. Mechanical properties of SLM 316L stainless steel plate before and after exposure to elevated temperature. *Constr. Build. Mater.* **2024**, *444*, 137786. [CrossRef]
14. Chen, M.T.; Gong, Z.; Zhang, T.; Zuo, W.; Zhao, Y.; Zhao, O.; Wang, Z. Mechanical behavior of austenitic stainless steels produced by wire arc additive manufacturing. *Thin Walled Struct.* **2023**, *196*, 111455. [CrossRef]
15. Liverani, E.; Toschi, S.; Ceschini, L.A.; Fortunato, A. Effect of selective laser melting (SLM) process parameters on microstructure and mechanical properties of 316L austenitic stainless steel. *Mater. Process. Technol.* **2017**, *249*, 255–263. [CrossRef]
16. Marattukalam, J.J.; Karlsson, D.; Pacheco, V.; Beran, P.; Wiklund, U.; Jansson, U.; Sahlberg, M. The effect of laser scanning strategies on texture, mechanical properties and site-specific grain orientation in selective laser melted 316L. *Mater. Des.* **2020**, *193*, 108852. [CrossRef]
17. Zhang, A.; Wu, W.; Wu, M.; Liu, Y.; Zhang, Y.; Wang, Q. Influence of laser power on mechanical properties and pitting corrosion behavior of additively manufactured 316L stainless steel by laser powder bed fusion (L-PBF). *Laser Technol.* **2024**, *176*, 110886. [CrossRef]
18. Qiu, C.; Kindi, M.A.; Aladawi, S.; Hatmi, I.A. A comprehensive study on microstructure and tensile behaviour of a selectively laser melted stainless steel. *Sci. Rep.* **2018**, *8*, 7785. [CrossRef]
19. Montero-Sistiaga, M.L.; Godino-Martinez, M.K.; Boschmans, J.; Kruth, P.; Van Humbeeck, J.; Vanmeensel, K. Microstructure evolution of 316L produced by HP-SLM (high power selective laser melting). *Addit. Manuf.* **2018**, *23*, 402–410. [CrossRef]
20. Wan, Y.C.; Xu, S.Y.; Liu, C.M.; Gao, Y.H.; Jiang, S.N.; Chen, Z.Y. Enhanced strength and corrosion resistance of Mg-Gd-Y-Zr alloy with ultrafine grains. *Mater Lett.* **2018**, *213*, 213–224. [CrossRef]
21. Sun, J.J.; Liu, Y.N.; Zhu, Y.T.; Lian, F.L. Super-strong dislocation-structured high-carbon martensite steel. *Sci. Rep.* **2017**, *7*, 6596. [CrossRef]
22. Zhang, Y.; Fang, Y.; Kim, M.K.; Duan, Z.; Yuan, Q.; Oh, E.; Suhr, J. In-situ TiCxNy nanoparticle reinforced crack-free CoCrFeNi medium-entropy alloy matrix nanocomposites with high strength and ductility via laser powder bed fusion. *Compos. Part B-Eng.* **2024**, *273*, 111237. [CrossRef]
23. Kim, W.R.; Bang, G.B.; Park, J.H.; Lee, T.W.; Lee, B.S.; Yang, S.M.; Kim, G.H.; Lee, K.H.; Kim, G. Microstructural study on a Fe-10Cu alloy fabricated by selective laser melting for defect-free process optimization based on the energy density. *J. Mater. Res. Technol.* **2020**, *9*, 12834–12839. [CrossRef]
24. Li, J.; Qu, H.; Bai, J. Grain boundary engineering during the laser powder bed fusion of TiC/316L stainless steel composites: New mechanism for forming TiC-induced special grain boundaries. *Acta. Mater.* **2022**, *226*, 117605. [CrossRef]
25. King, W.E.; Barth, H.D.; Castillo, V.M.; Gallegos, G.F.; Gibbs, J.W.; Hahn, D.E.; Kamath, C.; Rubenchik, A.M. Observation of keyhole-mode laser melting in laser powder-bed fusion additive manufacturing. *J. Mater. Process. Technol.* **2014**, *214*, 2915–2925. [CrossRef]
26. Bai, S.; Perevoshchikova, N.; Sha, Y.; Wu, X. The effects of selective laser melting process parameters on relative density of the AlSi10Mg parts and suitable procedures of the archimedes method. *Appl. Sci.* **2019**, *9*, 583. [CrossRef]
27. Sun, M.; Chen, S.Y.; Wei, M.W.; Liang, J.; Liu, C.S.; Wang, M. Microstructure and properties of high power-SLM 24CrNiMoY alloy steel at different laser energy density and tempering temperature. *Powder Metall* **2021**, *64*, 1–7. [CrossRef]
28. Khairallah, S.A.; Anderson, A. Mesoscopic simulation model of selection laser melting of stainless steel power. *J. Mater. Res. Technol.* **2014**, *214*, 2627–2636. [CrossRef]
29. Cui, X.; Zhang, S.; Wang, C.; Zhang, C.H.; Chen, J.; Zhang, J.B. Microstructure and fatigue behavior of a laser additive manufactured 12CrNi2 low alloy steels. *Mater. Sci. Eng. A* **2020**, *772*, 138685. [CrossRef]
30. Zinoviev, A.; Zinovieva, O.; Ploshikhin, V.; Romanova, V.; Balokhonov, R. Evolution of grain structure during laser additive manufacturing. Simulation by a cellular automata method. *Mater. Des.* **2016**, *106*, 321–329. [CrossRef]
31. Liang, Q.; Junru, L.; Pengfei, Z.; Leiying, W.; Guojun, M. Effect of Niobium Microalloying on the Mechanical Properties in Grade R4 Mooring Chain Steel. *Metall. Mater. Trans.* **2024**, *55*, 695–705.
32. Mainier, F.B.; Canto, M.O.; Meiriño, M.J. Characteristics of Grade R4 Steel Manufactured by Ingot Casting and Used in the Production of Offshore Mooring Chains. *Trans. Marit. Sci.* **2023**, 12. [CrossRef]
33. Augustin, A.; Huilgol, P.; Udupa, K.R.; Bhat, K.U. Effect of current density during electrodeposition on microstructure and hardness of textured Cu coating in the application of antimicrobial Al touch surface. *J. Mech. Behav. Biomed.* **2016**, *63*, 352–360. [CrossRef] [PubMed]
34. AlMangour, B.; Grzesiak, D.; Cheng, J.; Ertas, Y. Thermal behavior of the molten pool, microstructural evolution and tribological performance during selective laser melting of TiC/316L stainless steel nanocomposites: Experimental and simulation methods. *J. Mater. Res. Technol.* **2018**, *257*, 288–301. [CrossRef]

35. Sang Gun, L.; Massoud, K.; Jungho, L. Role of quenching method on cooling rate and microstructure of steels: Variations in coolant and its flow arrangement. *Int. J. Heat Mass Transf.* **2022**, *189*, 122702.
36. Chaolin, T.; Qian, L.; Xiling, Y.; Lequn, C.; Jinlong, S.; Fern Lan, N.; Yuchan, L.; Tao, Y.; Youxiang, C.; Chain, T.L.; et al. Machine Learning Customized Novel Material for Energy-Efficient 4D Printing. *Adv. Sci.* **2023**, *10*, 2206607.

metals

MDPI

Review

Additive Manufacturing of Biodegradable Metallic Implants by Selective Laser Melting: Current Research Status and Application Perspectives

Anna Gracheva, Igor Polozov * and Anatoly Popovich

Institute of Mechanical Engineering, Materials, and Transport, Peter the Great St. Petersburg Polytechnic University, Polytechnicheskaya, 29, St. Petersburg 195251, Russia; gracheva_am@spbstu.ru (A.G.); director@immet.spbstu.ru (A.P.)
* Correspondence: polozov_ia@spbstu.ru

Abstract: Biodegradable metallic implants represent a paradigm shift in implantology, eliminating secondary removal surgeries through predictable controlled degradation. This review systematizes current achievements in selective laser melting (SLM) of biodegradable metals (Mg, Fe, Zn), analyzing how processing parameters influence microstructure, mechanical properties, and degradation kinetics. Key findings demonstrate that SLM-produced Mg alloys achieve bone-matching modulus (40–45 GPa) with moderate degradation (1–3 mm/year); Fe-based systems provide superior strength (400–600 MPa) but slower degradation (0.1–0.5 mm/year); while Zn alloys offer intermediate properties. Design strategies for porous/lattice structures enhancing osseointegration and enabling property gradients are discussed. Major challenges include controlling degradation kinetics, optimizing SLM parameters for reactive metals, standardizing testing methodologies, and regulatory harmonization. This comprehensive analysis provides systematic guidelines for material selection and process optimization, establishing a foundation for developing next-generation personalized biodegradable implants.

Keywords: biodegradable metals; selective laser melting; additive manufacturing; magnesium alloys; iron alloys; zinc alloys; medical implants; personalized medicine

1. Introduction

Modern implantology is transitioning to biodegradable implants that are replaced by body tissues. According to a Grand View Research report, the global biodegradable implant market reached USD 5.35 billion in 2022 and is projected to grow to USD 13.1 billion by 2030 with a CAGR of 11.8% [1] This growth is driven by the advantages of biodegradable implants, including elimination of removal surgeries, reduced risk of long-term complications, and more physiological load distribution during healing [2].

Traditional metallic implants (titanium, stainless steel, cobalt-chromium alloys) possess high strength and biocompatibility, but their elastic modulus (100–210 GPa) significantly exceeds that of bone (5–23 GPa) [3,4]. This mismatch causes stress shielding, bone resorption around the implant, loosening, and the need for revisions [5]. Additionally, the permanent presence of the implant may be accompanied by metal ion release, corrosion, inflammatory reactions, and complications [6].

Biodegradable metallic alloys offer an innovative solution by providing temporary mechanical support with precisely controlled degradation kinetics that synchronizes with tissue regeneration, ultimately eliminating the need for implant removal surgeries. Alloys based on magnesium (Mg), iron (Fe), and zinc (Zn) are of particular interest, possessing unique properties and potential for various clinical applications [7].

Magnesium alloys are characterized by low density, elastic modulus close to bone tissue, and relatively rapid degradation capability (3–6 months) [8]. The main corrosion product is magnesium hydroxide, which gradually transforms into water-soluble compounds and is safely excreted from the body. Magnesium is an essential microelement involved in more than 300 biochemical reactions in the body, ensuring a favorable biological response [9].

Iron alloys feature high strength and toughness, making them suitable for implants experiencing significant loads. However, their slow in vivo degradation rate (more than 1 year) limits clinical application, stimulating the development of new compositions with accelerated corrosion [10].

Zinc alloys occupy an intermediate position in terms of mechanical properties and degradation rate between magnesium and iron alloys, demonstrating moderate corrosion rate (3–12 months) and satisfactory strength [11]. Zinc is an important microelement involved in regulating immune function and wound healing, making Zn alloys promising candidates for biodegradable implants [12].

Traditional manufacturing methods (casting, forging, machining) are limited in creating complex geometries and personalized implants. Additive manufacturing (AM), particularly selective laser melting (SLM), opens unique possibilities for overcoming these limitations, allowing the creation of implants with precise geometry, controlled porosity, and individualized properties [13].

The SLM process is based on layer-by-layer deposition of metal powder using a high-energy laser beam that selectively melts and solidifies material according to a digital 3D model [14]. This technology provides unprecedented design freedom, including the creation of complex lattice structures that mimic trabecular bone architecture, promoting osseointegration and controlled implant degradation.

Despite significant progress in developing biodegradable alloys and their additive manufacturing technologies, several unresolved problems remain that require further research. The main challenges include control of degradation rate [15], balance of mechanical properties, technological aspects of working with reactive metals [16], and certification issues for clinical application.

Figure 1 demonstrates the process of temporary functioning of a biodegradable implant and the dynamic relationship between implant degradation and bone tissue regeneration over time.

The purpose of this review is to provide a comprehensive and critical analysis of the current state of research in the field of additive manufacturing of biodegradable metallic implants, with a specific focus on the selective laser melting process. Unlike previous reviews that typically address either biodegradable metals or additive manufacturing separately, this work uniquely integrates both aspects while offering a comparative analysis of Mg, Fe, and Zn-based systems. This review makes several novel contributions to the field by (1) systematizing data on the influence of SLM parameters on microstructure, mechanical properties, and degradation rate of different biodegradable metal systems; (2) critically evaluating porous structure design strategies and their influence on biological response; (3) providing side-by-side comparison of key properties across different biodegradable metal families; and (4) identifying promising research directions to overcome existing limitations for successful clinical translation. The insights presented here will guide researchers, engineers, and clinicians in developing next-generation personalized biodegradable implants with optimized properties and controlled degradation behavior.

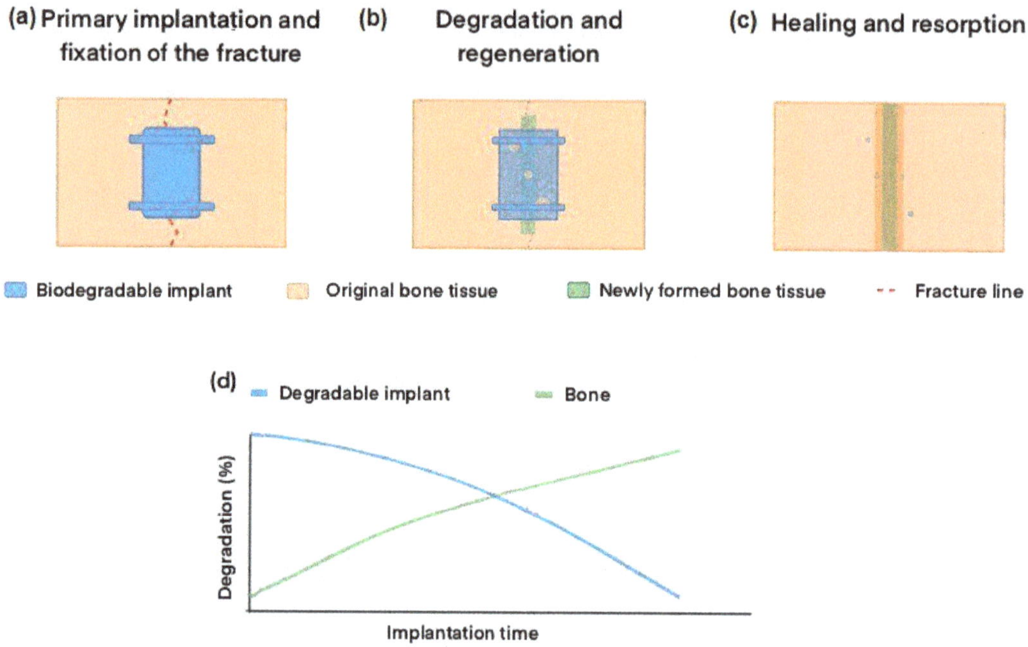

Figure 1. Schematic of biodegradable implant temporary functioning: (**a**) Initial implantation and fracture fixation. (**b**) Gradual implant degradation and new bone tissue formation. (**c**) Complete fracture healing and implant resorption. (**d**) Graph showing the relationship between implant degradation and forming bone during rehabilitation.

2. Additive Manufacturing of Biodegradable Metals: Process Overview

2.1. Technology of the Selective Laser Melting Process

Selective laser melting (SLM) is a promising additive manufacturing (AM) technology for biodegradable metallic implants, based on layer-by-layer melting of metal powder by a high-energy laser. SLM belongs to Powder Bed Fusion (PBF) methods, forming three-dimensional objects through the sequential creation of two-dimensional layers [17].

The SLM process includes the following main stages (Figure 2):

1. Deposition of a thin layer of metal powder (typically 20–100 μm) on the build platform using a recoater blade or roller mechanism.
2. Selective melting of the powder by a laser beam according to the geometry of the current cross-section of the model.
3. Lowering the build platform by the height of one layer.
4. Applying a new layer of powder and repeating the melting process.
5. Sequential repetition of these stages until the complete formation of the product.

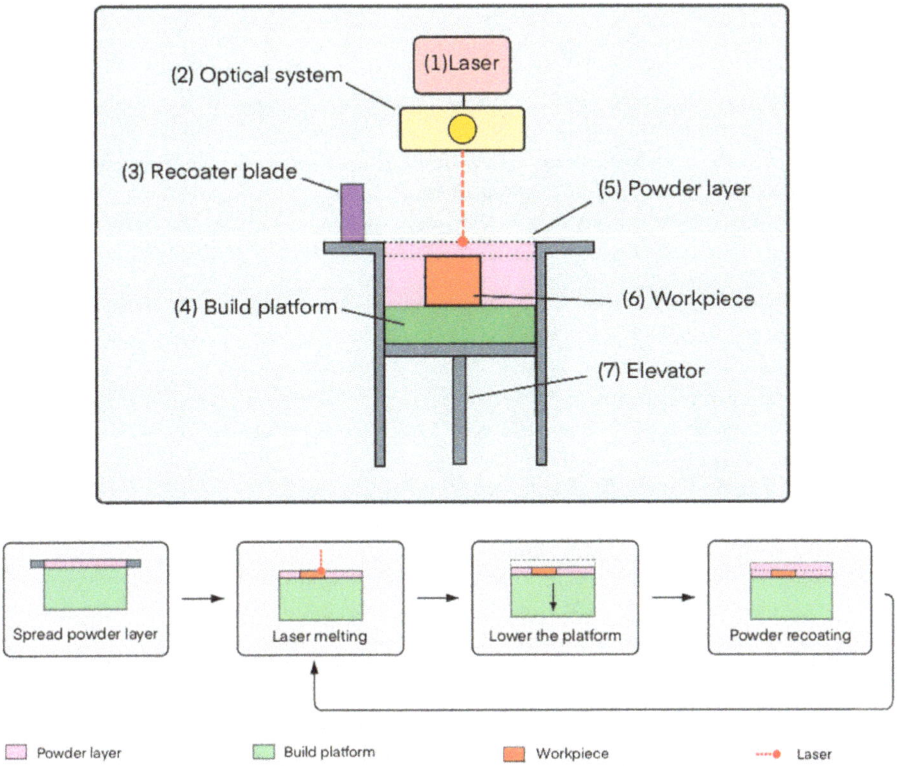

Figure 2. Schematic representation of the main components of an SLM setup and process stages: (1) laser, (2) focusing and scanning system, (3) powder delivery device, (4) build platform, (5) powder layer, (6) finished part, (7) elevator. The diagram also shows the sequential stages: powder layer deposition, laser beam scanning, platform lowering, and application of a new layer.

2.2. Features of Working with Biodegradable Metals

Working with biodegradable metals in the SLM process has a number of specific features due to their physicochemical properties and requirements for final products. These features must be considered for the successful production of implants with specified characteristics.

2.2.1. Reactivity and Temperature Regime

Biodegradable metals (magnesium, zinc, iron) possess high reactivity and specific thermal properties, complicating their use in SLM processes.

Magnesium and zinc readily oxidize, forming surface oxide films that degrade melting quality and increase product porosity. Studies show that low boiling points of magnesium (1090 °C) and melting of zinc (419.5 °C) create a risk of material evaporation during laser exposure, which can disrupt process stability [18]. In contrast, iron requires significant energy input, but selective evaporation of alloying elements is possible, leading to changes in alloy composition [19].

Thus, processing biodegradable metals by SLM requires strict control of the atmosphere and precise adjustment of laser parameters.

2.2.2. Requirements for Process Parameters and Control of Structural-Phase Composition

Selective laser melting of biodegradable metals requires precise control of parameters that determine optimal characteristics within a narrow "process window." For magnesium alloys, energy density is critical, ensuring complete melting without evaporation [20] and forming a fine-grained structure that increases strength but potentially accelerates corrosion [21]. For zinc alloys, cooling rate is important, affecting grain size and strength [22]. Iron alloys require high laser power while controlling scanning speed to minimize residual stresses and cracks [23], and SLM parameters allow regulation of phase ratios (γ and ε-martensite) to control biodegradation [24]. Rapid cooling in SLM leads to the formation of non-equilibrium structures with high defect density. SLM of biodegradable metals requires a comprehensive approach that considers the specifics of each alloy and the required properties.

2.2.3. Post-Processing, Sterilization and Degradation Rate Control

SLM implants require specific post-processing to ensure properties and safety. Traditional heat treatment for Mg alloys may be ineffective due to oxidation risk. Micro-arc and plasma electrolytic oxidation, biocompatible coatings (hydroxyapatite, polylactide) are applied to improve corrosion resistance and biocompatibility [25].

Degradation rate management is a key task. For this purpose, alloying (to increase corrosion resistance) [26], surface modification (creation of protective coatings) [27], porosity management (regulation of surface area) [28], and microstructure regulation (phase composition, grain size through SLM parameters) [29] are used. Combining these approaches allows creating implants with predetermined degradation rates.

Critically important is the choice of sterilization method, as biodegradable metals can be sensitive to aggressive environments and high temperatures. Studies show that gamma radiation may be the preferred sterilization method for magnesium implants, as it minimally affects their properties [27].

2.3. Key Process Parameters and Their Influence

The quality and performance characteristics of biodegradable implants manufactured by SLM are largely determined by process parameters. Optimization of these parameters allows achieving the required density, microstructure, mechanical properties, and material degradation rate.

2.3.1. Laser Power

Laser power is one of the key parameters determining the energy input in the SLM process. The optimal power depends on the thermophysical properties of the material and should ensure complete powder melting without excessive overheating and evaporation.

Studies show that for magnesium alloys such as AZ91D, the optimal power range is 100–200 W [30]. Incomplete powder melting and porous structure formation occur when laser power falls below 100 W. Conversely, powers exceeding 200 W cause intensive magnesium evaporation, resulting in chemical composition changes and keyhole defect formation [31].

For iron alloys, which have a higher melting temperature, increased laser power (200–400 W) is required. Ewald et al. (2021) showed that a power of at least 200 W is necessary to achieve optimal density and mechanical properties of the Fe-Mn-Al-Ni alloy (SLM Solutions Group AG, Lübeck, Germany) [32].

Zinc alloys, conversely, require lower power (80–150 W) due to zinc's low melting temperature. Precise power control is critically important to prevent overheating and excessive evaporation [33].

2.3.2. Scanning Speed

The scanning speed of the laser beam affects the interaction time between the laser and material and, consequently, the penetration depth, melt pool size, and cooling rate.

For magnesium alloys, the optimal scanning speed is in the range of 400–800 mm/s. At lower speeds, excessive material overheating occurs, while at higher speeds, incomplete powder melting occurs [34].

Iron alloys allow higher scanning speeds (600–1200 mm/s) due to their increased heat resistance. Studies by Xie et al. (2024) showed that the optimal scanning speed for the Fe–Mn–Al–C alloy is about 800 mm/s (SLM Solutions, Germany) [35].

Zinc alloys are characterized by lower scanning speeds (300–600 mm/s) due to the need for precise control of thermal exposure [36].

2.3.3. Layer Thickness and Hatch Distance

Layer thickness and distance between scanning tracks determine the spatial distribution of laser energy and the degree of overlap between adjacent melt pools.

The optimal layer thickness for most biodegradable metals is 20–50 μm. Thinner layers provide better accuracy and surface quality but increase build time. Thicker layers increase productivity but can lead to incomplete penetration and reduced mechanical properties [37].

The distance between scanning tracks should ensure sufficient overlap of adjacent melt pools (typically 20–30%) to form a monolithic structure. Typical values are 60–120 μm depending on the material type and laser spot diameter [38].

2.3.4. Scanning Strategy

The scanning strategy (laser beam path scheme) affects heat distribution, residual stresses, and microstructure. Common strategies include raster scanning (in one direction or with rotation between layers), island (chessboard) strategy, and contour scanning. For metals prone to oxidation and evaporation (Mg, Zn), a strategy with rotation of direction between layers is preferable, providing uniform heat distribution and reducing property anisotropy [39].

2.3.5. Energy Density

Energy density is an integral parameter that combines the influence of laser power, scanning speed, hatch distance, and layer thickness. It is calculated using Equation (1),

$$E = \frac{P}{v \cdot h \cdot d} \tag{1}$$

where E is the volumetric energy density (J/mm^3), P is the laser power (W), v is the scanning speed (mm/s), h is the hatch distance (mm), d is the layer thickness (mm).

For each type of biodegradable metal, there is an optimal energy density range that provides a balance between complete powder melting and defect minimization. For magnesium alloys, this range is 120–150 J/mm^3, for iron alloys—150–200 J/mm^3, for zinc alloys—100–130 J/mm^3 [40].

As shown in Figure 3, the quality of SLM processing of ZK60 alloy (SLM 125 H L, SLM Solutions, Lübeck, Germany) significantly depends on laser energy density. With excessive energy (area I), magnesium evaporation and melt splashing occur, leading to an uneven surface. Insufficient energy (areas III–IV) causes incomplete powder melting and defect formation. Optimal formation (area V) is achieved at 40–50 W and 500–800 mm/s, providing homogeneous samples with high dimensional accuracy. These results demonstrate the critical influence of energy density on the morphology and quality of

SLM products. As seen from the results, porosity shows a non-monotonic dependence on scanning speed at fixed laser power. The lowest porosity is achieved at a medium power of 50 W, whereas extreme values (40 and 60 W) lead to a significant increase in defects. At low power and high scanning speed, connected pores form due to capillary instability and insufficient track overlap, causing the balling effect. Conversely, the combination of high power and low scanning speed provokes the formation of large spherical pores due to the keyhole effect, associated with vapor trapping in a deep melt pool. Increasing the input energy exacerbates this effect, leading to increased porosity.

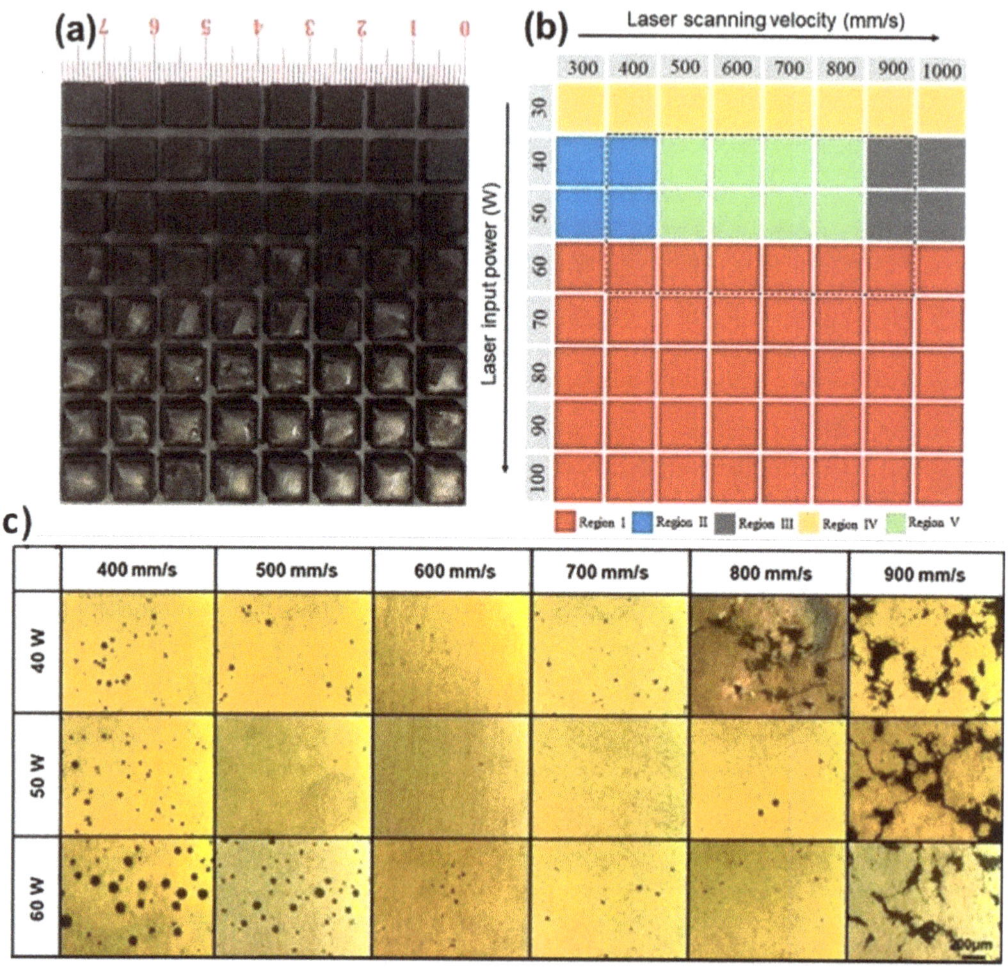

Figure 3. SLM processing optimization for ZK60 magnesium alloy: (a) macro-morphology in xy plane showing different processing regions, (b) processing parameter map with laser power (40–60 W) and scanning speed (400–900 mm/s), (c) porosity analysis and optical microscopy images corresponding to the marked region in (b), demonstrating the relationship between energy density and part quality. The dotted box in (b) outlines the region corresponding to the samples shown in (c), reprint with permission from Ref. [41], 2021, Elsevier.

2.4. Requirements for Powder Materials for Additive Manufacturing

The quality of metal powders for selective laser melting of biodegradable implants is determined by a set of interrelated characteristics. Optimal parameters include particle size distribution in the range of 15–63 μm (narrow distribution, D10 = 20–25 μm, D50 = 30–40 μm, D90 = 50–60 μm) for a balance between accuracy and flowability. The morphology of particles is critically important, with preference given to spherical particles with a roundness coefficient of at least 0.85 [42]. The chemical purity of powders must meet the strict requirements of international standards ISO 13485 [43] and ASTM F3049 [44].

Various methods are used for powder production, each with its own advantages. Different production methods are used: gas atomization (sphericity and purity), plasma spheroidization (shape correction), mechanical alloying (complex alloys), and hydride–dehydride method (high-purity powders).

For specific metals, specialized technological solutions are recommended. Magnesium powders are optimally obtained by gas atomization in argon followed by spheroidization. Iron powders are preferably produced by vacuum plasma atomization, while for zinc powders, gas atomization with controlled cooling is most suitable. Compliance with these parameters ensures reproducibility of the SLM process and controlled biodegradation characteristics of implants.

2.5. Problems in Additive Manufacturing of Biodegradable Alloys

Despite significant progress in the field of additive manufacturing of biodegradable metallic implants, there are a number of unresolved problems limiting their widespread clinical application.

2.5.1. Degradation Rate Control

One of the key problems is ensuring a predictable and controlled degradation rate of the implant, corresponding to the rate of tissue regeneration. Too rapid degradation leads to loss of mechanical integrity and adverse reactions, while too slow degradation negates the advantages of biodegradation. The complexity of control is exacerbated by differences between in vivo and in vitro, influence of biological factors [45], and patient response [46].

The three main biodegradable metal systems exhibit significantly different degradation kinetics under physiological conditions. Magnesium-based implants typically degrade at rates of 1.0–3.0 mm/year [47], with pure magnesium showing the highest degradation rate (up to 3 mm/year), while alloyed systems demonstrate more moderate rates. Iron-based materials show considerably slower degradation, typically in the range of 0.1–0.5 mm/year [48], with pure iron exhibiting rates as low as 0.008 mm/year in certain environments. Zinc alloys occupy an intermediate position with degradation rates between 0.2–0.5 mm/year [49].

Several key metallurgical and environmental factors influence degradation behavior. The microstructural features of additively manufactured metals—including grain size, phase distribution, and dislocation density—significantly impact corrosion mechanisms. Finer grain structures with increased grain boundary area, typical of SLM-produced materials, generally accelerate degradation compared to coarse-grained counterparts. Phase composition plays an equally critical role; for instance, in Fe-Mn alloys, ε-martensite phases degrade faster than austenitic γ-phases, while in magnesium alloys, the presence of secondary phases like Mg_2Ca or $Mg_{17}Sr_2$ can create galvanic couples that accelerate local corrosion.

Environmental factors such as pH, ion concentration, protein adsorption, and fluid flow dynamics profoundly affect degradation kinetics in vivo. The body's dynamic environment, with variations in mechanical loading, cellular activity, and local chemistry, creates significant complexity in predicting long-term degradation behavior. Static in vitro tests frequently underestimate the degradation rates observed in vivo due to these dynamic physiological factors.

The processing route—particularly for additively manufactured implants—introduces additional variables affecting degradation. SLM processing parameters directly influence microstructural features and residual stresses, which in turn affect corrosion susceptibility. For example, higher cooling rates typically produce finer microstructures with higher dislocation densities that can accelerate corrosion, while processing-induced porosity can lead to localized corrosion phenomena.

2.5.2. Manufacturing Difficulties

Additive manufacturing of biodegradable metals is associated with a number of technological challenges. The high reactivity of magnesium and zinc requires special storage conditions and modifications to SLM systems [50]. Material evaporation leads to smoke and condensate formation, contaminating laser optics and reducing process efficiency [51]. Rapid cooling creates residual stresses, leading to deformation and reduced strength.

2.5.3. Corrosion Fatigue and Stress Corrosion Cracking

The combination of cyclic loads and corrosive environment creates a risk of corrosion fatigue and stress corrosion cracking (SCC) in biodegradable implants [52]. This problem is especially critical for implants working under cyclic loading conditions (e.g., bone plates, screws, cardiovascular stents). Studies have shown that Al-free Mg alloys such as ZX50, developed for biodegradable implants, are susceptible to stress corrosion cracking in m-SBF at a strain rate of 10^{-7} s^{-1}, resulting in significant loss of plasticity (from 21% in air to 3.8% in m-SBF) [52].

Zinc alloys are characterized by increased sensitivity to SCC, especially in the presence of chlorides, which requires special attention when designing implants for long-term functioning [53].

2.6. Lattice Structures Manufactured by Additive Methods in Tissue Engineering

Additive manufacturing opens unique opportunities for creating complex three-dimensional lattice structures that mimic the architecture of natural tissues and provide optimal biological integration of implants.

2.6.1. Types of Lattice Structures

Depending on design and geometry, lattice structures of biodegradable implants are divided into several types. Regular lattice structures (Figure 4a) use repeating cells (cubic, tetrahedral, etc.) for predictable and customizable mechanical properties [54]. Triply periodic minimal surfaces (TPMS) (Figure 4b), such as Schwarz and gyroid structures, provide smooth transitions and good permeability [55]. Voronoi structures (Figure 4c) create stochastic structures and show enhanced cell proliferation [56]. Biomimetic structures recreate the architecture of natural tissues such as bone [57]. Functionally gradient structures (Figure 4d) change parameters (porosity, cell size) to create zones with different properties [57].

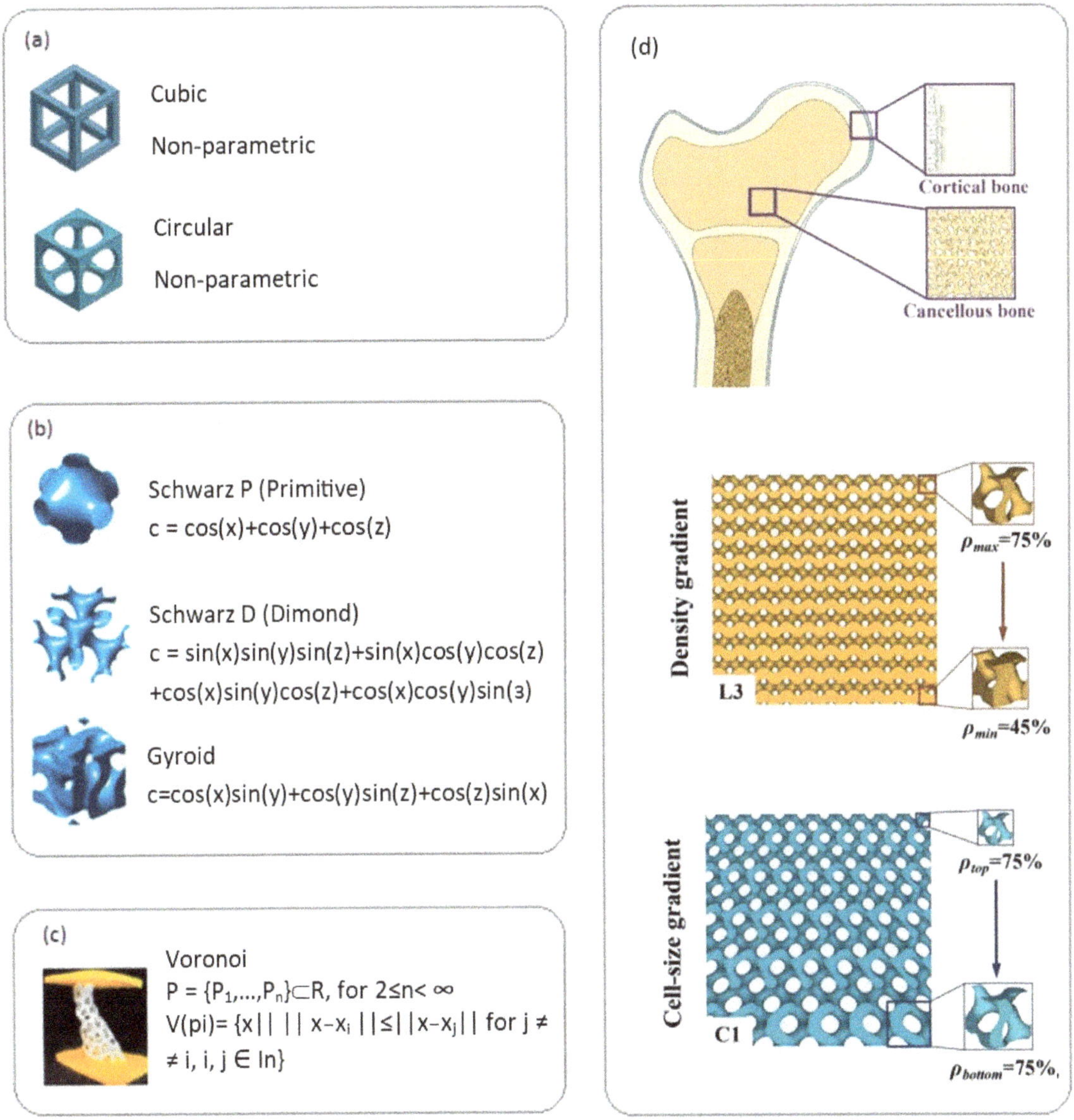

Figure 4. Main types of lattice structures for biodegradable implant materials: (**a**) cubic and circular, reprint with permission from Ref. [58], 2024, Elsevier. (**b**) Triply periodic minimal surfaces—Schwarz P, Schwarz D, gyroid, reprint with permission from Ref. [58], 2024, Elsevier, (**c**) Voronoi structure reprint with permission from Ref. [58], 2024, Elsevier, (**d**) schematic diagram depicting natural bone structure with gradient porous structures and gyroid CGPS structures with cell size gradient, reprint with permission from Ref. [57], 2024, Elsevier.

2.6.2. Influence of Lattice Structure Parameters on Mechanical Properties and Biodegradation

The mechanical properties of lattice structures made from biodegradable metals are determined by a combination of material properties and geometric parameters of the lattice. Key factors include porosity, as Schwarz lattices with 42% porosity demonstrated the highest combination of strength and plasticity [59], cell size (affects mechanics/permeability/cell integration; optimally above 300 μm for osseointegration [60]), strut thickness, and topology (determines stiffness/strength [59]). Higher porosity increases surface area, accelerating degradation. For magnesium alloys, interconnected pores promote uniform corrosion, whereas isolated pores lead to localized pitting corrosion [61]. Lattice structures made from biodegradable metals demonstrate improved biological response compared to monolithic implants [62].

Having established the fundamental principles and technical aspects of additive manufacturing for biodegradable metals, we now turn our attention to specific material systems. The following sections examine three primary biodegradable metal families—magnesium, iron, and zinc alloys—in detail. Each section analyzes the unique characteristics, challenges, and opportunities presented by these materials when processed through selective laser melting. We begin with magnesium alloys, which offer the closest mechanical match to bone tissue and have seen the most extensive clinical adoption among biodegradable metals.

3. Magnesium Alloys in Selective Laser Melting

3.1. Historical Overview and Basic Properties

The history of magnesium applications in medicine began in 1878 when Edward C. Huse first used magnesium wire to stop bleeding [63], and in 1900, Payr proposed magnesium plates for joint restoration [64]. Widespread clinical application began later, with CE marking obtained for the DREAMS magnesium stent (Biotronik, Bülach, Switzerland) and MAGNEZIX® screw (Syntellix AG, Hannover, Germany) [65,66]. Modern developments focus on creating new alloys with improved properties and implementing technologies such as selective laser melting, due to the unique characteristics of magnesium that make it promising for biodegradable implants:

1. Mechanical properties—magnesium's elastic modulus (41–45 GPa) is significantly closer to cortical bone (5–23 GPa) than traditional implant materials (titanium alloys—110–120 GPa, stainless steel—200–210 GPa), reducing the risk of stress shielding effects [67].
2. Biocompatibility—magnesium is an essential element for the human body, participating in more than 300 biochemical reactions, including protein synthesis and energy metabolism regulation. The daily requirement of an adult for magnesium is 300–400 mg [68].
3. Osteogenic properties—magnesium ions released during implant degradation stimulate the proliferation and differentiation of osteoblasts, promoting bone tissue formation [69].
4. Degradation in physiological conditions—magnesium undergoes electrochemical corrosion in biological environments with the formation of magnesium hydroxide and hydrogen: $Mg + 2H_2O \rightarrow Mg(OH)_2 + H_2$. Magnesium corrosion products are non-toxic and gradually dissolve or are excreted from the body [70].

The use of magnesium alloys as biodegradable implants is associated with several limitations. First, pure magnesium is characterized by a high corrosion rate (up to 3 mm/year), which can lead to premature loss of mechanical integrity of the implant and the formation of gas cavities [71]. Second, magnesium alloys possess relatively low strength, specifically, their yield strength is lower than that of titanium alloys and stainless steel, which limits their application under high load conditions [72]. Third, technological difficulties in processing and additive manufacturing are related to the high reactivity and low boiling point of magnesium [73]. To overcome these limitations, special magnesium alloys with improved properties are being developed, and their additive manufacturing processes are being optimized.

3.2. *Influence of Alloying Elements*

Magnesium alloying plays a key role in modifying its properties for biomedical applications, allowing control of degradation rate, mechanical characteristics, and biological response. The choice of alloying elements for biodegradable magnesium alloys is determined not only by their influence on material properties but also by biocompatibility, absence of toxicity, and potential therapeutic effect [74]. Table 1 summarizes the influence of alloying elements on the properties of magnesium alloys obtained by SLM.

Table 1. Influence of alloying elements on the properties of magnesium alloys obtained by SLM.

Alloying Element	Optimal Content	Effect on Mechanical Properties	Effect on Degradation Rate	Biological Effects
Calcium (Ca)	0.3–1.0%	Increased yield strength, grain refinement [75]	Improved corrosion resistance at content up to 0.5%, decreased at higher content [76]	Stimulation of bone tissue formation, activation of osteoblasts [77]
Zinc (Zn)	1.0–5.0%	Increases tensile strength by 30–40%, decreases plasticity [78]	Improved corrosion resistance at content up to 3–4%, deterioration at higher content	Stimulation of osteoblast proliferation, antibacterial action
Strontium (Sr)	0.5–2.0%	Moderate strength increase, improved plasticity [79]	Decreased corrosion resistance with increasing content [80]	Stimulation of osteogenesis, inhibition of osteoclast activity [80]
Rare Earth Elements (Gd, Y, Nd)	2.0–4.0%	Significant strength increase, improved fatigue characteristics [81]	Substantial improvement in corrosion resistance [82]	Depend on the specific element, Gd and Y show osteoinductive properties [83]
Manganese (Mn)	0.2–0.8%	Moderate effect on strength, reduced property anisotropy [84]	Improved corrosion resistance due to binding Fe impurities [84]	Participation in connective tissue formation, enzymatic processes [85]
Zirconium (Zr)	0.3–0.6%	Grain refinement, improved microstructure homogeneity, reduced anisotropy [51]	Moderate improvement in corrosion resistance [86]	Low cytotoxicity, neutral biological action [86]

3.2.1. Calcium (Ca)

Calcium is one of the most important minerals in human bone tissue, playing a key role in its strength and ability to recover after damage. In magnesium alloys, calcium forms the intermetallic compound Mg_2Ca, which is distributed predominantly along grain boundaries and affects mechanical properties [87]. The addition of calcium promotes grain refinement, leading to increased strength and plasticity [88,89]. However, calcium content should be controlled, usually less than 1% [80], as exceeding this value can deteriorate the material's corrosion resistance. Studies show that the Mg-2Zn-0.2Mn alloy with the addition of 0.36 wt.% Ca has optimal corrosion resistance due to the formation of a dense protective film and limited amount of secondary phase. At higher Ca content (0.76% and 1.10%), significant formation of secondary phase is observed, causing intensive galvanic corrosion [90]. In the SLM process, calcium affects the microstructure and properties of magnesium alloys. Modern studies of SLM Mg-Ca alloys demonstrate the possibility of achieving relative density above 99% with optimized process parameters (laser power 150–200 W, scanning speed 400 mm/s) [91].

3.2.2. Zinc (Zn)

Zinc, an important microelement for the human body [92–94], participates in solid solution and dispersion strengthening in magnesium alloys, refines grain, and forms the MgZn phase, improving mechanical properties and corrosion resistance at contents up to 5% [95]. However, when exceeding 5–7% Zn, corrosion resistance decreases due to galvanic corrosion, and the microstructure consists of α-Mg and MgZn distributed along grain boundaries. In the SLM process, Mg-Zn alloys demonstrate a fine-grained structure, and the addition of Ca increases strength to 182 MPa and elongation to 9.1% [96].

3.2.3. Strontium (Sr)

Strontium is an important element for bone tissue growth and strengthening. It stimulates osteoblast differentiation while slowing osteoclast activity, also improving calcium absorption by bone tissue [97]. In magnesium alloys, strontium forms secondary phases that improve mechanical properties and corrosion resistance [98]. Binary Mg-Sr alloys predominantly consist of α-Mg and $Mg_{17}Sr_2$ phases [98]. Increasing strontium content (0.5–2.5%) enhances tensile and compressive strength. However, the corrosion resistance of extruded Mg-Sr alloys decreases with increasing strontium content due to galvanic corrosion [3]. In SLM, strontium is of interest for biodegradable implants with osteogenic properties. Adding 0.5–1.0% Sr to magnesium alloys processed by SLM improves the biological response without significantly deteriorating technological properties [99].

3.2.4. Rare Earth Elements (REE)

Rare earth elements (REE) such as gadolinium (Gd), yttrium (Y), neodymium (Nd), and cerium (Ce) improve the mechanical properties and corrosion resistance of magnesium alloys. For biomedical purposes, Gd and Y are of greatest interest [100]. The commercial alloy WE43 (Mg-4Y-3RE-0.5Zr), originally developed for the aerospace industry, has found application in biomedicine due to its optimal properties. This alloy is characterized by enhanced corrosion resistance and mechanical strength, as well as relatively good processability in additive manufacturing [101].

3.2.5. Manganese (Mn) and Zirconium (Zr)

Manganese and zirconium modify the microstructure and neutralize impurities in magnesium alloys. Manganese improves corrosion resistance by binding iron and forming Al-Mn compounds [102]. Zirconium is a powerful grain modifier, promoting the formation of fine-grained structure [102]. In the context of SLM, adding 0.4–0.8% Mn to magnesium alloys helps improve surface quality and reduce porosity of the resulting products [85]. Zirconium, despite its advantages as a structure modifier, can create problems in laser melting due to its high melting temperature (1855 °C) and tendency to oxidize [103]. Recent research shows the promise of the Mg-10Zn-0.8Ca-0.5Zr alloy for the SLM process [104]. The powder of this alloy, obtained by mechanical alloying, is characterized by a homogeneous structure and relatively spherical particle shape. Optimization of SLM parameters allows obtaining samples with high strength, integrity, and minimal porosity (0.64%) at low laser energy density (138 J/mm^3) [104].

3.3. *Features of the SLM Process for Magnesium Alloys*

Selective laser melting of magnesium alloys is associated with a number of specific problems due to the physicochemical properties of magnesium. Understanding these features and developing approaches to overcome them are key to successful additive manufacturing of biodegradable implants from magnesium alloys.

3.3.1. Reactivity and Thermal Properties

Magnesium has high chemical activity, leading to intensive oxidation when heated during the SLM process with the formation of MgO, increasing the risk of ignition [105]. To prevent oxidation, sealed chambers with inert atmosphere (argon/nitrogen, $O_2 < 10$ ppm), evacuation (10^{-3}–10^{-5} mbar), and oxygen content control (<500 ppm for AZ91D alloy) [106] are used. The low boiling point of magnesium (1090 °C) causes evaporation, pore formation, and changes in alloy composition, while high thermal conductivity (156 W/(m·K)) promotes rapid cooling, leading to fusion defects [107]. These problems are solved by optimizing laser parameters (low power, high scanning speed), powder preheating, and vapor filtration. MgO oxide films (t_{plan} = 2852 °C) on powder particles deteriorate fusion, promoting inclusion formation [108]. To minimize them, powder pre-treatment, increased laser energy, introduction of modifying additives (e.g., Ca), and optimization of scanning modes are applied.

3.3.2. Optimization of SLM Parameters for Magnesium Alloys

For successful additive manufacturing of magnesium alloys, it is necessary to optimize SLM parameters: laser power (100–200 W, for AZ91D optimally about 150 W) (see Figure 5); scanning speed (high, 400–800 mm/s, for AZ91D optimally about 600–650 mm/s); layer thickness (30 μm); scanning strategy (rotation of direction between layers); energy density (70–120 J/mm^3) [106].

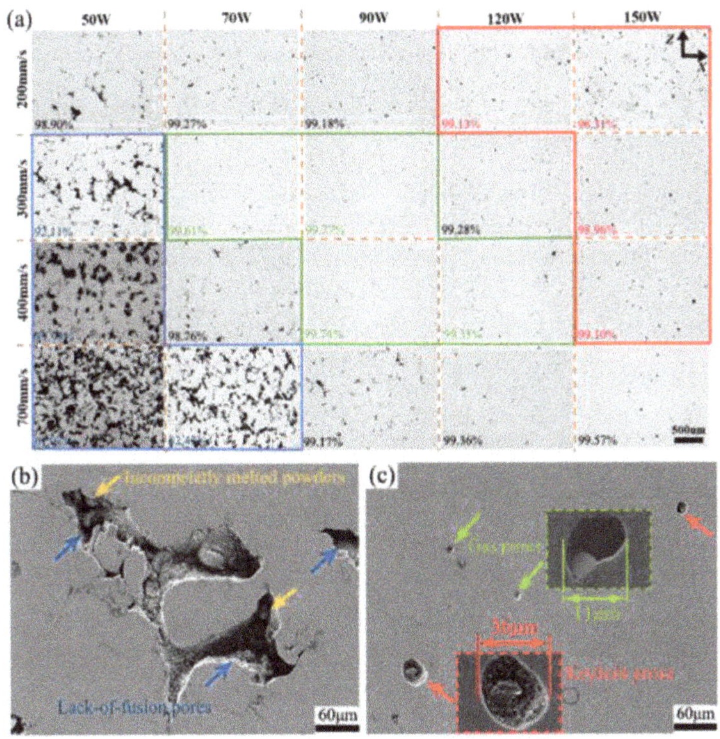

Figure 5. Typical optical micrographs of vertical planes of AZ91D manufactured at various processing parameters, values marked in the lower left corner represent the corresponding densification (**a**), SEM micrograph of lack of fusion pores (**b**), gas pores, keyhole-type pores and corresponding local magnifications (**c**), Reprint from [106].

Nopová et al. (2023) investigated the influence of SLM parameters on the quality and properties of AZ91D alloy [109]. It was established that optimal process parameters (laser power = 180 W, scanning speed = 612.5 mm/s, hatch distance = 0.133 mm, and layer thickness = 0.05 mm) provide relative density over 99%, low porosity, and homogeneous microstructure. Mechanical tests showed high values of yield strength (181 MPa) and tensile strength (305 MPa), with elongation to failure of 5.2%, which is more than twice the literature data for cast material [109].

Analysis of the microstructure of SLM-manufactured AZ91D showed the dependence of pore morphology on processing parameters (see Figure 3). Lack of fusion areas, keyhole areas, and transition areas were identified, which is consistent with previous studies [18,32,41]. Defects were classified as lack of fusion pores (low energy density), gas pores, and keyhole-type pores (high energy density). Optimal parameters provide densification >99.77%, with predominance of small gas pores associated with Mg and Zn evaporation [110] and argon entrapment.

For SLM WE43 yields two primary defect types: lack-of-fusion (LOF) at low power/high speed, and porosity at high power/low speed. LOF is attributed to insufficient energy density for complete powder melting, while porosity results from metal vapor entrapment at higher energy densities, exacerbated by the high vapor pressure of magnesium. A scan speed of 1000 mm/s and a laser power of 100 W resulted in optimal density (99.88%) [111].

3.4. Microstructure and Mechanical Properties

Additive manufacturing of magnesium alloys ensures the formation of a microstructure significantly different from that obtained by traditional methods (casting, extrusion). These differences are due to the unique crystallization conditions in SLM, characterized by extremely high cooling rates and directional heat removal [93].

3.4.1. Features of SLM-Magnesium Alloys Microstructure

The microstructure of SLM-magnesium alloys is characterized by a fine-grained structure of 5–20 μm, which is significantly smaller than in cast alloys (50–150 μm). The fine-grained structure provides enhanced mechanical properties. It also has a layered morphology ("fish-scale"), growth texture (preferential orientation of crystallographic directions parallel to the build direction), non-equilibrium phase distribution, microsegregation, and high density of crystal lattice defects. Studies have shown that SLM material is characterized by smaller and more uniformly distributed grains, with less anisotropy compared to traditional production methods [51].

3.4.2. Mechanical Properties of SLM-Magnesium Alloys

The mechanical properties of magnesium alloys obtained by SLM usually exceed those of similar materials manufactured by traditional methods, especially in terms of strength. This improvement is due to the fine-grained structure, solid solution strengthening, and high dislocation density.

Table 2 presents the mechanical properties of various magnesium alloys obtained by SLM compared to traditional production methods.

Table 2. Mechanical properties of magnesium alloys obtained by various methods.

Alloy	Production Method	Yield Strength (MPa)	Tensile Strength (MPa)	Elongation (%)	Source
WE43	SLM	296.3 ± 2.5	308.0 ± 1.0	12.2 ± 1.4	[112]
WE43	Extrusion	284.4 ± 0.9	306.6 ± 0.5	22.4 ± 3.6	[112]
WE43	Casting	145.4 ± 6.6	189.2 ± 9.2	4.4 ± 0.6	[112]
WE43	Rolling + aging	310 ± 3	357 ± 4	2.9 ± 0.1	[113]
AZ91D	SLM	181	305	5.2	[109]
AZ91D	Casting	160	220	5.8	[106]
GZ112K	SLM	252	275	4.3	[33]

As seen from the table, SLM materials demonstrate increased values of yield strength and tensile strength compared to cast analogues. Particularly impressive improvement is observed for the WE43 alloy, where the yield strength of SLM material (296.3 MPa) is more than twice the value for cast material (145.4 MPa). This is explained by a combination of fine-grained structure, solid solution strengthening, and dispersion strengthening by secondary phases [112].

Of particular interest are the fatigue properties of magnesium alloys obtained by SLM, as they are critically important for implants working under cyclic loading conditions. Studies show that the fatigue limit of SLM materials is 40–45% of the tensile strength, which is comparable to extruded materials and exceeds the values for cast analogues (30–35%) [114].

3.4.3. Effect of Heat Treatment

Heat treatment effectively modifies the microstructure and properties of SLM-magnesium alloys. Applied treatments include: stress relief annealing, homogenization annealing, T4 treatment (solid solution strengthening), and T6 treatment (combination of solid solution strengthening and artificial aging). For example, T4 treatment of WE43 alloy increases plasticity with a moderate decrease in yield strength [115]. Heat treatment of magnesium alloys requires precautions due to their tendency to oxidize, such as protective atmospheres or coatings.

3.5. Biodegradation and Biocompatibility

The rate and mechanism of biodegradation of magnesium alloys obtained by SLM play a key role in their functionality as temporary implants. The unique microstructure and properties of SLM materials significantly affect the corrosion process in the physiological environment.

3.5.1. Mechanism of Biodegradation of Magnesium Alloys

Biodegradation of magnesium alloys in physiological conditions is based on electrochemical corrosion, which can be described by the following reactions:

$$\text{Anodic reaction: } Mg \rightarrow Mg^{2+} + 2e^-$$

$$\text{Cathodic reaction: } 2H_2O + 2e^- \rightarrow H_2 + 2OH^-$$

$$\text{Hydroxide formation: } Mg^{2+} + 2OH^- \rightarrow Mg(OH)_2$$

$$\text{Overall reaction: } Mg + 2H_2O \rightarrow Mg(OH)_2 + H_2$$

In a physiological environment containing chloride ions, the formation of a protective $Mg(OH)_2$ layer can be disrupted due to the formation of soluble $MgCl_2$:

$$Mg(OH)_2 + 2Cl^- \rightarrow MgCl_2 + 2OH^-$$

Biodegradation of magnesium alloys in a living organism represents a more complex process, including interaction with proteins, cells, and enzymes. Protein adsorption on the implant surface can either accelerate or slow down corrosion depending on the type of proteins and environmental conditions [116].

3.5.2. Influence of Microstructure on Biodegradation

The microstructure of SLM-magnesium alloys significantly affects the rate and nature of their biodegradation. Fine-grained structure (5–20 μm) increases the area of grain boundaries, which can accelerate corrosion [117]. Non-equilibrium phase distribution and extended solubility limits of alloying elements in SLM materials change their electrochemical behavior—for example, in WE43 alloy, uniform distribution of rare earth elements contributes to the formation of a more stable protective film [118]. High residual stresses characteristic of the SLM process accelerate corrosion due to the mechano-chemical effect, but heat treatment can reduce their negative impact. Present microporosity creates conditions for localized corrosion due to limited mass transfer and local pH changes [119], and the crystallographic texture forming during SLM leads to anisotropy of corrosion behavior, as different crystallographic planes of magnesium have different electrochemical activity [120].

3.5.3. Comparison of Biodegradation Rate of SLM and Traditional Materials

Studies show that the biodegradation rate of SLM-magnesium alloys is higher than that of analogues obtained by traditional methods. For example, for SLM-WE43 alloy, the in vitro corrosion rate is 2.6 ± 1.9 versus 1.0 ± 0.5 mm/year for cast material [121]. This is associated with fine-grained structure, microporosity, and residual stresses, but the corrosion rate can be regulated by optimizing SLM parameters, post-processing, and surface modification.

As evident from the comparative data in Table 3, SLM-produced magnesium alloys consistently demonstrate higher corrosion rates compared to their traditionally manufactured counterparts, with increases ranging from 60% to 160% depending on the specific alloy composition and testing environment. This accelerated degradation can be attributed to several microstructural features unique to the SLM process: (1) the significantly finer grain size (5–20 μm compared to 50–150 μm in cast alloys) increases the total grain boundary area, creating more sites for preferential corrosion; (2) the presence of processing-induced microporosity creates localized areas for corrosion initiation; and (3) high residual stresses from rapid solidification promote stress-assisted corrosion mechanisms.

Table 3. Comparison of biodegradation rates of magnesium alloys produced by different methods.

Alloy	Manufacturing Method	Testing Environment	Corrosion Rate (mm/Year)	Reference
WE43	SLM	SBF, in vitro	2.6 ± 1.9	[121]
WE43	Casting	SBF, in vitro	1.0 ± 0.5	[121]
AZ91D	SLM	3.5% NaCl	1.68	[106]
AZ91D	Casting	3.5% NaCl	0.89	[106]
Mg-Ca	SLM	HBSS	1.35–1.81	[91]
Mg-Ca	Casting	HBSS	0.72–0.96	[91]
Pure Mg	SLM	PBS	2.28	[115]
Pure Mg	Casting	PBS	1.02	[115]
WE43	SLM, T4 treated	SBF, in vitro	1.8 ± 0.7	[41,122]
ZK60	SLM	HBSS	1.47	[41]
ZK60	Extrusion	HBSS	0.83	[41]

The data also reveals that post-processing treatments, particularly heat treatment, can substantially reduce this accelerated degradation. For instance, T4 heat treatment of SLM-WE43 reduced its corrosion rate from 2.6 to 1.8 mm/year by relieving residual stresses and homogenizing the microstructure. This modulation capability highlights the potential to tailor degradation rates to specific clinical requirements through controlled processing and post-processing strategies.

Testing environments significantly impact measured degradation rates, with more complex physiological solutions (containing proteins and various ions) generally resulting in different corrosion behavior compared to simple saline solutions. This underscores the importance of standardized testing protocols when comparing degradation performance across different studies and manufacturing methods.

3.5.4. Methods of Controlling Biodegradation Rate

A complex of methods is applied to regulate the biodegradation rate of SLM-magnesium implants. Alloy composition optimization (addition of RE, Ca, Zn) improves corrosion properties. Heat treatment (e.g., T4 regime for WE43) reduces the corrosion rate by relieving stresses [122]. Surface coatings are effective: MAO (reduces corrosion by 5–10 times) [123], PEO (creates regulated porous layers) [124], and biopolymer coatings. Additionally, porosity is controlled by adjusting SLM parameters, and gradient structures are created for zonal degradation [125]. These approaches allow precise tuning of implant dissolution rate.

3.5.5. Biocompatibility of SLM-Magnesium Implants

The biocompatibility of magnesium alloys obtained by SLM is determined by their chemical composition, microstructure, and surface topography. In vitro studies confirm their good cytocompatibility: they support adhesion, proliferation, and differentiation of osteoblasts, stem cells, and endothelial cells [126].

Porous lattice structures (60–70% porosity, 600–800 μm) provide optimal conditions for cell migration and bone tissue formation [57]. Figure 6 shows the Characterization of degradation products on the periphery of scaffolds. In vivo studies on animal models demonstrate active osseointegration and gradual replacement of the implant with bone tissue [127]. First clinical trials of individualized implants made of WE43 show their successful integration and controlled degradation [128].

Thus, SLM-magnesium alloys have high potential for clinical application due to the combination of biocompatibility, osteoinductive properties, and regulated degradation rate.

3.6. Clinical Applications and Prospects

SLM–magnesium implants represent a promising direction in orthopedics, craniomaxillofacial and cardiovascular surgery due to the unique possibilities of additive manufacturing and personalization. In orthopedics, they can be used for fracture fixation (screws, plates, rods), providing gradual load transfer and stimulation of osteogenesis [129], as well as bone scaffolds with optimized geometry and osteochondral implants with varying porosity. Clinical studies of MAGNEZIX® screws demonstrate effectiveness comparable to titanium analogues [130]. In craniomaxillofacial surgery, SLM–magnesium finds application in skull reconstruction (personalized implants), maxillofacial reconstruction (implants for defect restoration with improved osseointegration), and as experimental dental implants with controlled degradation rate [131]. In cardiovascular surgery, biodegradable stents (providing temporary support for vessels) and occluders for closing defects are being investigated [131]. Further research is directed towards developing new alloys with optimized properties, surface functionalization [131], creating bioactive implants [28], hybrid structures, improving SLM technology, and integration into personalized medicine platforms. In conclusion, SLM–magnesium alloys present significant potential in the field of biodegradable implants, opening new possibilities through additive manufacturing and magnesium biocompatibility.

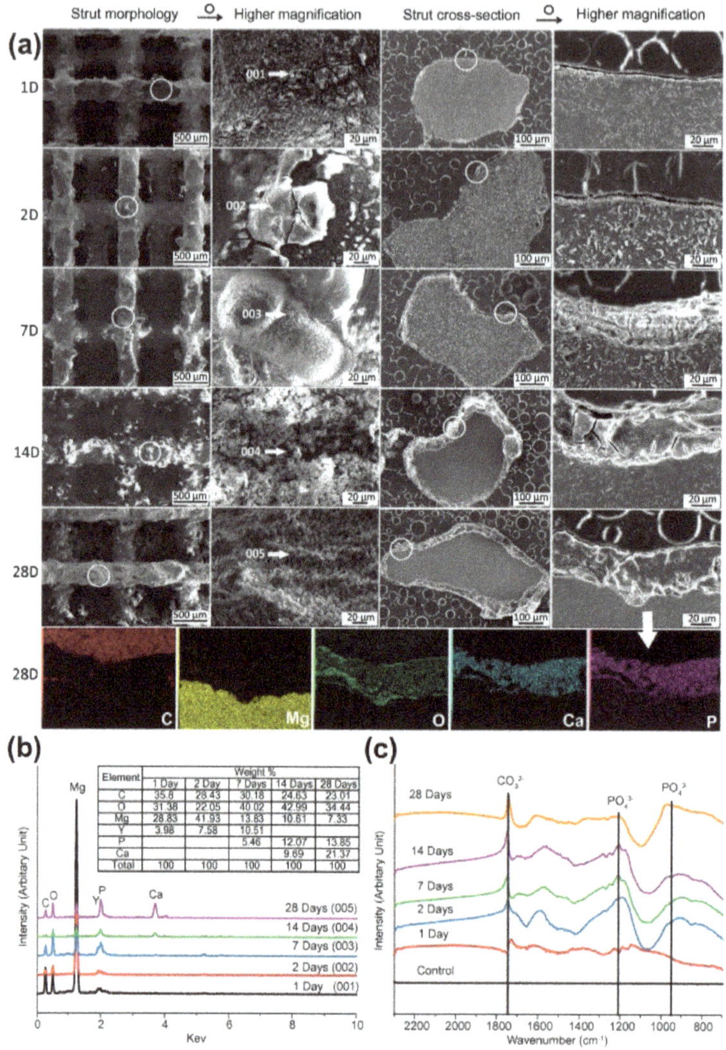

Figure 6. Characterization of degradation products on the periphery of scaffolds: (**a**) SEM images and EDS mapping, (**b**) EDS analysis, and (**c**) Fourier Transform Infrared Spectroscopy analysis, reproduced from [28], with permission from Elsevier, 2018.

While magnesium alloys offer excellent biocompatibility and mechanical similarity to bone, their relatively rapid degradation may be unsuitable for applications requiring prolonged structural support. Iron-based alloys present an alternative with superior strength and slower degradation kinetics, making them potentially valuable for load-bearing applications where longer-term support is necessary. The following section explores how selective laser melting can be applied to iron-based biodegradable systems, addressing both the opportunities and challenges presented by these high-strength alloys.

4. Iron-Based Alloys in Selective Laser Melting

4.1. Historical Overview and Basic Properties

Iron represents a promising material for biodegradable implants due to its combination of high strength, plasticity, excellent biocompatibility, and the possibility of manufacturing complex structures (stents, foils, metal foams) [38]. However, its clinical application is limited by an extremely slow degradation rate: experimental studies in pigs showed that iron stents in the descending aorta maintained structural integrity for 12 months without significant signs of corrosion [10]. Although the natural metabolism of iron in the body ensures good biocompatibility, its excessively low degradation rate makes a pure iron implant unsuitable for temporary clinical application, stimulating the development of methods to accelerate its corrosion.

Iron and its alloys are promising for creating biodegradable implants due to their combination of mechanical and biological properties. Their high strength and stiffness provide the necessary load-bearing capacity in orthopedic applications, while adjustable porosity allows approximating mechanical characteristics to those of natural bone [132,133]. An important advantage is the biocompatibility of iron, which is naturally metabolized in the body, minimizing the risk of undesirable immune reactions [131].

However, the use of iron as a biodegradable material is associated with several problems:

1. Low degradation rate—the corrosion rate of pure iron in physiological conditions is 0.1–0.5 mm/year, which is significantly lower than clinically acceptable values for temporary implants (0.5–2.0 mm/year) [134].
2. Formation of insoluble corrosion products—iron corrosion products (predominantly oxides and hydroxides) have low solubility and can accumulate around the implant, creating a diffusion barrier that further slows down corrosion [135].
3. Potential toxicity—Iron ions, especially when exceeding certain concentrations, can have a cytotoxic effect. Studies have shown that high levels of iron ions can reduce cell proliferation rate and affect metabolic activity [132].

To overcome these limitations, especially the low degradation rate, intensive research is being conducted on the development of new iron-based alloys and optimization of their production technologies, including selective laser melting.

4.2. Influence of Alloying Elements

Alloying of iron plays a key role in modifying its properties for biomedical applications, allowing control of degradation rate, mechanical characteristics, and biological response (Table 4). The choice of alloying elements for biodegradable iron alloys is determined not only by their influence on material properties but also by biocompatibility, absence of toxicity, and potential therapeutic effect.

Table 4. Influence of alloying elements on the properties of iron alloys obtained by SLM.

Alloying Element	Optimal Content	Effect on Mechanical Properties	Effect on Degradation Rate	Biological Effects
Manganese (Mn)	20–35%	Increased strength and plasticity, change in phase composition (γ, $\gamma + \varepsilon$), reduced magnetic properties [136]	Acceleration of corrosion compared to pure Fe [137]	Essential microelement, participates in metabolism [138]

Table 4. *Cont.*

Alloying Element	Optimal Content	Effect on Mechanical Properties	Effect on Degradation Rate	Biological Effects
Carbon (C)	0.5–1.2%	Significant strength increase [139], austenitic phase stabilization, TRIP effect [140]	Can accelerate or slow down corrosion depending on microstructure [141]	-
Silicon (Si)	3–6%	Increased yield strength, shape memory effect [142]	Significant acceleration of corrosion (by 2–3 times) [143]	Biocompatible with various cell types [144]
Calcium (Ca)	0.5–2.0%	-	Substantial acceleration of corrosion [145]	Stimulation of osteogenesis [146], important element for bone tissue
Magnesium (Mg)	0.5–2.0%	Increased hardness [147]	Significant acceleration of corrosion [145]	-
Palladium (Pd)	0.5–1.0%	Increased strength and corrosion rate [148]	Acceleration of corrosion by 3–4 times through microgalvanic effect [149]	Acceptable cytotoxicity [148]
Copper (Cu)	1.0–4.0%	Moderate reduction in mechanical properties [150]	Acceleration of corrosion [150]	Antimicrobial action [151]

4.2.1. Manganese (Mn)

Manganese (Mn) is a promising alloying element for biodegradable iron alloys, affecting phase composition, mechanical properties, and degradation rate. At 20–30% Mn, austenitic (γ, FCC) or austenitic-martensitic ($\gamma + \varepsilon$, FCC + HCP) structures form, and at >30%—fully austenitic structures [139]. Mn improves strength and enhances plasticity; for example, Fe-30Mn alloys demonstrate yield strength of 200–250 MPa and tensile strength of 430–550 MPa with elongation of 30–40%. Fabricated porous biodegradable SLM Fe-30Mn scaffolds showed mechanical adaptability, biocompatibility, and osseointegration in vivo (48 weeks) [28]. However, in the context of SLM, high reactivity and low evaporation temperature of Mn create technological challenges, such as selective evaporation of Mn and deviation of product composition.

4.2.2. Carbon (C)

Carbon (C) is an important alloying element for iron alloys, affecting mechanical properties and microstructure. It strengthens iron (0.5–1.2% C increases yield strength and strength) [139], stabilizes the austenitic phase, and affects corrosion resistance [141]. Fe-Mn-C possesses high strength and plasticity due to austenitic structure and TRIP effect, as well as accelerated degradation in vitro [140]. In SLM, it is necessary to control the formation of carbides that affect product properties.

4.2.3. Silicon (Si)

Silicon (Si) is an effective alloying element that accelerates the degradation of iron alloys: it increases corrosion rate [141], promotes the formation of duplex ε and γ phases (Fe-28Mn-6Si demonstrates 80% higher degradation rate) [143], improves strength by 70% compared to the base alloy (Fe-30Mn) and improves strain hardening capacity while maintaining good plasticity [144], and at certain compositions induces shape memory effect [142]. Optimal Si content (3–5%) for balance of properties and degradation; Fe-30Mn-5Si showed a degradation rate of 0.80 mm/year [152].

4.2.4. Calcium (Ca) and Magnesium (Mg)

Calcium (Ca) and magnesium (Mg) are biologically active elements that, while accelerating the degradation of iron alloys, can positively affect bone tissue regeneration. They significantly increase corrosion rate [145], and the released Ca^{2+} and Mg^{2+} ions can stimulate osteogenesis and angiogenesis [146]. Ca and Mg also affect microstructure and, consequently, alloy properties. Hong et al. (2016) showed that Fe-Mn-Ca/Mg alloys obtained by jet 3D printing accelerate degradation and promote osteoinduction and osteoconduction [153]. In SLM, it is necessary to consider the high reactivity of Ca and Mg and optimize process parameters.

4.2.5. Palladium (Pd) and Copper (Cu)

Palladium (Pd) and copper (Cu) accelerate the degradation of iron alloys and impart additional properties. Minor addition of Pd (0.5–1.0%) to Fe-10Mn increases degradation rate fourfold [149]. Cu-containing alloys possess antimicrobial properties [151], while Pd and Cu can improve mechanical characteristics. Schinhammer et al. (2013) showed that the combination of Mn, C, Pd optimizes degradation rate and mechanical properties [154]. In SLM, differences in melting temperatures and thermophysical properties of Pd and Cu should be considered.

4.3. *Features of the SLM Process for Iron Alloys*

Selective laser melting of iron alloys for biodegradable implants has a number of features related to both the physicochemical properties of the material and the requirements for final products.

4.3.1. Technological Features of SLM for Iron Alloys

Iron alloys, particularly Fe-Mn systems, possess a number of specific properties affecting the SLM process:

1. High melting temperature requires higher laser power and energy density for complete powder melting [155].
2. Selective evaporation of alloying elements—In the Fe-Mn SLM process, evaporation of large amounts of Mn will lead to Mn mass loss, defect formation, and chemical composition changes in the final product [156].
3. High coefficient of thermal expansion—austenitic Fe-Mn alloys have a relatively high coefficient of thermal expansion [157], which can lead to significant thermal stresses and deformations during SLM.
4. Oxidation—although iron alloys are less reactive than magnesium or zinc alloys, they are still subject to oxidation at high temperatures, especially alloys containing Mn, which requires working in a protective atmosphere [158].

4.3.2. Optimization of SLM Parameters for Iron Alloys

Key SLM parameters requiring optimization for iron alloys are:

1. Laser power—higher power (200–400 W) is usually required for effective melting of iron alloys compared to magnesium and zinc alloys. Donik et al. (2021) showed that for the Fe-Mn alloy, the optimal power is 250–300 W [159].
2. Scanning speed—relatively high scanning speeds (600–1200 mm/s) are applied for iron alloys, allowing reduction in laser interaction time and minimization of selective evaporation of alloying elements. The optimal speed for Fe-Mn alloys is about 800 mm/s [159].

3. Hatch distance—for iron alloys, a distance of 70–100 μm is usually used, providing sufficient overlap of tracks for forming a monolithic structure. Donik et al. (2021) used 80 μm [159].
4. Layer thickness—typical values for Fe-alloys are 20–40 μm, providing a balance between productivity and quality of the resulting products.
5. Scanning strategy—to minimize thermal stresses and property anisotropy, a strategy with rotation of scanning direction between layers (usually by 67° or 90°) is commonly applied [160].
6. Energy density—for iron alloys, the optimal energy density is usually 60–150 J/mm^3 [161,162].
7. Platform preheating temperature—preheating the platform to 500 °C allows reducing thermal gradients and residual stresses, as well as improving product quality [163].

4.3.3. Influence of SLM Parameters on Phase Composition and Microstructure of Fe-Mn Alloys

SLM parameters significantly affect phase formation and microstructure of Fe-Mn alloys, determining their mechanical properties and degradation rate. Energy density influences the formation of austenitic (γ) and ε-martensitic phases, with lower densities potentially reducing Mn losses [24]. The formation of γ and ε phases is directly related to Mn content [159]. Thermal effect determines the formation of columnar or equiaxed grains [24]. Rapid cooling leads to non-equilibrium structures with Mn segregation at melt pool boundaries. Donik et al. (2021) established the dependence of ε-phase content on laser energy density, allowing optimization of phase ratio and microstructure for controlled degradation [159].

4.3.4. Post-Processing of SLM Products from Iron Alloys

Post-processing is an important stage in the production of biodegradable implants from iron alloys, allowing modification of microstructure, improvement in mechanical properties, and control of degradation rate:

1. Heat treatment—various heat treatment regimes can be applied to reduce residual stresses, homogenize microstructure, and modify phase composition. For Fe-Mn alloys, Mn oxides have several transformations in the temperature range from 700 to 1000 °C, from which heat treatment regimes are often selected [24,164].
2. Hot isostatic pressing (HIP)—this method allows eliminating residual porosity and improving mechanical properties of SLM products.
3. Surface modification controls degradation and improves biocompatibility: electrochemical polishing, passivation, coating application [165,166]
4. Mechanical processing—to achieve the necessary dimensional accuracy and surface quality, finish mechanical processing can be applied. However, for complex porous structures, traditional mechanical processing methods are often inapplicable, requiring the use of specialized approaches such as electric discharge machining or chemical etching.

Figure 7 shows EBSD ε-phase content from 81% (20.5% Mn) to 71% (24.8% Mn). Mn content affects the $\varepsilon - \gamma$ phase transformation, with carbon also playing an important role (Mesquita et al., our previous studies). Despite increasing Mn content in the raw material to 53%, only 32% Mn was achieved in the SLM sample, a dual-phase structure ($\varepsilon + \gamma$) remained due to manufacturing issues [159].

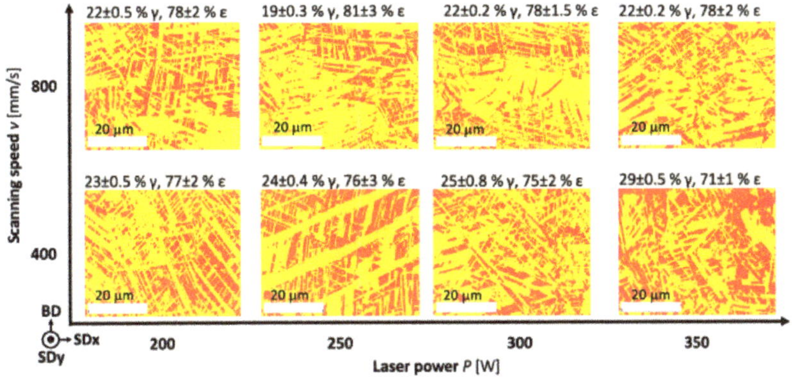

Figure 7. Phase maps of ε phase—yellow, and γ phase—red for SLM samples produced at different process parameters, reprint from [159].

4.4. Microstructure and Mechanical Properties

The microstructure of iron alloys obtained by SLM differs significantly from that formed by traditional production methods, due to the unique crystallization conditions during laser melting.

4.4.1. Features of SLM-Iron Alloys Microstructure

The microstructure of Fe-Mn alloys obtained by SLM is characterized by a number of features. Phase composition depends on Mn content and SLM parameters: γ-phase (austenite) forms at high Mn content, ε-phase (martensite) at medium content, and α′-phase (martensite) at low Mn content or deformation [159,167]. SLM typically leads to the formation of a fine-grained structure (10–50 μm) with preferential grain orientation, the shape of which depends on process parameters. Microporosity, hot cracks, and residual stresses may be present in SLM materials. The composition of the iron alloy 1.2% C, 32% Mn provides an austenitic structure regardless of method, but SLM parameters affect microstructure, element distribution, grain boundaries, mechanical properties, and corrosion [168].

4.4.2. Mechanical Properties of SLM-Iron Alloys

The mechanical properties of iron alloys obtained by SLM are determined by a combination of chemical composition, phase composition, microstructure, and presence of defects. Table 5 presents the mechanical properties of various Fe alloys obtained by SLM compared to traditional production methods.

Table 5. Mechanical properties of iron alloys obtained by various methods as well as 316L.

Alloy	Production Method	Yield Strength (MPa)	Tensile Strength (MPa)	Elongation (%)	Source
Fe-30Mn-1.2C-1Si	SLM	599 ± 27	843 ± 23	24 ± 4	[169]
Fe-30Mn-1.2C-1Si	Casting	341 ± 6	767 ± 51	30 ± 1	[169,170]
Fe-30Mn	PM	134 ± 5	216 ± 12	11 ± 1	[171]
Fe–30Mn	Casting	124	366	55	[172]
Fe–Mn–Si	SLM	325.8 ± 20.3	863.2 ± 10.7	11.0 ± 4.6	[164]
FeMn	SLM	331 ± 8	553 ± 10	3.8 ± 1	[24]
316L	SLM	638 ± 22	674 ± 9	30 ± 3	[173]

These results highlight the potential of Fe-based biodegradable alloys, particularly those produced via SLM, for applications requiring high strength, though improvements in elongation are needed for broader use in biomedical applications.

4.4.3. Special Mechanical Effects in Fe-Mn Alloys

Fe-Mn alloys, especially with the addition of C and Si, can demonstrate a number of unique mechanical effects that may be useful for biomedical applications:

1. TRIP effect (Transformation-Induced Plasticity)—plasticity induced by phase transformation. In Fe-Mn-C alloys with predominantly austenitic structure, mechanical deformation can cause martensite formation, leading to enhanced plasticity and strengthening [174].
2. TWIP effect (Twinning-Induced Plasticity)—plasticity induced by twinning. In high-manganese alloys (Fe-Mn with Mn content 25–35%), deformation occurs predominantly through the twinning mechanism, providing high plasticity and strengthening [174].
3. Shape memory effect—some Fe-Mn-Si alloys demonstrate shape memory effect associated with reversible martensitic transformation $\gamma \leftrightarrow \varepsilon$. This effect can be used to create implants with functional properties, for example, self-expanding stents [175].
4. Superelasticity—Fe-Mn-Si-Al alloys can demonstrate superelastic behavior similar to NiTi alloys, but with better biocompatibility and biodegradability [176].

These effects can be enhanced or modified by optimizing SLM parameters and subsequent heat treatment. For example, by controlling cooling rate and thermal cycles during the SLM process, the ratio of γ and ε phases can be regulated and, consequently, the expression of TRIP/TWIP effects and shape memory effect.

4.4.4. Fatigue Characteristics and Durability

For temporary implants, especially those working under cyclic loading conditions (such as orthopedic fixators or cardiovascular stents), fatigue characteristics and durability are critically important:

1. Fatigue limit—for Fe-Mn alloys obtained by SLM, the fatigue limit is usually 40–45% of the tensile strength, which corresponds to 330 MPa for alloys with tensile strength of 839 MPa [177].
2. Corrosion fatigue—high fatigue strength (70% of yield strength in air, 65% in r-SBF) due to the plasticity of iron and slow degradation is shown. Cyclic loading accelerated iron degradation, but iron remains a promising bioactive bone implant [178].
3. Microstructure influence—fatigue characteristics strongly depend on microstructure and presence of defects. Fine-grained structure with uniform phase distribution usually provides better fatigue strength [179].
4. Residual stresses—characteristic for SLM, residual stresses can significantly reduce fatigue strength. Heat treatment for stress relief (heat treatment at 600–700 °C [180]).

4.5. Biodegradation and Biocompatibility

The rate and mechanism of biodegradation of iron alloys obtained by SLM play a key role in their functionality as temporary implants. The unique microstructure and properties of SLM materials significantly affect the corrosion process in the physiological environment.

4.5.1. Mechanism of Biodegradation of Iron Alloys

Biodegradation of iron alloys in physiological conditions is based on electrochemical corrosion, which can be described by the following reactions:

$$\text{Anodic reaction: Fe} \rightarrow \text{Fe}^{2+} + 2\text{e}^-$$

$$\text{Cathodic reaction: O}_2 + 2\text{H}_2\text{O} + 4\text{e}^- \rightarrow 4\text{OH}^-$$

$$\text{Hydroxide formation: Fe}^{2+} + 2\text{OH}^- \rightarrow \text{Fe(OH)}_2$$

$$\text{Further oxidation: 4Fe(OH)}_2 + \text{O}_2 + 2\text{H}_2\text{O} \rightarrow 4\text{Fe(OH)}_3$$

$Fe(OH)_3$ can then dehydrate to form various forms of iron oxides (Fe_2O_3, Fe_3O_4).

Unlike magnesium alloys, during corrosion of iron alloys, a significant amount of gaseous hydrogen is usually not formed, since the main cathodic reaction is oxygen reduction [181].

An important feature of iron alloy corrosion is the formation of insoluble corrosion products (iron oxides and hydroxides), which form a diffusion barrier slowing down further corrosion. This explains the relatively low degradation rate of pure iron in vivo and the need for alloying to accelerate corrosion [182].

In a physiological environment, the corrosion process is further complicated by interaction with proteins, cells, and enzymes. Protein adsorption on the implant surface can either accelerate or slow down corrosion depending on the type of proteins and environmental conditions [183].

4.5.2. Influence of Microstructure on Biodegradation

The microstructure of SLM-iron alloys affects biodegradation, determined by grain size (fine-grainedness increases the area of electrochemically active boundaries, phase composition (different electrochemical activity of γ, ε, α' phases, where ε is more active than γ) [184], distribution of alloying elements, residual stresses (accelerate corrosion mechano-chemically). SLM-Fe-Mn alloys corrode faster than traditional ones, which is related to process parameters and microstructure [185].

4.5.3. Comparison of Biodegradation Rate of SLM and Traditional Materials

The biodegradation rate of SLM-iron alloys is usually higher than that of analogues manufactured traditionally. They compared the corrosion rate of pure iron obtained by laser melting, which was 13 times higher than that of cast iron [186]. For example, the corrosion rate of the Fe-Mn alloy (SLM) was 0.22 mm/year versus 0.008 mm/year for pure iron (cast) [187]. This is explained by fine-grained structure, non-uniform distribution of alloying elements, microporosity, and residual stresses.

4.5.4. Methods of Controlling Biodegradation Rate

To optimize the biodegradation rate of SLM-iron implants, the following are used: alloying; phase composition control (regulation of γ and ε phase ratio [159]); creation of galvanic pairs (introduction of Pd, Cu, Ag [188,189]); surface modification [165,166]; porous structure design.

4.5.5. Biocompatibility of SLM-Iron Implants

The biocompatibility of iron alloys obtained by SLM is determined by a combination of factors, including chemical composition, microstructure, surface topography, and degradation products. In vitro studies show that SLM materials based on Fe-Mn usually demonstrate good cytocompatibility, supporting adhesion, proliferation, and differentiation of various cell types, including osteoblasts, mesenchymal stem cells, and endothelial cells [190].

The influence of iron ions released during implant degradation on cell metabolism depends on their concentration. At low and moderate concentrations, Fe^{2+} ions can stimulate proliferation and differentiation of osteoblasts, whereas high concentrations can cause oxidative stress and cytotoxicity [191].

Of particular interest is the influence of alloying elements on biological response. Manganese at low concentrations is necessary for normal cell functioning, however, its excess can cause neurotoxic effects. Calcium and magnesium, on the contrary, have a positive effect on osteogenesis and angiogenesis, stimulating bone tissue formation [192].

In vivo studies on animal models (rats, rabbits, pigs) confirm the biocompatibility of SLM-iron implants. Good integration with surrounding tissues, minimal inflammatory reaction, and gradual degradation with replacement by new tissue are observed [193].

An important aspect of biocompatibility is control of local concentration of degradation products. The porous structure of SLM implants promotes uniform distribution of ions and prevents their local accumulation in toxic concentrations [125].

In Figure 8, the SEM showed degradation of struts with the formation of white products. By day 28, the surface is covered with loose products. Degradation is non-uniform: products in the center are thinner and denser than on the periphery. The composition of products (C, O, P, Ca, Fe) differs: more P and Ca on the periphery. By day 28, crystalline structures (spherical and feather-like) form in the center, without P. On the periphery, the composition is similar to day 7, but with increased O and decreased Fe [125].

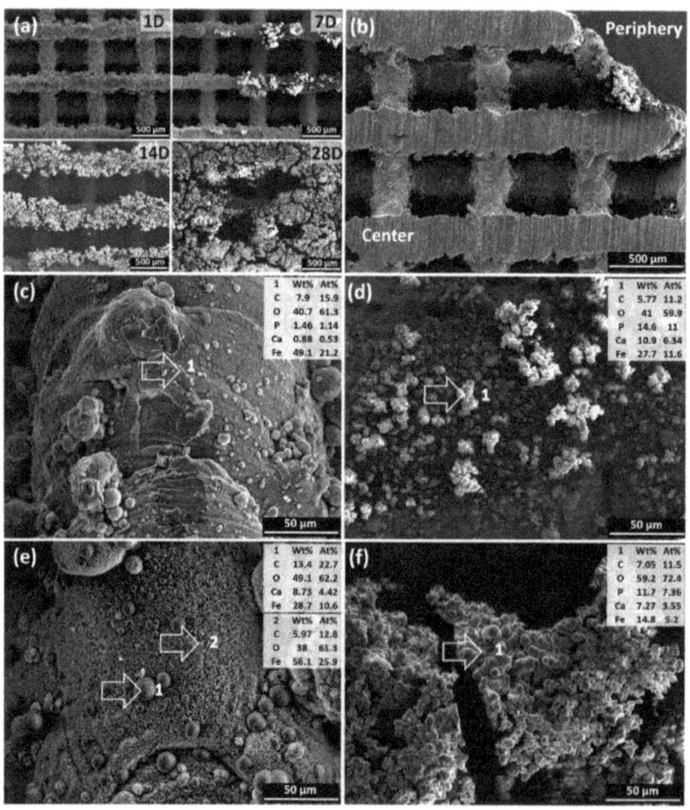

Figure 8. SEM and EDS analyses of degradation products from the scaffold periphery to the center: (**a**) degradation products on the periphery at different immersion time points, (**b**) cross section of the scaffolds after 7-day immersion, (**c**) degradation products in the center and (**d**) on the periphery after 7-day immersion, (**e**) degradation products in the center and (**f**) on the periphery after 28-day immersion. The numbers 1 and 2 indicates the spot where EDS analysis was performed [125].

4.6. Clinical Applications and Prospects

SLM-iron implants are promising in cardiovascular, orthopedic, and reconstructive surgery due to their mechanical properties and controlled degradation rate. In cardiovascular surgery, they are used for biodegradable stents (coronary, peripheral), the design of which is optimized using SLM [194]. Clinical studies demonstrate good biocompatibility, and the Fe-35Mn stent showed minimal inflammatory reaction and gradual degradation [195]. In orthopedics, SLM-iron is applied for fracture fixation (screws, plates, rods), optimizing design with SLM, for bone scaffolds, providing osseointegration and controlled degradation, and for intervertebral implants. Further research is directed toward developing new alloys, functionally gradient materials, bioactive implants, composite materials, and improving SLM technology. In conclusion, SLM–iron alloys are a promising direction in developing biodegradable implants, combining strength, biocompatibility, and controlled degradation. SLM opens possibilities for creating personalized structures with optimized properties.

Having examined both magnesium alloys with their rapid degradation profiles and iron alloys with their high strength but slow corrosion rates, we now focus on zinc-based systems. Zinc alloys occupy an intermediate position between magnesium and iron in terms of both mechanical properties and degradation behavior, potentially offering a balanced solution for certain biomedical applications. The following section investigates how selective laser melting can be applied to zinc alloys to create biodegradable implants with tailored properties.

5. Zinc Alloys in Selective Laser Melting

5.1. Historical Overview and Basic Properties

Zinc occupies a unique position among biodegradable metals, demonstrating intermediate properties between magnesium and iron, possessing a degradation rate of 0.2 mm/year [196] and good biocompatibility of corrosion products [197]. However, the application of zinc is limited by low mechanical strength [198], brittleness, technological complexities, and creep at body temperature [199], which requires the development of special zinc alloys and optimization of their production technologies, including selective laser melting.

5.2. Influence of Alloying Elements

Alloying of zinc plays a key role in improving its properties for biomedical applications, allowing enhanced mechanical strength, improved plasticity, and controlled degradation rate. The choice of alloying elements for biodegradable zinc alloys is determined not only by their influence on material properties but also by biocompatibility, absence of toxicity, and potential therapeutic effect.

5.2.1. Magnesium (Mg)

Magnesium is a promising alloying element for zinc alloys, affecting: mechanical properties (significantly increases strength) [200], microstructure [201], degradation rate (depends on concentration) [201], and biocompatibility. Studies have shown that the optimal magnesium content is 1–2%, providing a balance between improved mechanical properties, controlled degradation rate, and good processability in SLM [202]. Corrosion rates of Zn alloys are approximately 14–30 μm/year [203], these values are comparable to bone regeneration rate. The results of the study suggest that magnesium addition contributes to enhanced cytocompatibility of the Zn-3Cu alloy [204].

5.2.2. Calcium (Ca)

Calcium, being a biocompatible element, plays an important role in bone tissue formation and regeneration [205]. In zinc alloys, calcium moderately increases strength, contributes to grain refinement, and formation of a more homogeneous microstructure [206], and typically accelerates zinc corrosion. Calcium ions released during alloy degradation stimulate proliferation and differentiation of osteoblasts, promoting bone tissue formation [207]. Due to these properties, Zn-Ca alloys exhibit pronounced osteogenic properties and are considered promising for creating temporary implants used in orthopedics and maxillofacial surgery.

5.2.3. Strontium (Sr)

Strontium, structurally and chemically similar to calcium, has a positive effect on bone tissue formation [208]. Strontium moderately increases the strength of zinc (yield strength 130–230 MPa, tensile strength 220–360 MPa for Zn-(0.5–2%)Sr alloys [209]) through solid solution strengthening, promotes grain refinement and formation of intermetallic phases affecting mechanical properties and corrosion behavior. Strontium typically accelerates zinc corrosion due to the formation of microgalvanic pairs [210]. It stimulates osteoblasts and inhibits osteoclasts, and strontium ranelate is used for osteoporosis treatment [211]. Studies of Zn-Sr alloys for additive manufacturing are promising for orthopedic applications.

5.2.4. Copper (Cu)

Copper, an essential microelement with antibacterial properties, significantly increases zinc's strength and plasticity (yield strength 225 ± 9 MPa, tensile strength 330 ± 12 MPa, Zn3Cu [212]). Degradation rates of Zn-xCu alloys in c-SBF solution are at a relatively low level, ranging from 22.1 ± 4.7 to 33.0 ± 1.0 μm/year [213], with a trend towards a slight increase in corrosion rate with increasing copper content compared to pure zinc. The study [213] found that the best antibacterial properties are exhibited by alloys with copper content above 2%.

5.2.5. Silver (Ag)

Silver, due to its pronounced antibacterial properties, is of interest as an alloying element for biodegradable implants. Silver moderately increases zinc strength by forming solid solutions and intermetallic compounds [214]. Zn-Ag alloys corrode faster than pure Zn; silver typically accelerates zinc corrosion through the galvanic effect [215]. The bone marrow cavities of rat femur models were implanted and inoculated with S. aureus and E. coli; results showed that the Zn-Ag alloy exhibited optimal antibacterial properties [216].

5.2.6. Multi-Component Alloys

To optimize properties, multi-component zinc alloys are being developed: Zn-Cu-Ag (improved mechanical and antibacterial properties [217]); Zn-Mg-Sr (strengthening and hemocompatibility [218]); Zn-Li-Mg (improved plasticity, strength, and degradation rate [219]). Multi-component alloys expand optimization possibilities but require strict control of composition and microstructure to ensure reproducibility and biocompatibility.

5.3. *Features of the SLM Process for Zinc Alloys*

Selective laser melting of zinc alloys is associated with a number of specific problems and requires careful optimization of process parameters to ensure high quality of the resulting products.

5.3.1. Technological Features and Optimization of SLM Parameters for Zinc Alloys

Zinc and its alloys possess specific properties affecting SLM: low melting temperature (419.5 °C) requires control of energy parameters; high vapor pressure creates a risk of evaporation and defects [220]; high thermal conductivity (116 W/(m·K)) promotes rapid heat removal; high reflectivity reduces laser energy absorption efficiency; easy oxidation leads to oxide film formation [221].

5.3.2. Optimization of SLM Parameters for Zinc and Zinc Alloys

For SLM of zinc and zinc alloys, the following are optimized: laser power (pure Zn 50 W, for Zn-10Mg optimally 70 W [222]); scanning speed (pure Zn 700 mm/s, for Zn-10Mg optimally 600 mm/s [222]); hatch distance (600–800 μm [222]); layer thickness (30 μm [222]); scanning strategy (with 67° rotation between layers [223]); platform preheating temperature (120 °C [223]). Optimization of these parameters allows obtaining zinc and zinc alloy products with desired mechanical properties, density, and microstructure necessary for specific biomedical applications.

5.3.3. Influence of SLM Parameters on Microstructure of Zinc Alloys

SLM parameters affect the microstructure of zinc alloys and, consequently, their mechanical properties and degradation rate: high cooling rates lead to fine-grained structure (5–30 μm, with average grain size of 27.1 μm) [223]; metastable phases may form (e.g., Mg_2Zn_{11} in Zn-Mg alloys) [223]; porosity control is critical (high density >99% or controlled porosity is achieved) [224].

5.3.4. Post-Processing of SLM Products from Zinc Alloys

Post-processing is important for biodegradable implants from zinc alloys, as it allows modifying microstructure, improving mechanical properties, and controlling degradation rate. Heat treatment is applied to relieve internal stresses, improve microstructure, and mechanical properties. Heat treatment helps stabilize the material and reduce porosity [225]. Despite growing interest in additive manufacturing of biodegradable zinc implants by selective laser melting (SLM), issues related to post-processing of such products to improve their mechanical properties, biocompatibility, and degradation kinetics remain understudied, highlighting the need for further systematic research in this area.

5.4. Microstructure and Mechanical Properties

The microstructure and, consequently, mechanical properties of zinc alloys obtained by SLM significantly differ from those of similar materials manufactured by traditional methods, due to the unique crystallization conditions during laser melting.

5.4.1. Features of SLM Zinc and Zinc Alloys Microstructure

The microstructure of SLM-obtained pure zinc is characterized by equiaxed grains and lamellar structure with grain size of 8–10 μm, as well as the presence of oxides at grain boundaries, which is explained by the low temperature gradient and inhibiting effect of oxides on grain growth [226].

Figure 9 shows the microstructure of SLM Zn-xMg alloy samples, where Zn-xMg consists mainly of HCP phases (fewer slip systems), with Mg_2Zn_{11} precipitates being BCC, which reduces plasticity. Zn-1Mg shows fine α-Zn with Mg precipitates/oxides at boundaries (grain size ~2 μm). With increasing Mg, Mg_2Zn_{11} increases. In pure Zn, only α-Zn is present (size ~6 μm). In Zn-2Mg, even finer α-Zn grains are observed. Zn-5Mg consists of $MgZn_2$(~1 μm) and α-Zn + Mg_2Zn_{11} [227]. Zn-xMg consists mainly of HCP phases (fewer slip systems), with Mg_2Zn_{11} precipitates being BCC, which reduces plasticity [227].

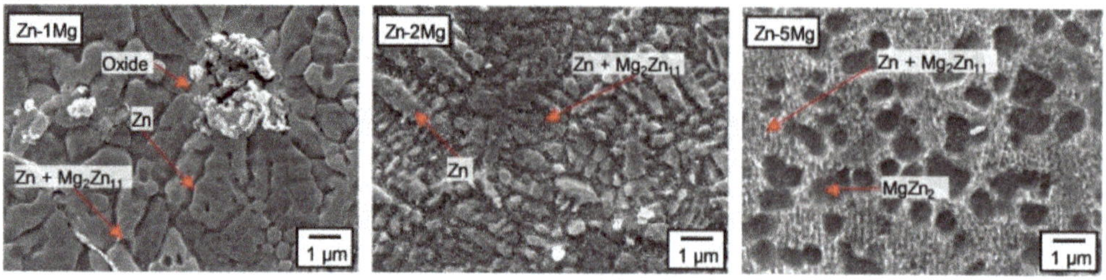

Figure 9. Microstructure of SLM Zn-xMg alloy samples, reprint from [227].

5.4.2. Mechanical Properties of SLM–Zinc Alloys

The mechanical properties of zinc alloys obtained by SLM usually significantly exceed those of similar materials manufactured by traditional methods, especially in terms of strength indicators. Table 6 presents the mechanical properties of various zinc alloys obtained by SLM compared to traditional production methods.

Table 6. Mechanical properties of zinc alloys obtained by various methods.

Alloy	Production Method	Yield Strength (MPa)	Tensile Strength (MPa)	Elongation (%)	Source
Pure Zn	SLM	84	95	11.7	[226]
Pure Zn	Casting	17	20	0.2	[46]
Zn-2Mg	SLM	117	162	4.1	[228]
Zn-2Mg	Casting	-	154	2	[229]
Zn-3Cu	SLM	152	222	7.2	[228]
Zn-3Cu	Casting	64	84	1.3	[230]
Zn-1Mg-0.5Sr	Casting	130	209	2	[231]

Table 6 reveals that zinc-based biodegradable alloys demonstrate significantly enhanced mechanical properties when processed via SLM compared to traditional casting methods. Pure zinc processed by SLM exhibits remarkable improvements, with yield strength increasing from 17 MPa (casting) to 84 MPa (SLM)—nearly a five-fold enhancement. Similarly, tensile strength improves from 20 MPa to 95 MPa, while elongation increases dramatically from 0.2% to 11.7%, indicating substantially improved ductility.

Alloying zinc with magnesium and copper further enhances mechanical performance. Zn-3Cu (SLM) demonstrates the highest strength among the zinc alloys tested, achieving 152 MPa yield strength and 222 MPa tensile strength—representing 137% and 165% improvements over cast Zn-3Cu, respectively. The Zn-2Mg system shows more moderate but still significant improvements when processed by SLM. These results highlight the effectiveness of SLM processing in refining microstructure and enhancing the mechanical properties of zinc alloys, making them more viable candidates for load-bearing biodegradable implant applications. However, elongation values remain relatively low compared to other biodegradable metal systems, indicating that further alloy development and processing optimization are needed to improve ductility for broader biomedical applications.

5.4.3. Fatigue Characteristics and Creep

For temporary implants, fatigue characteristics and creep resistance are important. The fatigue strength of porous zinc in air is 70% of yield strength, and in simulated physiological environment (r-SBF)—80% [232]. This is associated with the formation of corrosion products that strengthen the connections of the structure struts. To minimize creep, it is recommended to use multi-component alloys with thermally stable phases [219], optimized heat treatment, and to consider the creep factor in design.

5.5. Biodegradation and Biocompatibility

The rate and mechanism of biodegradation of zinc alloys obtained by SLM play a key role in their functionality as temporary implants. The unique microstructure and properties of SLM materials significantly affect the corrosion process in the physiological environment.

5.5.1. Mechanism of Biodegradation of Zinc Alloys

Biodegradation of zinc alloys in physiological conditions is based on electrochemical corrosion, which can be described by the following reactions:

$$\text{Anodic reaction: } Zn \rightarrow Zn^{2+} + 2e^{-}$$

$$\text{Cathodic reaction: } O_2 + 2H_2O + 4e^{-} \rightarrow 4OH^{-}$$

$$\text{Hydroxide formation: } Zn^{2+} + 2OH^{-} \rightarrow Zn(OH)_2$$

$$\text{Further reactions: } Zn(OH)_2 \rightarrow ZnO + H_2O$$

An important feature of zinc corrosion is the formation of relatively dense and stable corrosion products (oxides, hydroxides, carbonates, phosphates), which can form a protective layer slowing down further corrosion. However, in a physiological environment, this protective layer is usually less stable than in atmospheric conditions, due to the presence of chloride ions, which can cause pitting corrosion [232].

In a living organism, the corrosion process is further complicated by interaction with proteins, cells, and enzymes. Protein adsorption on the implant surface can either accelerate or slow down corrosion depending on the type of proteins and environmental conditions [233].

5.5.2. Influence of Microstructure on Biodegradation

The microstructure of SLM–zinc alloys significantly affects biodegradation: fine-grainedness increases the area of grain boundaries, accelerating corrosion, but uniform distribution of alloying elements can compensate for this effect [230]; different phases have different electrochemical activity (intermetallics, such as $MgZn_2$, $CaZn_{13}$, accelerate corrosion [234]); non-uniform distribution of alloying elements creates microgalvanic pairs; porosity promotes localized corrosion; grain orientation affects corrosion anisotropy [235].

5.5.3. Comparison of Biodegradation Rate of SLM and Traditional Materials

SLM technology allows significantly reducing the corrosion rate of zinc when alloyed with magnesium (up to 0.09 mm/year for the Zn-3Mg alloy [236]), demonstrating a clear advantage of additive manufacturing over traditional casting, which is characterized by a higher corrosion rate for the same alloy (0.21 mm/year [237]). This is explained by fine-grained structure, larger area of grain boundaries, microgalvanic pairs, and residual stresses.

5.5.4. Methods of Controlling Biodegradation Rate

To optimize the biodegradation rate of SLM-zinc implants, the following are used: alloy composition optimization (alloying with Mg, Ca, Cu, Ag) [238]; heat treatment (reduces corrosion rate due to homogenization and stress reduction) [239]; surface modification (anodizing, micro-arc oxidation, application of biopolymer coatings) [240]; porous structure design.

5.5.5. Biocompatibility of SLM-Zinc Implants

The biocompatibility of SLM-zinc alloys is determined by chemical composition, microstructure, surface topography, and degradation products. Low and moderate concentrations of Zn^{2+} ions (10–50 μM) stimulate osteoblast proliferation and mineralization, while high concentrations (>100 μM) can be cytotoxic [241]. Alloying elements also affect biocompatibility: magnesium and calcium stimulate osteogenesis, copper possesses antibacterial and angiogenic properties, and silver enhances antibacterial effects. However, in vivo studies on animals are needed to confirm biocompatibility (good integration, minimal inflammation, controlled degradation with tissue replacement). Zn-based alloys (AMed Zn-Li, Zn-Se) effectively suppress osteosarcoma and promote bone regeneration [238].

5.6. Clinical Applications and Prospects

SLM–zinc implants are promising in cardiovascular surgery; the antibacterial properties of Zn-Ag alloys help prevent infection, a serious problem in implants made from degradable metal, making Zn-Ag alloys promising candidates for vascular stents [215]. Zn-Mg and Zn-Li alloys in vitro and in vivo stimulate bone cell proliferation and new bone formation, making them promising for bone implants, fracture fixation, and dentistry [215]. Future research is directed towards applying new materials (coating application [242], use of nanoparticles [243], etc.), improving SLM process parameters, and studying in vivo degradation. SLM-zinc alloys represent a promising but still at early stages of clinical application direction in creating biodegradable implants.

6. Comparative Analysis of Biodegradable Metallic Systems

The preceding sections have individually examined magnesium, iron, and zinc-based biodegradable metals processed through selective laser melting. Each system presents distinct advantages and limitations regarding mechanical properties, degradation behavior, biocompatibility, and processing challenges. To facilitate material selection for specific biomedical applications, it is essential to directly compare these three metal systems across key parameters. The following comparative analysis synthesizes the findings from previous sections to provide a comprehensive overview of the relative performance of these biodegradable metallic systems, highlighting their respective strengths and weaknesses for various implant applications. Table 7 shows the comparison of key characteristics of biodegradable metallic alloys.

Table 7. Comparison of key characteristics of biodegradable metallic alloys for SLM.

Characteristic	Magnesium Alloys	Iron Alloys	Zinc Alloys
Density (g/cm^3)	1.7–2	7.8–8.1	7.0–7.2
Elastic modulus (GPa)	40–45	190–210	80–110
Yield strength of SLM material (MPa)	180–300	400–600	80–200
Tensile strength of SLM material (MPa)	250–350	250–850	100–350
Typical degradation rate (mm/year)	1.0–3.0	0.1–0.5	0.2–0.5
Main alloying elements	Ca, Zn, REE	Mn, C, Si	Mg, Ca, Cu

Magnesium alloys demonstrate the most balanced properties for biomedical implants. Their elastic modulus (40–45 GPa) is close to bone tissue, minimizing the stress shielding effect. Degradation rate (1–3 mm/year) corresponds to healing times, and alloying (Ca, Zn, REE) allows controlling corrosion and mechanical properties. However, the high reactivity of magnesium and risk of excessive gas formation require careful optimization of SLM parameters, including protective atmosphere and platform preheating.

Iron alloys possess outstanding strength (yield strength 400–600 MPa) and plasticity, but their extremely slow degradation (0.1–0.5 mm/year) limits application. Alloying with Mn, C, and Si accelerates corrosion, and SLM provides a fine-grained structure, improving mechanical characteristics. However, the necessity of post-processing (e.g., HIP) and risk of oxide accumulation complicate their use.

Zinc alloys occupy an intermediate position: their degradation rate (0.2–0.5 mm/year) is closer to magnesium, and strength (180–280 MPa)—to iron. Alloying with Mg and Cu improves mechanical properties and antibacterial activity. SLM allows achieving high density (>99%), but the low melting temperature of zinc requires precise control of parameters (laser power 50–70 W, platform preheating to 120 °C) to minimize evaporation and porosity.

Table 8 presents the main SLM process parameters for various biodegradable materials.

Table 8. Comparison of key SLM process parameters for various biodegradable metals.

Parameter	Magnesium Alloys	Iron Alloys	Zinc Alloys
Optimal laser power	100–200 W	200–400 W	80–150 W
Scanning speed	400–800 mm/s	600–1200 mm/s	300–700 mm/s
Layer thickness	30–50 μm	20–40 μm	20–40 μm
Hatch distance	80–120 μm	70–100 μm	60–100 μm
Energy density	120–150 J/mm^3	60–150 J/mm^3	100–130 J/mm^3
Protective atmosphere requirements	High purity argon (<10 ppm O_2)	Argon (<100 ppm O_2)	High purity argon (<50 ppm O_2)
Main technological problems	Evaporation, oxidation, high reactivity	High melting temperature, selective evaporation of Mn	Low melting temperature, evaporation, smoke formation
Literature references	[30]	[153,154]	[216,217]

The comparative analysis of magnesium, iron, and zinc-based biodegradable systems highlights the significant progress made in additive manufacturing of metallic implants. However, numerous challenges still impede the widespread clinical application of these promising materials. The following section identifies key technological, biological, regulatory, and economic barriers that must be overcome, while also exploring emerging research directions that may address these limitations. Understanding these challenges and opportunities is crucial for advancing the field toward successful clinical translation of additively manufactured biodegradable implants.

7. Current Problems and Prospects

7.1. Current Development Problems

Despite progress, additive manufacturing of biodegradable metallic implants faces technological, biological, regulatory, and economic challenges. Key technological challenges include control of degradation rate (differing in vitro and in vivo) [7,244], achieving balance between mechanical strength and degradation [245,246], protection of powders from oxidation [16,247], as well as optimization of SLM parameters to minimize defects, residual stresses, and selective evaporation of alloying elements [248–252]. An important task remains the development of accurate models for predicting degradation in vivo, considering the interaction of various factors [253,254].

A serious problem is the insufficient understanding of long-term effects of biodegradable metal degradation products, especially for new alloys, exacerbated by the variability of biological response [89,90], potential toxicity of alloying elements [100], and immune response [101,102]. Degradation in vivo includes complex interaction of corrosion, mechanical stresses, and biological processes [7,103], as well as mechano-corrosion effects [104]. The key task is optimizing implant-tissue interaction [105–107].

Regulatory challenges include compliance with clinical trial requirements, obtaining manufacturing certification, and standardization, which is particularly problematic for additively manufactured implants due to the lack of specialized standards [7], complicating obtaining approvals [255]. Economic challenges are due to high equipment and material costs, scaling problems, and competition with traditional implants. Proposed solutions include interdisciplinary collaboration and development of new biocompatible materials. Successful transition to clinical application requires a comprehensive solution to these problems. Quality control of small-batch production also presents a challenge [256].

Figure 10 shows the relationship between the different categories of challenges in additive manufacturing of biodegradable implants.

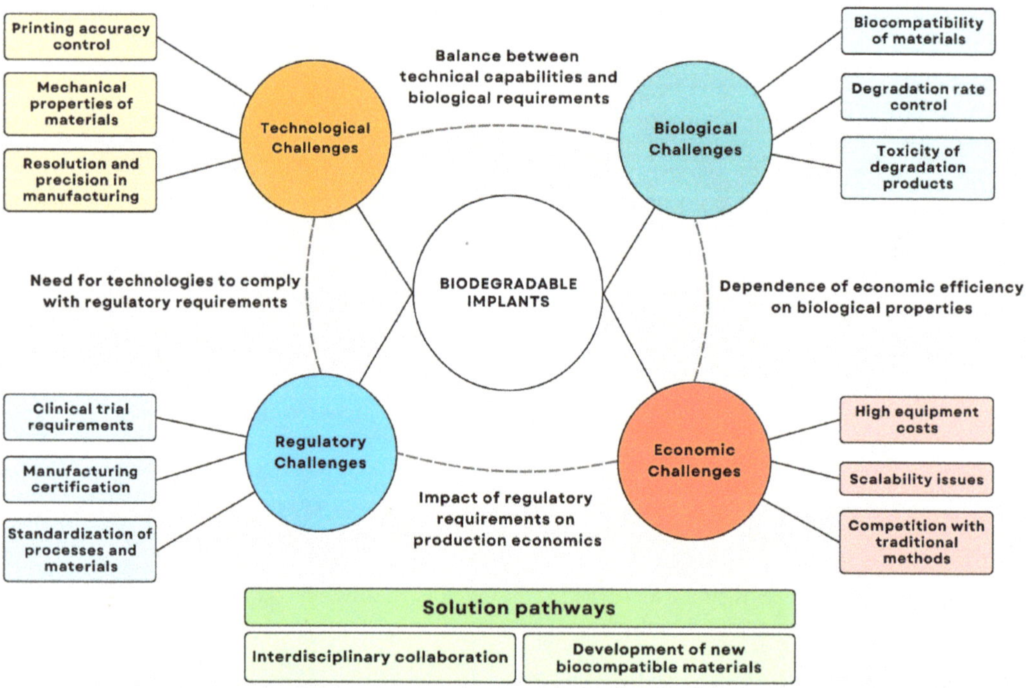

Figure 10. Comprehensive diagram illustrating the interrelationship between different categories of problems (technological, biological, regulatory, and economic) in the field of additive manufacturing of biodegradable implants. Includes examples of key challenges in each category and their mutual influence, as well as potential solutions.

7.2. Promising Research Directions

Additive manufacturing of biodegradable metallic implants is a rapidly developing field with the potential to overcome limitations and expand clinical applications. Main directions include: development of multi-component, composite, and functionally gradient materials with optimized properties [257–259]. An important direction is improvement of technological processes: use of hybrid additive manufacturing [260], improvement of SLM processes, and development of post-processing for controlling degradation and improving biological response [261,262]. Also considered are development of integrated models linking microstructure, properties, and degradation [263], application of AI for optimization and prediction [264], creation of digital twins for modeling implant behavior [265]. Development of methodology for fully personalized implants: development of personalized implants considering individual characteristics [266], minimally invasive solutions [267,268], and development of multifunctional implants combining mechanical support with controlled drug release and integrated sensors for monitoring the healing process [269].

8. Conclusions

Additive manufacturing of biodegradable metallic implants has emerged as a rapidly evolving, interdisciplinary field that leverages selective laser melting (SLM) technology to create personalized biomedical devices from magnesium, iron, and zinc alloys. This systematic review has demonstrated both the substantial progress achieved and the critical challenges that remain for successful clinical implementation.

Material science has advanced biodegradable alloys. Magnesium alloys (WE43, AZ91D, ZK60) are biocompatible, bone-like, but exhibit low degradation/strength balance. SLM boosts magnesium alloy degradation (by 60–160%) and strength (by up to 250%). Iron-based alloys (Fe-Mn, etc.) are strong, ductile, but degrade too slowly, which is why they require alloying. Zinc exhibits ideal degradation kinetics, improving mechanics via alloying.

SLM excels at complex geometries and controlled porosity, surpassing traditional methods. SLM's fine microstructure enhances mechanics but also speeds degradation. Topology optimization and biomimicry (TPMS) enable personalized implants with tailored property gradients.

However, additive manufacturing faces substantial limitations that impede widespread adoption. High equipment and material costs, strict powder quality requirements, and process complexity requiring extensive parameter optimization create significant barriers. Additionally, mandatory post-processing, batch-to-batch variability in small-scale production, and limited standardized testing protocols complicate quality control and certification for medical applications.

Despite clinical promise (e.g., MAGNEZIX® screws), widespread use remains limited due to some key challenges. Optimizing the materials' composition/microstructure for mechanics/degradation, as well as managing reactivity and controlling evaporation/stress during SLM is difficult. Additionally, long-term degradation effects and tissue responses are unclear, while regulatory criteria lack standardization. Finally, the high material costs currently lead to limited production.

Future AM biodegradable implants require the following: (1) multi-material and graded designs; (2) AI/ML-driven optimization; (3) interdisciplinary collaboration; (4) international regulatory harmonization; and (5) smart functionalities (drug delivery, sensors).

Prioritizing research and setting timelines is key. Short term (1–3 years): standardize tests, model degradation, optimize SLM process. Medium term (3–5 years): multi-material implants, real-time SLM monitoring, producing new alloys. Long term (5–10 years): automated manufacturing, "smart" implants, regulatory harmonization. This approach streamlines resource allocation.

AM biodegradable implants are transformative for personalized treatment. Understanding composition-structure-property-biology, standardizing evaluation, and interdisciplinary collaboration are key to overcoming limitations and achieving clinical adoption.

Author Contributions: Conceptualization, A.P. and I.P.; methodology, I.P.; formal analysis, A.G. and I.P.; investigation, A.G.; resources, A.P.; data curation, A.G.; writing—original draft preparation, A.G.; writing—review and editing, I.P. and A.P.; visualization, A.G.; supervision, I.P. and A.P.; project administration, A.P.; funding acquisition, A.P. All authors have read and agreed to the published version of the manuscript.

Funding: This research was supported by the Ministry of Science and Higher Education of the Russian Federation (agreement No. 075-15-2024-562).

Data Availability Statement: No new data were created or analyzed in this study.

Conflicts of Interest: The authors declare no conflict of interest.

Abbreviations

AI	Artificial Intelligence
AM	Additive Manufacturing
ASTM	American Society for Testing and Materials
BCC	Body-Centered Cubic
CAGR	Compound Annual Growth Rate
CE	Conformité Européenne
EBSD	Electron Backscatter Diffraction
FCC	Face-Centered Cubic
HBSS	Hank's Balanced Salt Solution
HCP	Hexagonal Close-Packed
HIP	Hot Isostatic Pressing
ISO	International Organization for Standardization
MAO	Micro-Arc Oxidation
ML	Machine Learning
OM	Optical Microscopy
PBF	Powder Bed Fusion
PBS	Phosphate Buffered Saline
PEO	Plasma Electrolytic Oxidation
PM	Powder Metallurgy
REE	Rare Earth Elements
SBF	Simulated Body Fluid
SCC	Stress Corrosion Cracking
SEM	Scanning Electron Microscopy
SLM	Selective Laser Melting
TPMS	Triply Periodic Minimal Surfaces
TRIP	Transformation-Induced Plasticity
TWIP	Twinning-Induced Plasticity

References

1. Grand View Research. Orthopedic Implants Market Size, Share & Trends Analysis Report By Product (Lower Extremity Implants, Spinal Implants, Dental Implants, Upper Extremity Implants), By End Use (Hospitals, Outpatient Specialties), By Region, And Segment Forecasts, 2025–2030. Available online: https://www.researchandmarkets.com/reports/5899491/orthopedic-implants-market-size-share-and-trends?srsltid=AfmBOoohp1h9BX4o9SoqB78gB-6JwVdlyL84saMdYnVP8u5KmxCDEnF1 (accessed on 22 April 2025).
2. Li, J.; Liu, Y.; Hermansson, L.; Soremark, R. Evaluation o biocompatibility of various ceramic powders with human fibroblasts in vitro. *Clin. Mater.* **1993**, *12*, 197–201. [CrossRef] [PubMed]
3. Li, H.; Wen, J.; Liu, Y.; He, J.; Shi, H.; Tian, P. Progress in Research on Biodegradable Magnesium Alloys: A Review. *Adv. Eng. Mater.* **2020**, *22*, 2000213. [CrossRef]
4. Li, J.; Qin, L.; Yang, K.; Ma, Z.; Wang, Y.; Cheng, L.; Zhao, D. Materials evolution of bone plates for internal fixation of bone fractures: A review. *J. Mater. Sci. Technol.* **2020**, *36*, 190–208. [CrossRef]

5. Joshi, M.G.; Advani, S.G.; Miller, F.; Santare, M.H. Analysis of a femoral hip prosthesis designed to reduce stress shielding. *J. Biomech.* **2000**, *33*, 1655–1662. [CrossRef] [PubMed]
6. Chen, J.; Tan, L.; Yu, X.; Etim, I.P.; Ibrahim, M.; Yang, K. Mechanical properties of magnesium alloys for medical application: A review. *J. Mech. Behav. Biomed. Mater.* **2018**, *87*, 68–79. [CrossRef]
7. Zheng, Y.F.; Gu, X.N.; Witte, F. Biodegradable metals. *Mater. Sci. Eng. R Rep.* **2014**, *77*, 1–34. [CrossRef]
8. Witte, F.; Fischer, J.; Nellesen, J.; Crostack, H.-A.; Kaese, V.; Pisch, A.; Beckmann, F.; Windhagen, H. In vitro and in vivo corrosion measurements of magnesium alloys. *Biomaterials* **2006**, *27*, 1013–1018. [CrossRef]
9. de Baaij, J.H.F.; Hoenderop, J.G.J.; Bindels, R.J.M. Magnesium in Man: Implications for Health and Disease. *Physiol. Rev.* **2015**, *95*, 1–46. [CrossRef]
10. Peuster, M.; Hesse, C.; Schloo, T.; Fink, C.; Beerbaum, P. Long-term biocompatibility of a corrodible peripheral iron stent in the porcine descending aorta. *Biomaterials* **2006**, *27*, 4955–4962. [CrossRef]
11. Kádár, C.; Gorejová, R.; Kubelka, P.; Oriňaková, R.; Orbulov, I.N. Mechanical and Degradation Behavior of Zinc-Based Biodegradable Metal Foams. *Adv Eng Mater* **2024**, *26*, 2301496. [CrossRef]
12. John, E.; Laskow, T.C.; Buchser, W.J.; Pitt, B.R.; Basse, P.H.; Butterfield, L.H.; Kalinski, P.; Lotze, M.T. Zinc in innate and adaptive tumor immunity. *J. Transl. Med.* **2010**, *8*, 118. [CrossRef] [PubMed]
13. Hüner, B.; Kıstı, M.; Uysal, S.; Uzgören, İ.N.; Özdoğan, E.; Süzen, Y.O.; Demir, N.; Kaya, M.F. An Overview of Various Additive Manufacturing Technologies and Materials for Electrochemical Energy Conversion Applications. *ACS Omega* **2022**, *7*, 40638–40658. [CrossRef] [PubMed]
14. Sing, S.L.; An, J.; Yeong, W.Y.; Wiria, F.E. Laser and electron-beam powder-bed additive manufacturing of metallic implants: A review on processes, materials and designs. *J. Orthop. Res.* **2016**, *34*, 369–385. [CrossRef]
15. Wadge, M.D.; McGuire, J.; Hanby, B.V.T.; Felfel, R.M.; Ahmed, I.; Grant, D.M. Tailoring the degradation rate of magnesium through biomedical nano-porous titanate coatings. *J. Magnes. Alloys* **2021**, *9*, 336–350. [CrossRef]
16. Qin, Y.; Wen, P.; Guo, H.; Xia, D.; Zheng, Y.; Jauer, L.; Poprawe, R.; Voshage, M.; Schleifenbaum, J.H. Additive manufacturing of biodegradable metals: Current research status and future perspectives. *Acta Biomater.* **2019**, *98*, 3–22. [CrossRef]
17. Soni, N.; Renna, G.; Leo, P. Advancements in Metal Processing Additive Technologies: Selective Laser Melting (SLM). *Metals* **2024**, *14*, 1081. [CrossRef]
18. Liu, S.; Yang, W.; Shi, X.; Li, B.; Duan, S.; Guo, H.; Guo, J. Influence of laser process parameters on the densification, microstructure, and mechanical properties of a selective laser melted AZ61 magnesium alloy. *J. Alloys Compd.* **2019**, *808*, 151160. [CrossRef]
19. Kruth, J.P.; Froyen, L.; Van Vaerenbergh, J.; Mercelis, P.; Rombouts, M.; Lauwers, B. Selective laser melting of iron-based powder. *J. Mater. Process Technol.* **2004**, *149*, 616–622. [CrossRef]
20. Li, K.; Ji, C.; Bai, S.; Jiang, B.; Pan, F. Selective laser melting of magnesium alloys: Necessity, formability, performance, optimization and applications. *J. Mater. Sci. Technol.* **2023**, *154*, 65–93. [CrossRef]
21. Gajanan, M.N.; Narendranath, S.; Satheesh Kumar, S.S. Effect of grain refinement on mechanical and corrosion behavior of AZ91 magnesium alloy processed by ECAE. *IOP Conf. Ser. Mater. Sci. Eng.* **2019**, *591*, 012015. [CrossRef]
22. Qin, Y.; Wen, P.; Xia, D.; Guo, H.; Voshage, M.; Jauer, L.; Zheng, Y.; Schleifenbaum, J.H.; Tian, Y. Effect of grain structure on the mechanical properties and in vitro corrosion behavior of additively manufactured pure Zn. *Addit. Manuf.* **2020**, *33*, 101134. [CrossRef]
23. Mugwagwa, L.; Dimitrov, D.; Matope, S.; Muvunz, R. Residual Stresses and Distortions in Selective Laser Melting—A Review. In Proceedings of the 17th Rapid Product Development Association of South Africa, Vanderbijlpark, South Africa, 2–4 November 2016.
24. Kraner, J.; Medved, J.; Godec, M. Thermodynamic Behavior of Fe-Mn and Fe-Mn-Ag Powder Mixtures during Selective Laser Melting. *Metals* **2021**, *11*, 234. [CrossRef]
25. Jo, J.-H.; Hong, J.-Y.; Shin, K.-S.; Kim, H.-E.; Koh, Y.-H. Enhancing biocompatibility and corrosion resistance of Mg implants via surface treatments. *J. Biomater. Appl.* **2012**, *27*, 469–476. [CrossRef]
26. Gao, C.; Li, S.; Liu, L.; Bin, S.; Yang, Y.; Peng, S.; Shuai, C. Dual alloying improves the corrosion resistance of biodegradable Mg alloys prepared by selective laser melting. *J. Magnes. Alloys* **2021**, *9*, 305–316. [CrossRef]
27. Seitz, J.; Collier, K.; Wulf, E.; Bormann, D.; Angrisani, N.; Meyer-Lindenberg, A.; Bach, F. The Effect of Different Sterilization Methods on the Mechanical Strength of Magnesium Based Implant Materials. *Adv. Eng. Mater.* **2011**, *13*, 1146–1151. [CrossRef]
28. Li, Y.; Zhou, J.; Pavanram, P.; Leeflang, M.A.; Fockaert, L.I.; Pouran, B.; Tümer, N.; Schröder, K.U.; Mol, J.M.C.; Weinans, H.; et al. Additively Manufactured Biodegradable Porous Magnesium. *Acta Biomater.* **2018**, *67*, 378–392. [CrossRef]
29. Han, H.-S.; Minghui, Y.; Seok, H.-K.; Byun, J.-Y.; Cha, P.-R.; Yang, S.-J.; Kim, Y.C. The modification of microstructure to improve the biodegradation and mechanical properties of a biodegradable Mg alloy. *J. Mech. Behav. Biomed. Mater.* **2013**, *20*, 54–60. [CrossRef] [PubMed]
30. Liu, S.; Guo, H. A Review of SLMed Magnesium Alloys: Processing, Properties, Alloying Elements and Postprocessing. *Metals* **2020**, *10*, 1073. [CrossRef]

31. Palanivel, S.; Nelaturu, P.; Glass, B.; Mishra, R.S. Friction stir additive manufacturing for high structural performance through microstructural control in an Mg based WE43 alloy. *Mater. Des. (1980–2015)* **2015**, *65*, 934–952. [CrossRef]
32. Ewald, F.C.; Brenne, F.; Gustmann, T.; Vollmer, M.; Krooß, P.; Niendorf, T. Laser Powder Bed Fusion Processing of Fe-Mn-Al-Ni Shape Memory Alloy—On the Effect of Elevated Platform Temperatures. *Metals* **2021**, *11*, 185. [CrossRef]
33. Deng, Q.; Wu, Y.; Luo, Y.; Su, N.; Xue, X.; Chang, Z.; Wu, Q.; Xue, Y.; Peng, L. Fabrication of high-strength Mg-Gd-Zn-Zr alloy via selective laser melting. *Mater. Charact.* **2020**, *165*, 110377. [CrossRef]
34. Zhang, W.; Wang, L.; Feng, Z.; Chen, Y. Research progress on selective laser melting (SLM) of magnesium alloys: A review. *Optik* **2020**, *207*, 163842. [CrossRef]
35. Xie, T.; Liu, J.; Xiao, L.; Xie, Y.; Qian, S.; Dai, Y.; Wu, J. Additive Manufacturing of Fe–Mn–Al–C Lightweight Steel by Laser Powder Bed Fusion: The Role of Laser Scanning Speed on Forming Quality, Microstructure and Properties. *J. Mater. Res. Technol.* **2024**, *33*, 6610–6621. [CrossRef]
36. Pola, A.; Tocci, M.; Goodwin, F.E. Review of Microstructures and Properties of Zinc Alloys. *Metals* **2020**, *10*, 253. [CrossRef]
37. Venezuela, J.; Dargusch, M.S. Addressing the slow corrosion rate of biodegradable Fe-Mn: Current approaches and future trends. *Curr. Opin. Solid. State Mater. Sci.* **2020**, *24*, 100822. [CrossRef]
38. Rabeeh, V.P.M.; Hanas, T. Progress in manufacturing and processing of degradable Fe-based implants: A review. *Prog. Biomater.* **2022**, *11*, 163–191. [CrossRef]
39. Bai, J.; Wang, Q.; Men, Z.; Chen, W.; Huang, H.; Ji, C.; Li, Y.; Wang, L.; Zhu, L.; Li, K.; et al. Generation Mechanism of Anisotropy in Mechanical Properties of WE43 Fabricated by Laser Powder Bed Fusion. *Micromachines* **2024**, *15*, 976. [CrossRef]
40. Mouzou, E.; Paternoster, C.; Tolouei, R.; Purnama, A.; Chevallier, P.; Dubé, D.; Prima, F.; Mantovani, D. In vitro degradation behavior of Fe–20Mn–1.2C alloy in three different pseudo-physiological solutions. *Mater. Sci. Eng. C* **2016**, *61*, 564–573. [CrossRef]
41. Wu, C.L.; Zai, W.; Man, H.C. Additive manufacturing of ZK60 magnesium alloy by selective laser melting: Parameter optimization, microstructure and biodegradability. *Mater. Today Commun.* **2021**, *26*, 101922. [CrossRef]
42. Choi, J.-P.; Shin, G.-H.; Lee, H.-S.; Yang, D.-Y.; Yang, S.; Lee, C.-W.; Brochu, M.; Yu, J.-H. Evaluation of Powder Layer Density for the Selective Laser Melting (SLM) Process. *Mater. Trans.* **2017**, *58*, 294–297. [CrossRef]
43. *ISO 13485:2016*; Medical Devices—Quality Management Systems—Requirements for Regulatory Purposes. International Organization for Standardization (ISO): Geneva, Switzerland, 2016.
44. ASTM International. *Standard Guide for Characterizing Properties of Metal Powders Used for Additive Manufacturing Processes*; ASTM: West Conshohocken, PA, USA, 2014.
45. Jana, A.; Das, M.; Balla, V.K. In vitro and in vivo degradation assessment and preventive measures of biodegradable Mg alloys for biomedical applications. *J. Biomed. Mater. Res. A* **2022**, *110*, 462–487. [CrossRef]
46. Pogorielov, M.; Husak, E.; Solodivnik, A.; Zhdanov, S. Magnesium-based biodegradable alloys: Degradation, application, and alloying elements. *Interv. Med. Appl. Sci.* **2017**, *9*, 27–38. [CrossRef] [PubMed]
47. Kawamura, N.; Nakao, Y.; Ishikawa, R.; Tsuchida, D.; Iijima, M. Degradation and Biocompatibility of AZ31 Magnesium Alloy Implants In Vitro and In Vivo: A Micro-Computed Tomography Study in Rats. *Materials* **2020**, *13*, 473. [CrossRef]
48. Li, Y.; Jahr, H.; Lietaert, K.; Pavanram, P.; Yilmaz, A.; Fockaert, L.I.; Leeflang, M.A.; Pouran, B.; Gonzalez-Garcia, Y.; Weinans, H.; et al. Additively manufactured biodegradable porous iron. *Acta Biomater.* **2018**, *77*, 380–393. [CrossRef]
49. Kabir, H.; Munir, K.; Wen, C.; Li, Y. Recent research and progress of biodegradable zinc alloys and composites for biomedical applications: Biomechanical and biocorrosion perspectives. *Bioact. Mater.* **2021**, *6*, 836–879. [CrossRef] [PubMed]
50. Manjhi, S.K.; Sekar, P.; Bontha, S.; Balan, A.S.S. Additive manufacturing of magnesium alloys: Characterization and post-processing. *Int. J. Lightweight Mater. Manuf.* **2024**, *7*, 184–213. [CrossRef]
51. Esmaily, M.; Zeng, Z.; Mortazavi, A.N.; Gullino, A.; Choudhary, S.; Derra, T.; Benn, F.; D'Elia, F.; Müther, M.; Thomas, S.; et al. A detailed microstructural and corrosion analysis of magnesium alloy WE43 manufactured by selective laser melting. *Addit. Manuf.* **2020**, *35*, 101321. [CrossRef]
52. Jafari, S.; Harandi, S.E.; Singh Raman, R.K. A review of stress-corrosion cracking and corrosion fatigue of magnesium alloys for biodegradable implant applications. *JOM Miner. Met. Mater. Soc.* **2015**, *67*, 1143–1153. [CrossRef]
53. Singh Raman, R.K.; Wen, C.; Löffler, J.F. Human Body-Fluid-Assisted Fracture of Zinc Alloys as Biodegradable Temporary Implants: Challenges, Research Needs and Way Forward. *Materials* **2023**, *16*, 4984. [CrossRef]
54. Ali, M.; Sajjad, U.; Hussain, I.; Abbas, N.; Ali, H.M.; Yan, W.M.; Wang, C.C. On the assessment of the mechanical properties of additively manufactured lattice structures. *Eng. Anal. Bound. Elem.* **2022**, *142*, 93–116. [CrossRef]
55. Feng, J.; Fu, J.; Lin, Z.; Shang, C.; Niu, X. Layered infill area generation from triply periodic minimal surfaces for additive manufacturing. *Comput.-Aided Des.* **2019**, *107*, 50–63. [CrossRef]
56. Liang, H.; Yang, Y.; Xie, D.; Li, L.; Mao, N.; Wang, C.; Tian, Z.; Jiang, Q.; Shen, L. Trabecular-like Ti-6Al-4V scaffolds for orthopedic: Fabrication by selective laser melting and in vitro biocompatibility. *J. Mater. Sci. Technol.* **2019**, *35*, 1284–1297. [CrossRef]

57. Chen, J.; Song, C.; Deng, Z.; Huang, J.; Han, C.; Yang, Y.; Wang, J.; Xu, K. Functional gradient design of additive manufactured gyroid tantalum porous structures: Manufacturing, mechanical behaviors and permeability. *J. Manuf. Process* **2024**, *125*, 202–216. [CrossRef]
58. Ataollahi, S. A review on additive manufacturing of lattice structures in tissue engineering. *Bioprinting* **2023**, *35*, e00304. [CrossRef]
59. Dargusch, M.S.; Soro, N.; Demir, A.G.; Venezuela, J.; Sun, Q.; Wang, Y.; Abdal-hay, A.; Alali, A.Q.; Ivanovski, S.; Previtali, B.; et al. Optimising degradation and mechanical performance of additively manufactured biodegradable Fe–Mn scaffolds using design strategies based on triply periodic minimal surfaces. *Smart Mater. Med.* **2024**, *5*, 127–139. [CrossRef]
60. Karageorgiou, V.; Kaplan, D. Porosity of 3D biomaterial scaffolds and osteogenesis. *Biomaterials* **2005**, *26*, 5474–5491. [CrossRef]
61. Demir, C.D.; Bermingham, A.G.; Dargusch, M.J. Challenges and Opportunities in the Selective Laser Melting of Biodegradable Metals for Load-Bearing Bone Scaffold Applications. *Metall. Mater. Trans. A* **2020**, *51*, 3311–3334.
62. Bandyopadhyay, A.; Mitra, I.; Goodman, S.B.; Kumar, M.; Bose, S. Improving biocompatibility for next generation of metallic implants. *Prog. Mater. Sci. Pergamon* **2023**, *133*, 101053. [CrossRef]
63. Huse, E.C. A New Ligature. *Chic. Med. J. Exam.* **1878**, *37*, 171–172.
64. Rostock, P. Ist das Magnesium als Naht-und Schienungsmaterial für Knochenoperationen geeignet? *Arch. Orthop. Unfallchir.* **1937**, *38*, 486–492. [CrossRef]
65. Plaass, C.; Ettinger, S.; Sonnow, L.; Koenneker, S.; Noll, Y.; Weizbauer, A.; Reifenrath, J.; Claassen, L.; Daniilidis, K.; Stukenborg-Colsman, C.; et al. Early results using a biodegradable magnesium screw for modified chevron osteotomies. *J. Orthop. Res.* **2016**, *34*, 2207–2214. [CrossRef] [PubMed]
66. Windhagen, H.; Radtke, K.; Weizbauer, A.; Diekmann, J.; Noll, Y.; Kreimeyer, U.; Schavan, R.; Stukenborg-Colsman, C.; Waizy, H. Biodegradable magnesium-based screw clinically equivalent to titanium screw in hallux valgus surgery: Short term results of the first prospective, randomized, controlled clinical pilot study. *Biomed. Eng. Online* **2013**, *12*, 62. [CrossRef] [PubMed]
67. Cha, P.-R.; Han, H.-S.; Yang, G.-F.; Kim, Y.-C.; Hong, K.-H.; Lee, S.-C.; Jung, J.-Y.; Ahn, J.-P.; Kim, Y.-Y.; Cho, S.-Y.; et al. Biodegradability engineering of biodegradable Mg alloys: Tailoring the electrochemical properties and microstructure of constituent phases. *Sci. Rep.* **2013**, *3*, 2367. [CrossRef] [PubMed]
68. Strumińska-Parulska, D.; Strumińska-Parulska, D.; Moniakowska, A.; Block, K. Radiation safety of calcium and magnesium diet supplements. *J. Elem.* **2023**, *29*, 679–691. [CrossRef]
69. Wu, J.; Cheng, X.; Wu, J.; Chen, J.; Pei, X. The development of magnesium-based biomaterials in bone tissue engineering: A review. *J. Biomed. Mater. Res. B Appl. Biomater.* **2024**, *112*, e35326. [CrossRef]
70. Wang, C.; Sun, M.; Yang, C.; Wang, H.; Wang, J.; Mao, L.; Yang, Y.; Ying, T.; Chu, P.K.; Zeng, X. Degradation behavior of pure Mg in the physiological medium and growth mechanism of surface corrosion product films. *J. Magnes. Alloys* **2024**, *13*, 1523–1535. [CrossRef]
71. Mei, D.; Li, Y.; Tian, Y.; Zhang, Q.; Liu, M.; Zhu, S.; Wang, L.; Guan, S. The effect of selected corrosion inhibitors on localized corrosion of magnesium alloy: The expanded understanding of "inhibition efficiency". *Corros. Sci.* **2024**, *226*, 111650. [CrossRef]
72. Pan, H.; Qin, G.; Huang, Y.; Ren, Y.; Sha, X.; Han, X.; Liu, Z.-Q.; Li, C.; Wu, X.; Chen, H.; et al. Development of low-alloyed and rare-earth-free magnesium alloys having ultra-high strength. *Acta Mater.* **2018**, *149*, 350–363. [CrossRef]
73. Çam, G.; Günen, A. Challenges and Opportunities in the Production of Magnesium Parts by Directed Energy Deposition Processes. *J. Magnes. Alloys* **2024**, *12*, 1663–1686. [CrossRef]
74. Witte, F.; Kaese, V.; Haferkamp, H.; Switzer, E.; Meyer-Lindenberg, A.; Wirth, C.J.; Windhagen, H. In vivo corrosion of four magnesium alloys and the associated bone response. *Biomaterials* **2005**, *26*, 3557–3563. [CrossRef]
75. Jung, Y.-G.; Yang, W.; Kim, Y.J.; Kim, S.K.; Yoon, Y.-O.; Lim, H.; Kim, D.H. Effect of Ca addition on the microstructure and mechanical properties of heat-treated Mg-6.0Zn-1.2Y-0.7Zr alloy. *J. Magnes. Alloys* **2021**, *9*, 1619–1631. [CrossRef]
76. Chaudry, U.M.; Farooq, A.; bin Tayyab, K.; Malik, A.; Kamran, M.; Kim, J.-G.; Li, C.; Hamad, K.; Jun, T.-S. Corrosion behavior of AZ31 magnesium alloy with calcium addition. *Corros. Sci.* **2022**, *199*, 110205. [CrossRef]
77. Lin, X.; Saijilafu; Wu, X.; Wu, K.; Chen, J.; Tan, L.; Witte, F.; Yang, H.; Mantovani, D.; Zhou, H.; et al. Biodegradable Mg-based alloys: Biological implications and restorative opportunities. *Int. Mater. Rev.* **2023**, *68*, 365–403. [CrossRef]
78. Cai, S.; Lei, T.; Li, N.; Feng, F. Effects of Zn on microstructure, mechanical properties and corrosion behavior of Mg–Zn alloys. *Mater. Sci. Eng. C* **2012**, *32*, 2570–2577. [CrossRef]
79. Yu, F.; Guohua, W.; Chunquan, Z. Effect of Strontium on Mechanical Properties and Corrosion Resistance of AZ91D. *Materials Science Forum* **2007**, *546–549*, 567–570.
80. Ding, Y.; Wen, C.; Hodgson, P.; Li, Y. Effects of alloying elements on the corrosion behavior and biocompatibility of biodegradable magnesium alloys: A review. *J. Mater. Chem. B* **2014**, *2*, 1912–1933. [CrossRef]
81. Tang, C.; Wu, K.; Liu, W.; Feng, D.; Wang, X.; Miao, G.; Yang, M.; Liu, X.; Li, Q. Effects of Gd, Y Content on the Microstructure and Mechanical Properties of Mg-Gd-Y-Nd-Zr Alloy. *Metals* **2018**, *8*, 790. [CrossRef]
82. Yang, Y.; Ling, C.; Yang, M.; Yang, L.; Wang, D.; Peng, S.; Shuai, C. Selective Laser Melted Rare Earth Magnesium Alloy with High Corrosion Resistance. *Int. J. Bioprint* **2022**, *8*, 574. [CrossRef]

83. Shi, L.; Yan, Y.; Shao, C.; Yu, K.; Zhang, B.; Chen, L. The influence of yttrium and manganese additions on the degradation and biocompatibility of magnesium-zinc-based alloys: In vitro and in vivo studies. *J. Magnes. Alloys* **2024**, *12*, 608–624. [CrossRef]
84. Yu, Z.; Tang, A.; Li, C.; Liu, J.; Pan, F. Effect of manganese on the microstructure and mechanical properties of magnesium alloys. *Int. J. Mater. Res.* **2019**, *110*, 1016–1024. [CrossRef]
85. Zhou, Y.; Zhang, K.; Liang, Y.; Cheng, J.; Dai, Y. Selective Laser Melted Magnesium Alloys: Fabrication, Microstructure and Property. *Materials* **2022**, *15*, 7049. [CrossRef] [PubMed]
86. Mousavian, S.M.H.; Tabaian, S.H.; Badihehaghdam, M. Effects of zirconium addition on electrochemical and mechanical properties of Mg-3Zn-1Ca-1RE alloy. *Anti-Corros. Methods Mater.* **2020**, *67*, 583–591. [CrossRef]
87. Seong, J.W.; Kim, W.J. Development of biodegradable Mg–Ca alloy sheets with enhanced strength and corrosion properties through the refinement and uniform dispersion of the Mg2Ca phase by high-ratio differential speed rolling. *Acta Biomater.* **2015**, *11*, 531–542. [CrossRef]
88. Du, Y.Z.; Qiao, X.G.; Zheng, M.Y.; Wang, D.B.; Wu, K.; Golovin, I.S. Effect of microalloying with Ca on the microstructure and mechanical properties of Mg-6 mass%Zn alloys. *Mater. Des.* **2016**, *98*, 285–293. [CrossRef]
89. Erdmann, N.; Angrisani, N.; Reifenrath, J.; Lucas, A.; Thorey, F.; Bormann, D.; Meyer-Lindenberg, A. Biomechanical testing and degradation analysis of MgCa0.8 alloy screws: A comparative in vivo study in rabbits. *Acta Biomater.* **2011**, *7*, 1421–1428. [CrossRef]
90. Zuo, Y.B.; Fu, X.; Mou, D.; Zhu, Q.F.; Li, L.; Cui, J.Z. Study on the role of Ca in the grain refinement of Mg–Ca binary alloys. *Mater. Res. Innov.* **2015**, *19*, S1-94–S1-97. [CrossRef]
91. Ahmadi, M.; Tabary, S.A.A.B.; Rahmatabadi, D.; Ebrahimi, M.S.; Abrinia, K.; Hashemi, R. Review of selective laser melting of magnesium alloys: Advantages, microstructure and mechanical characterizations, defects, challenges, and applications. *J. Mater. Res. Technol.* **2022**, *19*, 1537–1562. [CrossRef]
92. Nagata, M.; Lönnerdal, B. Role of zinc in cellular zinc trafficking and mineralization in a murine osteoblast-like cell line. *J. Nutr. Biochem.* **2011**, *22*, 172–178. [CrossRef]
93. Xi, H.; Jihua, C.; Hongge, Y.; Bin, S.; Guanghao, Z.; Chongming, M. Effects of minor Sr addition on microstructure and mechanical properties of the as-cast Mg–4.5Zn–4.5Sn–2Al-based alloy system. *J. Alloys Compd.* **2013**, *579*, 39–44. [CrossRef]
94. Peng, Q.; Li, X.; Ma, N.; Liu, R.; Zhang, H. Effects of backward extrusion on mechanical and degradation properties of Mg–Zn biomaterial. *J. Mech. Behav. Biomed. Mater.* **2012**, *10*, 128–137. [CrossRef]
95. Shuai, C.; He, C.; Feng, P.; Guo, W.; Gao, C.; Wu, P.; Yang, Y.; Bin, S.; Feng, P. Biodegradation mechanisms of selective laser-melted Mg-xAl-Zn alloy: Grain size and intermetallic phase. *Virtual Phys. Prototyp.* **2018**, *13*, 59–69. [CrossRef]
96. Zhang, B.; Hou, Y.; Wang, X.; Wang, Y.; Geng, L. Mechanical properties, degradation performance and cytotoxicity of Mg–Zn–Ca biomedical alloys with different compositions. *Mater. Sci. Eng. C* **2011**, *31*, 1667–1673. [CrossRef]
97. Atkins, G.J.; Welldon, K.J.; Halbout, P.; Findlay, D.M. Strontium ranelate treatment of human primary osteoblasts promotes an osteocyte-like phenotype while eliciting an osteoprotegerin response. *Osteoporos. Int.* **2009**, *20*, 653–664. [CrossRef]
98. Zhao, C.; Pan, F.; Zhang, L.; Pan, H.; Song, K.; Tang, A. Microstructure, mechanical properties, bio-corrosion properties and cytotoxicity of as-extruded Mg-Sr alloys. *Mater. Sci. Eng. C* **2017**, *70*, 1081–1088. [CrossRef]
99. Bornapour, M.; Celikin, M.; Cerruti, M.; Pekguleryuz, M. Magnesium implant alloy with low levels of strontium and calcium: The third element effect and phase selection improve bio-corrosion resistance and mechanical performance. *Mater. Sci. Eng. C* **2014**, *35*, 267–282. [CrossRef]
100. Li, H.; Wang, P.; Lin, G.; Huang, J. The role of rare earth elements in biodegradable metals: A review. *Acta Biomater.* **2021**, *129*, 33–42. [CrossRef]
101. Wang, L.; Hu, J.; Guo, E.; Li, Y.; Kang, H.; Feng, Y. Microstructure and Mechanical Properties of Thick-Walled WE43A Alloy Fabricated by Wire Arc Additive Manufacturing. *J. Mater. Eng. Perform.* **2024**. [CrossRef]
102. Davis, J.R. (Ed.) Magnesium and Magnesium Alloys. In *Alloying*; ASM International: Almere, The Netherlands, 2001; pp. 432–453. [CrossRef]
103. Cox, B. Oxidation of Zirconium and its Alloys. In *Advances in Corrosion Science and Technology*; Springer: Boston, MA, USA, 1976; pp. 173–391.
104. Hendea, R.E.; Raducanu, D.; Claver, A.; García, J.A.; Cojocaru, V.D.; Nocivin, A.; Stanciu, D.; Serban, N.; Ivanescu, S.; Trisca-Rusu, C.; et al. Biodegradable Magnesium Alloys for Personalised Temporary Implants. *J. Funct. Biomater.* **2023**, *14*, 400. [CrossRef]
105. Ng, C.C.; Savalani, M.; Man, H.C. Fabrication of Magnesium Using Selective Laser Melting Technique. *Rapid Prototyp. J.* **2011**, *17*, 479–490.
106. Li, X.; Fang, X.; Wang, S.; Wang, S.; Zha, M.; Huang, K. Selective laser melted AZ91D magnesium alloy with superior balance of strength and ductility. *J. Magnes. Alloys* **2023**, *11*, 4644–4658. [CrossRef]
107. Liu, S.H.; Liu, X.J.; Liu, B.; Liu, L.M.; Jin, W.Z.; Hu, X.J. Effect of some alloying elements on boiling point of magnesium. *Mater. Sci. Technol.* **2005**, *21*, 735–738. [CrossRef]

108. Lun Sin, S.; Elsayed, A.; Ravindran, C. Inclusions in magnesium and its alloys: A review. *Int. Mater. Rev.* **2013**, *58*, 419–436. [CrossRef]
109. Nopová, K.; Jaroš, J.; Červinek, O.; Pantělejev, L.; Gneiger, S.; Senck, S.; Koutný, D. Processing of AZ91D Magnesium Alloy by Laser Powder Bed Fusion. *Appl. Sci.* **2023**, *13*, 1377. [CrossRef]
110. Wei, K.; Wang, Z.; Zeng, X. Influence of element vaporization on formability, composition, microstructure, and mechanical performance of the selective laser melted Mg–Zn–Zr components. *Mater. Lett.* **2015**, *156*, 187–190. [CrossRef]
111. Ji, C.; Li, K.; Liao, R.; Li, Z.; Yin, B.; Wen, P.; Jiang, B.; Murr, L.E.; Pan, F. Tensile creep mechanisms of laser powder bed fused WE43 alloy with heterogeneous microstructure: Evolution in dislocations and precipitates. *J. Mater. Sci. Technol.* **2025**, *238*, 209–229. [CrossRef]
112. Zumdick, N.A.; Jauer, L.; Kersting, L.C.; Kutz, T.N.; Schleifenbaum, J.H.; Zander, D. Additive manufactured WE43 magnesium: A comparative study of the microstructure and mechanical properties with those of powder extruded and as-cast WE43. *Mater. Charact.* **2019**, *147*, 384–397. [CrossRef]
113. Neh, K.; Ullmann, M.; Kawalla, R. Twin-Roll-Casting and Hot Rolling of Magnesium Alloy WE43. *Procedia. Eng.* **2014**, *81*, 1553–1558. [CrossRef]
114. Cosma, C.; Kessler, J.; Gebhardt, A.; Campbell, I.; Balc, N. Improving the Mechanical Strength of Dental Applications and Lattice Structures SLM Processed. *Materials* **2020**, *13*, 905. [CrossRef]
115. Hu, Q.; Chen, C.; Liu, M.; Chang, C.; Yan, X.; Dai, Y. Improved corrosion resistance of magnesium alloy prepared by selective laser melting through T4 heat treatment for biomedical applications. *J. Mater. Res. Technol.* **2023**, *27*, 813–825. [CrossRef]
116. Höhn, S.; Virtanen, S.; Boccaccini, A.R. Protein adsorption on magnesium and its alloys: A review. *Appl. Surf. Sci.* **2019**, *464*, 212–219. [CrossRef]
117. Yao, H.; Wen, J.-B.; Xiong, Y.; Lu, Y.; Huttula, M. Microstructure Evolution in Mg-Zn-Zr-Gd Biodegradable Alloy: The Decisive Bridge Between Extrusion Temperature and Performance. *Front. Chem.* **2018**, *6*, 71. [CrossRef] [PubMed]
118. Mraied, H.; Wang, W.; Cai, W. Influence of chemical heterogeneity and microstructure on the corrosion resistance of biodegradable WE43 magnesium alloys. *J. Mater. Chem. B* **2019**, *7*, 6399–6411. [CrossRef] [PubMed]
119. Yang, L.; Zhou, X.; Curioni, M.; Pawar, S.; Liu, H.; Fan, Z.; Scamans, G.; Thompson, G. Corrosion Behavior of Pure Magnesium with Low Iron Content in 3.5 wt% NaCl Solution. *J. Electrochem. Soc.* **2015**, *162*, 362–368. [CrossRef]
120. Yuwono, J.A.; Birbilis, N.; Williams, K.S.; Medhekar, N.V. Electrochemical stability of magnesium surfaces in an aqueous environment. *J. Phys. Chem. C Am. Chem. Soc.* **2016**, *120*, 26922–26933. [CrossRef]
121. Lovašiová, P.; Lovaši, T.; Kubásek, J.; Jablonská, E.; Msallamová, Š.; Michalcová, A.; Vojtěch, D.; Suchý, J.; Koutný, D.; Ghassan Hamed Alzubi, E. Biodegradable WE43 Magnesium Alloy Produced by Selective Laser Melting: Mechanical Properties, Corrosion Behavior, and In-Vitro Cytotoxicity. *Metals* **2022**, *12*, 469. [CrossRef]
122. Ling, C.; Li, Q.; Zhang, Z.; Yang, Y.; Zhou, W.; Chen, W.; Dong, Z.; Pan, C.; Shuai, C. Influence of heat treatment on microstructure, mechanical and corrosion behavior of WE43 alloy fabricated by laser-beam powder bed fusion. *Int. J. Extrem. Manuf.* **2023**, *6*, 015001. [CrossRef]
123. Shi, X.; Wang, Y.; Li, H.; Zhang, S.; Zhao, R.; Li, G.; Zhang, R.; Sheng, Y.; Cao, S.; Zhao, Y.; et al. Corrosion resistance and biocompatibility of calcium-containing coatings developed in near-neutral solutions containing phytic acid and phosphoric acid on AZ31B alloy. *J. Alloys Compd.* **2020**, *823*, 153721. [CrossRef]
124. Dou, J.; Wang, J.; Lu, Y.; Chen, C.; Yu, H.; Ma, R.L.-W. Bioactive MAO/CS composite coatings on Mg-Zn-Ca alloy for orthopedic applications. *Prog. Org. Coat.* **2021**, *152*, 106112. [CrossRef]
125. Li, Y.; Jahr, H.; Pavanram, P.; Bobbert, F.S.L.; Puggi, U.; Zhang, X.Y.; Pouran, B.; Leeflang, M.A.; Weinans, H.; Zhou, J.; et al. Additively manufactured functionally graded biodegradable porous iron. *Acta Biomater.* **2019**, *96*, 646–661. [CrossRef]
126. Charyeva, O.; Dakischew, O.; Sommer, U.; Heiss, C.; Schnettler, R.; Lips, K.S. Biocompatibility of magnesium implants in primary human reaming debris-derived cells stem cells in vitro. *J. Orthop. Traumatol.* **2016**, *17*, 63–73. [CrossRef]
127. Nourisa, J.; Zeller-Plumhoff, B.; Helmholz, H.; Luthringer-Feyerabend, B.; Ivannikov, V.; Willumeit-Römer, R. Magnesium ions regulate mesenchymal stem cells population and osteogenic differentiation: A fuzzy agent-based modeling approach. *Comput. Struct. Biotechnol. J.* **2021**, *19*, 4110–4122. [CrossRef] [PubMed]
128. Siring, J.; Cökelek, A.; Mohnfeld, N.; Wester, H.; Behrens, B.A. Evaluating the Degradation of WE43 for Implant Applications: Optical and Mechanical Insights. *Appl. Sci.* **2025**, *15*, 3300. [CrossRef]
129. Jayasathyakawin, S.; Ravichandran, M.; Naveenkumar, R.; Radhika, N.; Ismail, S.O.; Mohanavel, V. Recent advances in magnesium alloys for biomedical applications: A review. *Mater. Today Commun.* **2025**, *42*, 111239. [CrossRef]
130. Coheña-Jiménez, M.; Prieto-Domínguez, R.; Pérez-Belloso, A.J.; Muriel-Sánchez, J.M.; Gómez-Carrión, Á.; Montaño-Jiménez, P. Comparison of Resorbable and Non-Resorbable Osteosynthesis Material in Hallux Surgery: A Systematic Review. *Life* **2023**, *13*, 2018. [CrossRef] [PubMed]
131. Zhang, T.; Wang, W.; Liu, J.; Wang, L.; Tang, Y.; Wang, K. A review on magnesium alloys for biomedical applications. *Front. Bioeng. Biotechnol.* **2022**, *10*, 953344. [CrossRef] [PubMed]

132. Wegener, B.; Sichler, A.; Milz, S.; Sprecher, C.; Pieper, K.; Hermanns, W.; Jansson, V.; Nies, B.; Kieback, B.; Müller, P.E.; et al. Development of a novel biodegradable porous iron-based implant for bone replacement. *Sci. Rep.* **2020**, *10*, 9141. [CrossRef]
133. Salama, M.; Vaz, M.F.; Colaço, R.; Santos, C.; Carmezim, M. Biodegradable Iron and Porous Iron: Mechanical Properties, Degradation Behaviour, Manufacturing Routes and Biomedical Applications. *J. Funct. Biomater.* **2022**, *13*, 72. [CrossRef]
134. Shuai, C.; Li, S.; Peng, S.; Feng, P.; Lai, Y.; Gao, C. Biodegradable metallic bone implants. *Mater. Chem. Front.* **2019**, *3*, 544–562. [CrossRef]
135. Md Yusop, A.H.; Al Sakkaf, A.; Nur, H. Modifications on porous absorbable Fe-based scaffolds for bone applications: A review from corrosion and biocompatibility viewpoints. *J. Biomed. Mater. Res. B Appl. Biomater.* **2022**, *110*, 18–44. [CrossRef]
136. Hermawan, H.; Purnama, A.; Dube, D.; Couet, J.; Mantovani, D. Fe–Mn alloys for metallic biodegradable stents: Degradation and cell viability studies. *Acta Biomater.* **2010**, *6*, 1852–1860. [CrossRef]
137. Hermawan, H.; Alamdari, H.; Mantovani, D.; Dubé, D. Iron–manganese: New class of metallic degradable biomaterials prepared by powder metallurgy. *Powder Metall.* **2008**, *51*, 38–45. [CrossRef]
138. Li, L.; Yang, X. The Essential Element Manganese, Oxidative Stress, and Metabolic Diseases: Links and Interactions. *Oxid. Med. Cell. Longev.* **2018**, *2018*, 7580707. [CrossRef]
139. Seol, J.B.; Jung, J.E.; Jang, Y.W.; Park, C.G. Influence of carbon content on the microstructure, martensitic transformation and mechanical properties in austenite/ε-martensite dual-phase Fe–Mn–C steels. *Acta Mater.* **2013**, *61*, 558–578. [CrossRef]
140. Dai, Y.J.; Mi, Z.L. Influence of Carbon on Mechanical Behavior of Fe-Mn-C System Alloys. *Adv. Mat. Res.* **2014**, *941–944*, 1469–1472. [CrossRef]
141. Liu, B.; Zheng, Y.F. Effects of alloying elements (Mn, Co, Al, W, Sn, B, C and S) on biodegradability and in vitro biocompatibility of pure iron. *Acta Biomater.* **2011**, *7*, 1407–1420. [CrossRef] [PubMed]
142. Roman, A.M.; Cimpoeșu, R.; Pricop, B.; Cazacu, M.M.; Zegan, G.; Istrate, B.; Cocean, A.; Chelariu, R.; Moscu, M.; Bădărău, G.; et al. Investigations on the Degradation Behavior of Processed FeMnSi-xCu Shape Memory Alloys. *Nanomaterials* **2024**, *14*, 330. [CrossRef]
143. Xu, Z.; Hodgson, M.A.; Cao, P. Microstructure and degradation behavior of forged Fe-Mn-Si alloys. *Int. J. Mod. Phys. B. World Sci.* **2015**, *29*, 10–11. [CrossRef]
144. Fiocchi, J.; Lemke, J.N.; Zilio, S.; Biffi, C.A.; Coda, A.; Tuissi, A. The effect of Si addition and thermomechanical processing in an Fe-Mn alloy for biodegradable implants: Mechanical performance and degradation behavior. *Mater Today Commun.* **2021**, *27*, 102447. [CrossRef]
145. Hermawan, H.; Dubé, D.; Mantovani, D. Development of Degradable Fe-35Mn Alloy for Biomedical Application. *Adv. Mat. Res.* **2006**, *15–17*, 107–112.
146. Park, J.W.; Hanawa, T.; Chung, J.H. The relative effects of Ca and Mg ions on MSC osteogenesis in the surface modification of microrough Ti implants. *Int. J. Nanomed.* **2019**, *14*, 5697–5711. [CrossRef]
147. Alsakkaf, A.; Md Yusop, A.H.; Idris, H.; Hassan, A.G.; Iqbal, N.; Unal, R.; Sudin, I. Development of Fe-Mg alloys with intermediate degradation kinetics as potential biodegradable bone implants. *Mater. Today Commun.* **2024**, *41*, 110609. [CrossRef]
148. Čapek, J.; Msallamová, Š.; Jablonská, E.; Lipov, J.; Vojtěch, D. A novel high-strength and highly corrosive biodegradable Fe-Pd alloy: Structural, mechanical and in vitro corrosion and cytotoxicity study. *Mater. Sci. Eng. C* **2017**, *79*, 550–562. [CrossRef]
149. Schinhammer, M.; Hänzi, A.C.; Löffler, J.F.; Uggowitzer, P.J. Design strategy for biodegradable Fe-based alloys for medical applications. *Acta Biomater.* **2010**, *6*, 1705–1713. [CrossRef]
150. Wei, S.; Ma, Z.; Tan, L.; Chen, J.; Misra, R.D.K.; Yang, K. Effect of copper content on the biodegradation behavior of Fe-Mn-C alloy system. *Mater. Technol.* **2022**, *37*, 1109–1119. [CrossRef]
151. Paul, B.; Kiel, A.; Otto, M.; Gemming, T.; Hoffmann, V.; Giebeler, L.; Kaltschmidt, B.; Hütten, A.; Gebert, A.; Kaltschmidt, B.; et al. Inherent Antibacterial Properties of Biodegradable FeMnC(Cu) Alloys for Implant Application. *ACS Appl Bio Mater. Am. Chem. Soc.* **2024**, *7*, 839–852. [CrossRef] [PubMed]
152. Prokoshkin, S.; Pustov, Y.; Zhukova, Y.; Kadirov, P.; Karavaeva, M.; Prosviryakov, A.; Dubinskiy, S. Effect of Thermomechanical Treatment on Structure and Functional Fatigue Characteristics of Biodegradable Fe-30Mn-5Si (wt%) Shape Memory Alloy. *Materials* **2021**, *14*, 3327. [CrossRef]
153. Hong, D.; Chou, D.T.; Velikokhatnyi, O.I.; Roy, A.; Lee, B.; Swink, I.; Issaev, I.; Kuhn, H.A.; Kumta, P.N. Binder-jetting 3D printing and alloy development of new biodegradable Fe-Mn-Ca/Mg alloys. *Acta Biomater.* **2016**, *45*, 375–386. [CrossRef]
154. Schinhammer, M.; Steiger, P.; Moszner, F.; Löffler, J.F.; Uggowitzer, P.J. Degradation performance of biodegradable FeMnC(Pd) alloys. *Mater. Sci. Eng. C* **2013**, *33*, 1882–1893. [CrossRef]
155. Chen, D.; Wang, P.; Pan, R.; Zha, C.; Fan, J.; Liang, D.; Zhao, Y. Characteristics of Metal Specimens Formed by Selective Laser Melting: A State-of-the-Art Review. *J. Mater. Eng. Perform.* **2020**, *30*, 7073–7100. [CrossRef]
156. Block-Bolten, A.; Eagar, T.W. Metal vaporization from weld pools. *Metall. Trans. B* **1984**, *15*, 461–469. [CrossRef]
157. Cabañas, N.; Akdut, N.; Penning, J.; De Cooman, B.C. High-temperature deformation properties of austenitic Fe-Mn alloys. *Met. Mater. Trans. A Phys. Met. Mater. Sci.* **2006**, *37*, 3305–3315. [CrossRef]

158. Wang, F.; Tan, Q.; Liu, T.; Venezuela, J.; Shi, Z.; Hurley, S.; Ly, A.; Xu, C.; Erbulut, D.; Yin, J.; et al. Reassessing the Biodegradation Behavior of Additively Manufactured Pure Iron and Iron-Manganese Alloys. 2025. Available online: https://ssrn.com/abstract=5135577 (accessed on 22 April 2025).
159. Donik, Č.; Kraner, J.; Kocijan, A.; Paulin, I.; Godec, M. Evolution of the ε and γ phases in biodegradable Fe–Mn alloys produced using laser powder-bed fusion. *Sci. Rep.* **2021**, *11*, 19506. [CrossRef] [PubMed]
160. Zheng, Z.; Sun, B.; Mao, L. Effect of Scanning Strategy on the Manufacturing Quality and Performance of Printed 316L Stainless Steel Using SLM Process. *Materials* **2024**, *17*, 1189. [CrossRef] [PubMed]
161. Cao, Q.N.; Cherqaoui, A.; Henrique Michelin Beraldo, C.; Paternoster, C.; Gélinas, S.; Blais, C.; Mengucci, P.; Mantovani, D. Effect of Volumetric Energy Density on the Evolution of the Microstructure and the Degradation Behavior of 3D-Printed Fe-Mn-C Alloys from Water-Atomized Powders. *Metals* **2025**, *15*, 101. [CrossRef]
162. Gökhan, A.; Carluccio, D.; Bermingham, M.; Dargusch, M.; Gökhan Demir, A.; Caprio, L.; Previtali, B. Selective laser melting Fe and Fe-35Mn for biodegradable implants. *Int. J. Mod. Phys. B* **2020**, *34*, 2040034.
163. Mugwagwa, L.; Dimitrov, D.; Matope, S.; Yadroitsev, I. Influence of process parameters on residual stress related distortions in selective laser melting. *Procedia Manuf.* **2018**, *21*, 92–99. [CrossRef]
164. Nie, Y.; Yuan, B.; Liang, J.; Deng, T.; Li, X.; Chen, P.; Zhang, K.; Li, X.; Li, K.; Peng, H.; et al. Mechanical and functional properties of Fe–Mn–Si biodegradable alloys fabricated by laser powder bed fusion: Effect of heat treatment. *Mater. Sci. Eng. A* **2024**, *908*, 146725. [CrossRef]
165. de Andrade, L.M.; Paternoster, C.; Chevallier, P.; Gambaro, S.; Copes, F.; de Oliveira Sales, V.F.; Mantovani, D. Electropolishing Fe-based biodegradable metals for vascular applications: Impact on surface properties, corrosion and cell viability. RSC Applied Interfaces. *R. Soc. Chem.* **2025**, *2*, 420–438. [CrossRef]
166. Huang, S.; Ulloa, A.; Nauman, E.; Stanciu, L. Collagen Coating Effects on Fe–Mn Bioresorbable Alloys. *J. Orthop. Res.* **2020**, *38*, 523–535. [CrossRef]
167. Liu, J.; Xie, T.; Xie, Y.; Xiao, L.; Lin, Y.; Dai, Y.; Wu, J. Microstructure and Mechanical Properties of Fe–30Mn–9Al–C–3Ni Low-Density Steel Manufactured by Selective Laser Melting. *J. Mater. Res. Technol.* **2024**, *33*, 4280–4289. [CrossRef]
168. Jafarian, H.R.; Sabzi, M.; Mousavi Anijdan, S.H.; Eivani, A.R.; Park, N. The influence of austenitization temperature on microstructural developments, mechanical properties, fracture mode and wear mechanism of Hadfield high manganese steel. *J. Mater. Res. Technol.* **2021**, *10*, 819–831. [CrossRef]
169. Hufenbach, J.; Sander, J.; Kochta, F.; Pilz, S.; Voss, A.; Kühn, U.; Gebert, A. Effect of Selective Laser Melting on Microstructure, Mechanical, and Corrosion Properties of Biodegradable FeMnCS for Implant Applications. *Adv. Eng. Mater.* **2020**, *22*, 2000182. [CrossRef]
170. Shi, Z.-Z.; Gao, X.-X.; Zhang, H.-J.; Liu, X.-F.; Li, H.-Y.; Zhou, C.; Yin, Y.-X.; Wang, L.-N. Design biodegradable Zn alloys: Second phases and their significant influences on alloy properties. *Bioact. Mater.* **2020**, *5*, 210–218. [CrossRef] [PubMed]
171. Dehestani, M.; Trumble, K.; Wang, H.; Wang, H.; Stanciu, L.A. Effects of microstructure and heat treatment on mechanical properties and corrosion behavior of powder metallurgy derived Fe–30Mn alloy. *Mater. Sci. Eng. A* **2017**, *703*, 214–226. [CrossRef]
172. Liu, B.; Zheng, Y.F.; Ruan, L. In vitro investigation of Fe30Mn6Si shape memory alloy as potential biodegradable metallic material. *Mater. Lett.* **2011**, *65*, 540–543. [CrossRef]
173. Tolosa, I.; Garciandía, F.; Zubiri, F.; Zapirain, F.; Esnaola, A. Study of mechanical properties of AISI 316 stainless steel processed by "selective laser melting", following different manufacturing strategies. *Int. J. Adv. Manuf. Technol.* **2010**, *51*, 639–647. [CrossRef]
174. Grässel, O.; Krüger, L.; Frommeyer, G.; Meyer, L.W. High strength Fe–Mn–(Al, Si) TRIP/TWIP steels development—Properties—Application. *Int. J. Plast.* **2000**, *16*, 1391–1409. [CrossRef]
175. Sawaguchi, T.; Bujoreanu, L.G.; Kikuchi, T.; Ogawa, K.; Koyama, M.; Murakami, M. Mechanism of reversible transformation-induced plasticity of Fe–Mn–Si shape memory alloys. *Scr. Mater.* **2008**, *59*, 826–829. [CrossRef]
176. Omori, T.; Ando, K.; Okano, M.; Xu, X.; Tanaka, Y.; Ohnuma, I.; Kainuma, R.; Ishida, K. Superelastic Effect in Polycrystalline Ferrous Alloys. *Science* **2011**, *333*, 68–71. [CrossRef]
177. Liu, P.; Wu, H.; Liang, L.; Song, D.; Liu, J.; Ma, X.; Li, K.; Fang, Q.; Tian, Y.; Baker, I. Microstructure, mechanical properties and corrosion behavior of additively-manufactured Fe–Mn alloys. *Mater. Sci. Eng. A* **2022**, *852*, 143585. [CrossRef]
178. Li, Y.; Lietaert, K.; Li, W.; Zhang, X.Y.; Leeflang, M.A.; Zhou, J.; Zadpoor, A.A. Corrosion fatigue behavior of additively manufactured biodegradable porous iron. *Corros Sci.* **2019**, *156*, 106–116. [CrossRef]
179. Han, S.Z.; Choi, E.A.; Lim, S.H.; Kim, S.; Lee, J. Alloy design strategies to increase strength and its trade-offs together. *Prog. Mater. Sci.* **2021**, *117*, 100720. [CrossRef]
180. Shiomi, M.; Osakada, K.; Nakamura, K.; Yamashita, T.; Abe, F. Residual stress within metallic model made by selective laser melting process. *CIRP Ann. Manuf. Technol.* **2004**, *53*, 195–198. [CrossRef]
181. Gorejová, R.; Haverová, L.; Oriňaková, R.; Oriňak, A.; Oriňak, M. Recent advancements in Fe-based biodegradable materials for bone repair. *J. Mater. Sci.* **2019**, *54*, 1913–1947. [CrossRef]

182. Zhang, E.; Chen, H.; Shen, F. Biocorrosion properties and blood and cell compatibility of pure iron as a biodegradable biomaterial. *J. Mater. Sci. Mater. Med.* **2010**, *21*, 2151–2163. [CrossRef] [PubMed]
183. Williams, D.F. On the Mechanisms of Biocompatibility. *Biomaterials* **2008**, *29*, 2941–2953. [CrossRef] [PubMed]
184. Kadirov, P.; Zhukova, Y.; Gunderov, D.; Antipina, M.; Teplyakova, T.; Tabachkova, N.; Baranova, A.; Gunderova, S.; Pustov, Y.; Prokoshkin, S. Effect of Accumulative High-Pressure Torsion on Structure and Electrochemical Behavior of Biodegradable Fe-30Mn-5Si (wt.%) Alloy. *Crystals* **2025**, *15*, 351. [CrossRef]
185. Li, Y.; Jahr, H.; Zhou, J.; Zadpoor, A.A. Additively Manufactured Biodegradable Porous Metals. *Acta Biomater.* **2020**, *115*, 29–50. [CrossRef]
186. Zhao, Y.; Feng, J.; Yu, H.; Lin, W.; Li, X.; Tian, Y.; Zhao, M. Comparative Study on Biodegradation of Pure Iron Prepared by Microwave Sintering and Laser Melting. *Materials* **2022**, *15*, 1604. [CrossRef]
187. Cheng, J.; Zheng, Y.F. In vitro study on newly designed biodegradable Fe-X composites (X = W, CNT) prepared by spark plasma sintering. *J. Biomed. Mater. Res. B Appl. Biomater.* **2013**, *101*, 485–497. [CrossRef]
188. Rybalchenko, O.; Anisimova, N.; Martynenko, N.; Rybalchenko, G.; Kiselevskiy, M.; Tabachkova, N.; Shchetinin, I.; Raab, A.; Dobatkin, S. Structure Optimization of a Fe–Mn–Pd Alloy by Equal-Channel Angular Pressing for Biomedical Use. *Materials* **2022**, *16*, 45. [CrossRef] [PubMed]
189. Rivkin, B.; Akbar, F.; Otto, M.; Beyer, L.; Paul, B.; Kosiba, K.; Gustmann, T.; Hufenbach, J.; Medina-Sánchez, M. Remotely Controlled Electrochemical Degradation of Metallic Implants. *Small* **2024**, *20*, 2307742. [CrossRef]
190. Aufa, A.N.; Hassan, M.Z.; Ismail, Z.; Ramlie, F.; Jamaludin, K.R.; Md Daud, M.Y.; Ren, J. Current trends in additive manufacturing of selective laser melting for biomedical implant applications. *J. Mater. Res. Technol.* **2024**, *31*, 213–243. [CrossRef]
191. Chen, L.; Shen, Q.; Liu, Y.; Zhang, Y.; Sun, L.; Ma, X.; Song, N.; Xie, J. Homeostasis and metabolism of iron and other metal ions in neurodegenerative diseases. *Signal Transduct. Target. Ther.* **2025**, *10*, 31.
192. Hu, J.; Shao, J.; Huang, G.; Zhang, J.; Pan, S. In Vitro and In Vivo Applications of Magnesium-Enriched Biomaterials for Vascularized Osteogenesis in Bone Tissue Engineering: A Review of Literature. *J. Funct. Biomater.* **2023**, *14*, 326. [CrossRef] [PubMed]
193. Tai, C.-C.; Huang, Y.-M.; Liaw, C.-K.; Yang, K.-Y.; Ma, C.-H.; Huang, S.-I.; Huang, C.-C.; Tsai, P.-I.; Shen, H.-H.; Sun, J.-S.; et al. Biocompatibility and Biological Performance of Additive-Manufactured Bioabsorbable Iron-Based Porous Interference Screws in a Rabbit Model: A 1-Year Observational Study. *Int. J. Mol. Sci.* **2022**, *23*, 14626. [CrossRef]
194. Flege, C.; Vogt, F.; Höges, S.; Jauer, L.; Borinski, M.; Schulte, V.A.; Hoffmann, R.; Poprawe, R.; Meiners, W.; Jobmann, M.; et al. Development and characterization of a coronary polylactic acid stent prototype generated by selective laser melting. *J. Mater. Sci. Mater. Med.* **2013**, *24*, 241–255. [CrossRef]
195. Sing, N.B.; Mostavan, A.; Hamzah, E.; Mantovani, D.; Hermawan, H. Degradation behavior of biodegradable Fe35Mn alloy stents. *J. Biomed. Mater. Res. B Appl. Biomater.* **2015**, *103*, 572–577. [CrossRef] [PubMed]
196. Bowen, P.K.; Shearier, E.R.; Zhao, S.; Guillory, R.J.; Zhao, F.; Goldman, J.; Drelich, J.W. Biodegradable Metals for Cardiovascular Stents: From Clinical Concerns to Recent Zn-Alloys. *Adv. Health Mater.* **2016**, *5*, 1121–1140. [CrossRef]
197. Bowen, P.K.; Guillory, R.J.; Shearier, E.R.; Seitz, J.-M.; Drelich, J.; Bocks, M.; Zhao, F.; Goldman, J. Metallic zinc exhibits optimal biocompatibility for bioabsorbable endovascular stents. *Mater. Sci. Eng. C* **2015**, *56*, 467–472. [CrossRef]
198. Kong, L.; Heydari, Z.; Lami, G.H.; Saberi, A.; Baltatu, M.S.; Vizureanu, P. A Comprehensive Review of the Current Research Status of Biodegradable Zinc Alloys and Composites for Biomedical Applications. *Materials* **2023**, *16*, 4797. [CrossRef]
199. Zhu, S.; Wu, C.; Li, G.; Zheng, Y.; Nie, J.-F. Creep properties of biodegradable Zn-0.1Li alloy at human body temperature: Implications for its durability as stents. *Mater. Res. Lett.* **2019**, *7*, 347–353. [CrossRef]
200. Domínguez López, G.; Williams, P.L.; Llorca, J.; Echeverry-Rendón, M.; Echeverry-Rendon, M. Screening of Zn Alloys for Cardiovascular and Bone Applications: Effect of the Alloying Elements. Available online: https://ssrn.com/abstract=5027614 (accessed on 14 March 2025).
201. Imtiaz, H.; Riaz, M.; Anees, E.; Bashir, F.; Hussain, T. Biodegradable zinc–magnesium alloys for bone fixation: A study of their structural integrity, corrosion resistance, and mechanical properties. *Mater. Chem. Phys.* **2025**, *334*, 130429. [CrossRef]
202. Cao, X.; Wang, X.; Chen, J.; Geng, X.; Tian, H. 3D Printing of a Porous Zn-1Mg-0.1Sr Alloy Scaffold: A Study on Mechanical Properties, Degradability, and Biosafety. *J. Funct. Biomater.* **2024**, *15*, 109. [CrossRef]
203. Yang, H.; Jia, B.; Zhang, Z.; Qu, X.; Li, G.; Lin, W.; Zhu, D.; Dai, K.; Zheng, Y. Alloying design of biodegradable zinc as promising bone implants for load-bearing applications. *Nat. Commun.* **2020**, *11*, 401. [CrossRef] [PubMed]
204. Tang, Z.; Huang, H.; Niu, J.; Zhang, L.; Zhang, H.; Pei, J.; Tan, J.; Yuan, G. Design and characterizations of novel biodegradable Zn-Cu-Mg alloys for potential biodegradable implants. *Mater. Des.* **2017**, *117*, 84–94. [CrossRef]
205. Zhu, K.; Prince, R.L. Calcium and bone. *Clin. Biochem.* **2012**, *45*, 936–942. [CrossRef]
206. Čapek, J.; Kubásek, J.; Pinc, J.; Drahokoupil, J.; Čavojský, M.; Vojtěch, D. Extrusion of the biodegradable ZnMg0.8Ca0.2 alloy–The influence of extrusion parameters on microstructure and mechanical characteristics. *J. Mech. Behav. Biomed. Mater.* **2020**, *108*, 103796. [CrossRef]

207. Zhang, Z.; Jia, B.; Yang, H.; Han, Y.; Wu, Q.; Dai, K.; Zheng, Y. Biodegradable ZnLiCa ternary alloys for critical-sized bone defect regeneration at load-bearing sites: In vitro and in vivo studies. *Bioact. Mater.* **2021**, *6*, 3999–4013. [CrossRef] [PubMed]
208. Cabrera, W.E.; Schrooten, I.; De Broe, M.E.; D'Haese, P.C. Strontium and Bone. *J. Bone Miner. Res.* **1999**, *14*, 661–668. [CrossRef]
209. Mostaed, E.; Sikora-Jasinska, M.; Drelich, J.W.; Vedani, M. Zinc-based alloys for degradable vascular stent applications. *Acta Biomater.* **2018**, *71*, 1–23. [CrossRef]
210. Song, J.; Gao, Y.; Liu, C.; Chen, Z. The effect of Sr addition on the microstructure and corrosion behaviour of a Mg-Zn-Ca alloy. *Surf. Coat. Technol.* **2022**, *437*, 128328. [CrossRef]
211. Reginster, J. Strontium Ranelate in Osteoporosis. *Curr. Pharm. Des.* **2005**, *8*, 1907–1916. [CrossRef] [PubMed]
212. Palai, D.; Roy, T.; Prasad, P.S.; Hazra, C.; Dhara, S.; Sen, R.; Das, S.; Das, K. Influence of Copper on the Microstructural, Mechanical, and Biological Properties of Commercially Pure Zn-Based Alloys for a Potential Biodegradable Implant. *ACS Biomater. Sci. Eng. Am. Chem. Soc.* **2022**, *8*, 1443–1463. [CrossRef]
213. Tang, Z.; Niu, J.; Huang, H.; Zhang, H.; Pei, J.; Ou, J.; Yuan, G. Potential biodegradable Zn-Cu binary alloys developed for cardiovascular implant applications. *J. Mech. Behav. Biomed. Mater.* **2017**, *72*, 182–191. [CrossRef]
214. Čákyová, V.; Gorejová, R.; Kupková, M.; Sopčák, T.; Strečková, M.; Fáberová, M.; Özaltin, K.; Džupon, M.; Oriňaková, R. Study of Zn-Ag Alloys Prepared via Powder Metallurgy. 2025. Available online: https://ssrn.com/abstract=5115727 (accessed on 17 March 2025).
215. Xiao, X.; Liu, E.; Shao, J.; Ge, S. Advances on biodegradable zinc-silver-based alloys for biomedical applications. *J. Appl. Biomater. Funct. Mater.* **2021**, *19*, 22808000211062407. [CrossRef]
216. Xie, Z.; Yang, M.Z.; Xue, J.B.; Chen, Y.; Peng, L.J.; Cai, B.; Zhang, Y.Y. Antibacterial properties of biodegradable zinc alloys in vivo. Chinese Journal of Tissue Engineering Research. *Chin. J. Tissue Eng. Res.* **2019**, *23*, 2196–2201.
217. Di, T.; Xu, Y.; Liu, D.; Sun, X. Microstructure, Mechanical Performance and Anti-Bacterial Activity of Degradable Zn-Cu-Ag Alloy. *Metals* **2022**, *12*, 1444. [CrossRef]
218. Liu, X.; Sun, J.; Yang, Y.; Zhou, F.; Pu, Z.; Li, L.; Zheng, Y. Microstructure, mechanical properties, in vitro degradation behavior and hemocompatibility of novel Zn–Mg–Sr alloys as biodegradable metals. *Mater. Lett.* **2016**, *162*, 242–245. [CrossRef]
219. Waqas, M.; He, D.; Wu, X.; Tan, Z.; Shao, W.; Guo, X. Investigation on the preparation, microstructure, mechanical and degradation properties of laser additive manufactured Zn–Li–Mg alloy for bioresorbable application. *J. Mater. Res. Technol.* **2023**, *26*, 8509–8526. [CrossRef]
220. Meng, F.; Du, Y. Research Progress on Laser Powder Bed Fusion Additive Manufacturing of Zinc Alloys. *Materials* **2024**, *17*, 4309. [CrossRef] [PubMed]
221. Gao, W.; Li, Z.W.; Harikisun, R.; Chang, S.S. Zinc oxide films formed by oxidation of zinc under low partial pressure of oxygen. *Mater. Lett.* **2003**, *57*, 1435–1440. [CrossRef]
222. Waqas, M.; He, D.; Tan, Z.; Yang, P.; Gao, M.; Guo, X. A study of selective laser melting process for pure zinc and Zn10mg alloy on process parameters and mechanical properties. *Artic. Rapid Prototyp. J.* **2023**, *29*, 1923–1939. [CrossRef]
223. Cui, Z.; Zhang, Y.; Wang, J.; Zhou, L.; Hao, X.; Wang, W.; Li, W.; Cheng, W.; Chang, C. Comparison on microstructure and properties of Zn-10 Mg alloy prepared by SPS and SLM techniques. *Mater Lett.* **2025**, *389*, 138327. [CrossRef]
224. Cockerill, I.; Su, Y.; Sinha, S.; Qin, Y.X.; Zheng, Y.; Young, M.L.; Zhu, D. Porous zinc scaffolds for bone tissue engineering applications: A novel additive manufacturing and casting approach. *Mater. Sci. Eng. C* **2020**, *110*, 110738. [CrossRef]
225. Mahmood, M.A.; Chioibasu, D.; Rehman, A.U.; Mihai, S.; Popescu, A.C. Post-Processing Techniques to Enhance the Quality of Metallic Parts Produced by Additive Manufacturing. *Metals* **2022**, *12*, 77. [CrossRef]
226. Wang, C.; Hu, Y.; Zhong, C.; Lan, C.; Li, W.; Wang, X. Microstructural evolution and mechanical properties of pure Zn fabricated by selective laser melting. *Mater. Sci. Eng. A* **2022**, *846*, 143276. [CrossRef]
227. Voshage, M.; Megahed, S.; Schückler, P.G.; Wen, P.; Qin, Y.; Jauer, L.; Poprawe, R.; Schleifenbaum, J.H. Additive manufacturing of biodegradable Zn-xMg alloys: Effect of Mg content on manufacturability, microstructure and mechanical properties. *Mater. Today Commun.* **2022**, *32*, 103805. [CrossRef]
228. Yang, Y.; Yuan, F.; Gao, C.; Feng, P.; Xue, L.; He, S.; Shuai, C. A combined strategy to enhance the properties of Zn by laser rapid solidification and laser alloying. *J. Mech. Behav. Biomed. Mater.* **2018**, *82*, 51–60. [CrossRef]
229. Galib, R.H.; Sharif, A. Development of Zn-Mg Alloys as a Degradable Biomaterial. *Columbia Int. Publ. Adv. Alloys Compd.* **2016**, *1*, 1–7.
230. Dambatta, M.S.; Izman, S.; Kurniawan, D.; Hermawan, H. Processing of Zn-3Mg alloy by equal channel angular pressing for biodegradable metal implants. *J. King Saud. Univ. Sci.* **2017**, *29*, 455–461. [CrossRef]
231. Liu, X.; Sun, J.; Qiu, K.; Yang, Y.; Pu, Z.; Li, L.; Zheng, Y. Effects of alloying elements (Ca and Sr) on microstructure, mechanical property and in vitro corrosion behavior of biodegradable Zn–1. *5Mg alloy. J. Alloys Compd.* **2016**, *664*, 444–452. [CrossRef]
232. Li, Y.; Li, W.; Bobbert, F.S.L.; Lietaert, K.; Dong, J.H.; Leeflang, M.A.; Zhou, J.; Zadpoor, A.A. Corrosion fatigue behavior of additively manufactured biodegradable porous zinc. *Acta Biomater.* **2020**, *106*, 439–449. [CrossRef]
233. Eliaz, N. Corrosion of Metallic Biomaterials: A Review. *Materials* **2019**, *12*, 407. [CrossRef] [PubMed]

234. Sun, S.; Chen, W.; Wu, C.; Liu, Y.; Su, X.; Wang, J. Effect of calcium on microstructure and corrosion resistance of Zn-6Al-3Mg alloy coatings. *Mater. Today Commun.* **2025**, *46*, 112495. [CrossRef]
235. Chao, Z.; Wang, B.; Xu, C.; Li, Y. Study of grain orientation effect on the corrosion behavior of biocompatible magnesium alloy Mg–2Zn-0.5Ca. *Mater. Chem. Phys.* **2024**, *328*, 130039. [CrossRef]
236. Ning, J.; Ma, Z.-X.; Zhang, L.-J.; Wang, D.-P.; Na, S.-J. Effects of magnesium on microstructure, properties and degradation behaviors of zinc-based alloys prepared by selective laser melting. *Mater. Res. Express.* **2022**, *9*, 086511. [CrossRef]
237. Dambatta, M.S.; Izman, S.; Kurniawan, D.; Farahany, S.; Yahaya, B.; Hermawan, H. Influence of thermal treatment on microstructure, mechanical and degradation properties of Zn–3Mg alloy as potential biodegradable implant material. *Mater. Des.* **2015**, *85*, 431–437. [CrossRef]
238. Bandekian, S.; Baghbaderani, M.Z.; Drelich, J.W.; Sharif, S.; Ismail, A.F.; Bakhsheshi-Rad, H.R. Additive manufacturing of zinc-based biomaterials: Fabrication, performance and property evaluation. *J. Mater. Res. Technol.* **2025**, *36*, 5484–5508. [CrossRef]
239. Pham, D.N.; Hiromoto, S.; Kobayashi, E. Influences of Zinc Content and Solution Heat Treatment on Microstructure and Corrosion Behavior of Mg-Zn Binary Alloys. *Corrosion* **2021**, *77*, 323–338. [CrossRef]
240. Yuan, W.; Xia, D.; Wu, S.; Zheng, Y.; Guan, Z.; Rau, J.V. A review on current research status of the surface modification of Zn-based biodegradable metals. *Bioact. Mater.* **2022**, *7*, 192–216. [CrossRef] [PubMed]
241. Shearier, E.R.; Bowen, P.K.; He, W.; Drelich, A.; Drelich, J.; Goldman, J.; Zhao, F. In Vitro Cytotoxicity, Adhesion, and Proliferation of Human Vascular Cells Exposed to Zinc. *ACS Biomater. Sci. Eng. Am. Chem. Soc.* **2016**, *2*, 634–642. [CrossRef]
242. Du, C.; Zuo, K.; Ma, Z.; Zhao, M.; Li, Y.; Tian, S.; Lu, Y.; Xiao, G. Effect of Substrates Performance on the Microstructure and Properties of Phosphate Chemical Conversion Coatings on Metal Surfaces. *Molecules* **2022**, *27*, 6434. [CrossRef] [PubMed]
243. Guan, Z.; Linsley, C.S.; Pan, S.; Yao, G.; Wu, B.M.; Levi, D.S.; Li, X. Zn-Mg-WC Nanocomposites for Bioresorbable Cardiovascular Stents: Microstructure, Mechanical Properties, Fatigue, Shelf Life, and Corrosion. *ACS Biomater. Sci. Eng. Am. Chem. Soc.* **2022**, *8*, 328–339. [CrossRef] [PubMed]
244. Hermawan, H.; Dubé, D.; Mantovani, D. Developments in metallic biodegradable stents. *Acta Biomater.* **2010**, *6*, 1693–1697. [CrossRef]
245. Liu, Y.; Zheng, Y.; Chen, X.; Yang, J.; Pan, H.; Chen, D.; Wang, L.; Zhang, J.; Zhu, D.; Wu, S.; et al. Fundamental Theory of Biodegradable Metals—Definition, Criteria, and Design. *Adv. Funct. Mater.* **2019**, *29*, 1805402. [CrossRef]
246. Chen, Y.; Xu, Z.; Smith, C.; Sankar, J. Recent advances on the development of magnesium alloys for biodegradable implants. *Acta Biomater.* **2014**, *10*, 4561–4573. [CrossRef]
247. Zhang, E.; Yang, L. Microstructure, mechanical properties and bio-corrosion properties of Mg–Zn–Mn–Ca alloy for biomedical application. *Mater. Sci. Eng. A* **2008**, *497*, 111–118. [CrossRef]
248. Zhang, B.; Li, Y.; Bai, Q. Defect Formation Mechanisms in Selective Laser Melting: A Review. *Chin. J. Mech. Eng.* **2017**, *30*, 515–527. [CrossRef]
249. Gu, D.D.; Meiners, W.; Wissenbach, K.; Poprawe, R. Laser additive manufacturing of metallic components: Materials, processes and mechanisms. *Int. Mater. Rev.* **2012**, *57*, 133–164. [CrossRef]
250. Mercelis, P.; Kruth, J. Residual stresses in selective laser sintering and selective laser melting. *Rapid Prototyp. J.* **2006**, *12*, 254–265. [CrossRef]
251. Wang, D.; Yang, Y.; Liu, R.; Xiao, D.; Sun, J. Study on the designing rules and processability of porous structure based on selective laser melting (SLM). *J. Mater. Process Technol.* **2013**, *213*, 1734–1742. [CrossRef]
252. Yasa, E.; Kruth, J.-P. Microstructural investigation of Selective Laser Melting 316L stainless steel parts exposed to laser re-melting. *Procedia Eng.* **2011**, *19*, 389–395. [CrossRef]
253. Mellal, A.; Wiskott, H.W.A.; Botsis, J.; Scherrer, S.S.; Belser, U.C. Stimulating effect of implant loading on surrounding bone. *Clin. Oral. Implant. Res.* **2004**, *15*, 239–248. [CrossRef]
254. Jahr, H.; Li, Y.; Zhou, J.; Zadpoor, A.A.; Schröder, K.-U. Additively Manufactured Absorbable Porous Metal Implants–Processing, Alloying and Corrosion Behavior. *Front. Mater.* **2021**, *8*, 628633. [CrossRef]
255. Fazel-Rezai, R. *Biomedical Engineering–From Theory to Applications Edited by Reza Fazel-Rezai*; InTechOpen: London, UK, 2011; Volume 498.
256. Doll, K.; Doll, P.W.; Doll, C.; Doll, T. Importance of interdisciplinary training for successful development of innovative biomedical implants. *Curr. Dir. Biomed. Eng.* **2022**, *8*, 761–764. [CrossRef]
257. Khan, A.R.; Grewal, N.S.; Zhou, C.; Yuan, K.; Zhang, H.-J.; Jun, Z. Recent advances in biodegradable metals for implant applications: Exploring in vivo and in vitro responses. *Results Eng.* **2023**, *20*, 101526. [CrossRef]
258. Rashidy Ahmady, A.; Ekhlasi, A.; Nouri, A.; Haghbin Nazarpak, M.; Gong, P.; Solouk, A. High entropy alloy coatings for biomedical applications: A review. *Smart Mater. Manuf.* **2023**, *1*, 100009. [CrossRef]
259. Pompe, W.; Worch, H.; Epple, M.; Friess, W.; Gelinsky, M.; Greil, P.; Hempel, U.; Scharnweber, D.; Schulte, K. Functionally graded materials for biomedical applications. *Mater. Sci. Eng. A* **2003**, *362*, 40–60. [CrossRef]

260. Dilberoglu, U.M.; Gharehpapagh, B.; Yaman, U.; Dolen, M. Current trends and research opportunities in hybrid additive manufacturing. *Int. J. Adv. Manuf. Technol.* **2021**, *113*, 623–648. [CrossRef]
261. Gao, B.; Zhao, H.; Peng, L.; Sun, Z. A Review of Research Progress in Selective Laser Melting (SLM). *Micromachines* **2022**, *14*, 57. [CrossRef] [PubMed]
262. Li, J.; Fan, H.; Li, H.; Hua, L.; Du, J.; He, Y.; Jin, Y. Recent Advancements in the Surface Modification of Additively Manufactured Metallic Bone Implants. *Addit. Manuf. Front.* **2025**, *4*, 200195. [CrossRef]
263. Yi Wang, W.; Li, J.; Liu, W.; Liu, Z.-K. Integrated computational materials engineering for advanced materials: A brief review. *Comput. Mater. Sci.* **2019**, *158*, 42–48. [CrossRef]
264. Ng, W.L.; Goh, G.L.; Goh, G.D.; Ten, J.S.J.; Yeong, W.Y. Progress and Opportunities for Machine Learning in Materials and Processes of Additive Manufacturing. *Adv. Mater.* **2024**, *36*, e2310006. [CrossRef]
265. Diniz, P.; Grimm, B.; Garcia, F.; Fayad, J.; Ley, C.; Mouton, C.; Oeding, J.F.; Hirschmann, M.T.; Samuelsson, K.; Seil, R. Digital twin systems for musculoskeletal applications: A current concepts review. Knee Surgery, Sports Traumatology. *Arthroscopy* **2025**, *33*, 1892–1910.
266. Maintz, M.; Tourbier, C.; de Wild, M.; Cattin, P.C.; Beyer, M.; Seiler, D.; Honigmann, P.; Sharma, N.; Thieringer, F.M. Patient-specific implants made of 3D printed bioresorbable polymers at the point-of-care: Material, technology, and scope of surgical application. *3D Print. Med.* **2024**, *10*, 13. [CrossRef]
267. Merces, L.; Ferro, L.M.M.; Thomas, A.; Karnaushenko, D.D.; Luo, Y.; Egunov, A.I.; Zhang, W.; Bandari, V.K.; Lee, Y.; McCaskill, J.S.; et al. Bio-Inspired Dynamically Morphing Microelectronics toward High-Density Energy Applications and Intelligent Biomedical Implants. *Adv. Mater.* **2024**, *36*, e2313327. [CrossRef]
268. Xue, L.; Dai, S.; Li, Z. Biodegradable shape-memory block co-polymers for fast self-expandable stents. *Biomaterials* **2010**, *31*, 8132–8140. [CrossRef]
269. Kadhim, A.C.; Azzahrani, A.S.; Alsheikhy, A.A.; Resen, D.A. Smart Bandages with Integrated Sensors for Real-Time Monitoring of Wound Inflammation and Infection. *J. Opt.* 2023. [CrossRef]

metals MDPI

Article

High-Cycle Fatigue Characteristics of Aluminum/Steel Clinched and Resistance-Spot-Welded Joints Based on Failure Modes

Ákos Meilinger *, Péter Zoltán Kovács and János Lukács

Institute of Materials Science and Technology, Faculty of Mechanical Engineering and Informatics, University of Miskolc, 3515 Miskolc, Hungary; peter.kovacs@uni-miskolc.hu (P.Z.K.); janos.lukacs@uni-miskolc.hu (J.L.)
* Correspondence: akos.meilinger@uni-miskolc.hu

Abstract: Materials for lightweight vehicle structures play an increasingly important role in both economic and environmental terms; high-strength steels and aluminum alloys are suitable for this role. Resistance spot welding (RSW) and conventional clinching (CCL) methods can be used for joining vehicle bodies and can also be applied for aluminum/steel hybrid joints. Whereas vehicle structures are subjected to cyclic loading, damages can occur due to high-cycle fatigue (HCF) during long-term operation. Systematic HCF test results are rarely found in the literature, while HCF loading basically determines the lifetime of the hybrid joints. The base materials 5754-H22, 6082-T6, and DP600 were used for similar and hybrid RSW and CCL joints, and HCF tests were performed. The number of cycles-to-failure values and failure modes were studied and analyzed. Based on the experimental results, HCF design curves belonging to a 50% failure probability were calculated for all cases, and the curves were compared. Clear relationships were found between the failure modes and fatigue cycle numbers for both joining methods. Considering the steel/steel joints as a base, the load-bearing capacity of the hybrid joints is lower (48.7% and 73.0% for RSW, 35.0% and 38.7% for CCL) and it is even lower for the aluminum/aluminum joints (39.9% and 50.4% for RSW, 31.7% and 35.0% for CCL). With one exception, the load-bearing capacity of the CCL joints is higher than that of the RSW joints (156.1–108.3%).

Keywords: hybrid joining; aluminum/steel joining; resistance spot welding (RSW); conventional clinching (CCL); high-cycle fatigue (HCF); failure mode

1. Introduction

Different structures and equipment are made up of several different elements, the features (e.g., material quality, characteristic geometric dimensions, and surface area) of which may differ significantly. The construction of structures or equipment is inconceivable without the interconnection (joining) of the individual elements, and the correlation of joining technologies is achieved within the structure–material–manufacturing technology triangle. The diversity of structures, together with the economic and environmental demands and constraints, require continuous innovation in joining technologies [1–3], regardless of the size of the structures, but at different scales and/or levels [4].

Joining technologies play an exceptionally important role in all phases of a product's life—these phases include the design, development, fabrication, utilization, maintenance, and repair of a product as well as its recycling and disposal [5]. As a direct consequence, joining technologies and joining processes can be classified in several ways. They can be classified according to the size scale (e.g., micro/macro [4,5]), the combined materials (e.g., similar/dissimilar [6]), the main effect that creates the joint (e.g., heat/force and associated (plastic) deformation [2,7]), the variations within a process (e.g., clinching sheet materials [8]), and the geometry of the joint (continuous/spot), and the list is continuous.

Among the various structures, particular attention should be paid to vehicular/automotive structures, where a distinction can be made between those made of thicker plates (e.g., the chassis) and those made of thinner plates (e.g., the car body). Although not exclusively, plate thickness influences the joining technologies that can be used. Focusing on welding, fusion welding technologies may be preferred for thicker plates (e.g., trucks, heavy machinery), while solid-state welding may be preferred for thinner plates (e.g., cars). However, there are certainly processes (e.g., laser welding) that can be used regardless of the plate or sheet thicknesses. For thinner plates, there is a wide range of combined processes, where the combined formulation can have multiple meanings. It can be interpreted as a combination of several technologies (e.g., forming and welding), as a combination of different processes from one technology (e.g., root welding by the tungsten inert gas (TIG) welding method, the welding of additional layers by the metal inert gas (MIG) welding method, or laser MIG (or TIG) hybrid joining), or there are many variations of joints made with auxiliary or additional elements (e.g., rivets, screws) [9,10].

Joining technologies involving significant plastic deformation are typical for thinner plates or sheets; one of the characteristic technologies for car bodies is clinching. It is possible to clinch steel elements [11], steel and non-steel elements, or metallic elements [12], and both metallic and non-metallic elements [10]. As can be seen from the list, clinching technology can be used to join materials with significantly different physical properties (e.g., metal and fiber-reinforced plastic, or steel and aluminum alloy). In addition to the more typical two-plate joints, three-plate joints are also used [13].

One of the most typical material combinations for dissimilar material connections is the joining of steel to aluminum. The welding technologies used are summarized in [9,14], friction-based technologies are reviewed in [15], and details of bump or projection welding and ultrasonic welding processes are summarized in [16,17].

In the case of spot-like welding technologies, essentially resistance spot welding (RSW) and ultrasonic welding (UW) can be applied for hybrid joints [18–20]. In recent years, special process variants have emerged to improve the joint properties, such as resistance element or resistance rivet welding (REW or RRW) [21,22], metallic bump-assisted RSW (MBaRSW) [23], (high power) ultrasonic welding ((HP)USW) [24], and a combination of ultrasonic and resistance spot welding [25]. Micro RSW (MRSW) [26] and vaporizing foil actuators welding (VFAW) [27] have also been investigated as prospective technologies for aluminum/steel joining. These advanced joining methods frequently require additional materials or/and process steps; or rather, the process cycle is longer in comparison with the basic RSW cycle. Different spot welding process variants used for aluminum/steel joining, including their main characteristics, are summarized in Table 1. An overview of clinching technologies suitable for making aluminum/steel hybrid joints can be found in Table 2, where conventional clinching (CLL) [28] was selected as a basic technology. The material grades in both tables have retained the designations used in the original sources and, within each technology, the applied order corresponds to the order of the aluminum groups. It is worth noting that the differences between the (clinching) technologies listed in Table 2 are more significant than in the case of the (spot welding) technologies found in Table 1.

After comparing the data found in Tables 1 and 2, also with regard to the specific examples presented, two features should be highlighted. Firstly, for both technologies, the application of coatings on steel plates and auxiliary elements is common, which is clearly unrelated to the joining technologies, and secondly, the use of intermediate (third material) layers is less common in clinching technologies.

It is a well-known fact that car bodies are subjected to cyclic loads that fall within the range of high-cycle fatigue (HCF) or ultra-high-cycle fatigue (UHCF). Obviously, both the applied materials and the joints made with different technologies must withstand these loads.

Table 1. Spot welding-based joining technologies for aluminum/steel hybrid joints, including their main characteristics.

Joining Process	Aluminum Part		Steel Part		Additional Information	Source
	Material	Thickness (mm)	Material	Thickness (mm)		
RSW	A5052	1.0	A366/A366M-97	1.0	commercial steel cover plate (1.0 mm) on aluminum side	[29]
	A5052	1.0	SUS304	1.0	commercial steel cover plate (1.0 mm) on aluminum side	[29]
	A5052	1.5	DP600	1.2	pure Zn interlayer	[30]
	AW5754-H22	1.0	DP500	1.5	DP500 uncoated	[31]
	5182-O	2.0	SAE 1008	1.4	1050 clad SAE 1006 transition material	[32]
	6008-T6	1.5	H220YD	1.0	H220YD galvanised	[33]
	6008-T6	1.5	H220YD	1.0	H220YD galvanised; 4047 AlSi12 interlayer	[34]
	AA6016-T4	1.0	Interstitial-free steel (IFS)	0.7	IFS bare, galvanized (60 g/mm^2), galvannealed (40 g/mm^2, 60 g/mm^2)	[35]
	AA6022-T4	1.2	Rolled interstitial-free steel (IFS)	2.0	IFS hot-dipped galvanized	[36,37]
	AA6022-T4	1.2	Low-carbon steel (LCS)	2.0	LCS hot-dipped galvanized	[38]
	AA6022-T4	1.2	mild steel (MS)	2.0	MS hot-dipped galvanized	[39]
	X626	0.8	Low-carbon steel (LCS)	0.9; 2.0	LCS uncoated	[40]
	AA6022	1.2	Low-carbon steel (LCS)	0.9; 2.0	LCS uncoated	[40]
	AA6022	1.2	HSLA steel	1.2; 2.0	HSLA steel uncoated	[40]
	AA6062	1.2	HSLA steel	2.0	N/A	[41]
	AA6022	1.2	HSLA steel and CR780T *	0.65 and 1.4	N/A	[41]
	A6061	1.5	AISI-SAE 1005	1.5	pure Cu insert	[42]
	A6061	1.5	AISI-SAE 1005	1.5	pure Zn insert	[42]
	AA6061-T6	1.0	DP590	1.6	DP590 bare	[23]
	6063-T6	1.5	16Mn	1.0	16Mn uncoated	[43]
	Al6K32	1.0; 1.6	SGARC440	1.0; 1.4	SGARC uncoated; PT3000 (CrNi) process tape on aluminum side and PT1407 (steel) process tape on steel side	[44]
	Al6K32	1.0; 1.6	SGARC440	1.0; 1.4	SGARC Zn-coated; PT3000 (CrNi) process tape on aluminum side and PT1407 (steel) process tape on steel side	[44]
MBaRSW	AA6061-T6	1.0	DP590	1.6	DP590 bare; ER4043 printed bump	[23]

Table 1. *Cont.*

Joining Process	Aluminum Part		Steel Part		Additional Information	Source
	Material	Thickness (mm)	Material	Thickness (mm)		
MRSW	AA1100	0.4	SS301	0.2	circular low-carbon steel (LCS) interlayer (0.2 mm)	[26]
	AA1100	0.2	SS301	0.4	N/A	[45]
REW/RRW	Al5052-H32	1.0; 2.0; 3.0	DP780	1.2	DP780 Zn-coated; 20MnB4 solid steel element	[46]
	Al5182-O	1.0	22MnB5	1.2	22MnB5 Al-Si-coated; 20MnB4 solid steel element	[46]
	AW5754-H22	1.0	DP500	1.5	Q235 solid steel rivet on aluminum side	[31]
	EN AW-6016	1.2	22MnB5	1.5	20MnB4 solid steel rivet with numerically optimized geometry	[47]
	EN AW-6016	1.2	22MnB5	1.5	22MnB5 Al-Si-coated (AS150); 20MnB4 solid steel rivet with numerically optimized geometry and Zn-Ni-coated	[47]
	EN AW-6016-T4	1.2	22MnB5 and 22MnB5 *	1.5 and 1.0	both 22MnB5 Al-Si-coated (AS150); 20MnB4 solid steel rivet with numerically optimized geometry and Zn-Ni-coated	[47]
	EN AW-6016-T66	1.2	22MnB5 and 22MnB5 *	1.5 and 1.0	both 22MnB5 Al-Si-coated (AS150); 20MnB4 solid steel rivet with numerically optimized geometry and Zn-Ni-coated	[47]
	AA6061-T6	1.0	HS1300T	1.55	HS1300T Al-Si-coated; SWRCH16A solid steel rivet on aluminum side	[21]
	AA6061-T6	1.0	DP780 and press-hardened steel (PHS) *	1.2 and 1.55	DP780 Zn-coated and PHS Al-Si-coated; 35CrMo semi-tubular steel rivet on aluminum side	[22]
(HP)USW	6061-T6	1.5	AISI 304	1.5	AISI 304 uncoated	[48]
	6061-T6	1.5	ASTM A36	1.5	ASTM A36 uncoated	[48]
	Al-6011	0.93	DC04	0.97	DC04 uncoated	[24]
	Al-6011	0.93	DX53-ZF	0.97	DX53-ZF hard galvannealed Zn coating	[24]
	Al-6011	0.93	DX56-Z	0.75	DX56-Z soft hot-dipped Zn coating	[24]
	AA7075-T6	2.0	HSLA steel	1.2	AA7075-T6 clad with AA7072; HSLA steel hot-dip galvanized	[49]
USW + RSW	A6061-T6	1.0	AISI 1008	0.9	A6061-T6 insert (0.4 mm)	[25]
VFAW	5A06	1.8	SS321	4.0	3003 interlayer (1.02 mm)	[27]
	AA6111	2.5	HSLA 340	2.5	HSLA steel bare	[50]

* Three-sheet aluminum/steel dissimilar joint. N/A: not applicable.

Table 2. Clinching-based joining technologies for aluminum/steel hybrid joints, including their main characteristics.

Joining Process	Aluminum Part		Steel Part		Additional Information	Source
	Material	Thickness (mm)	Material	Thickness (mm)		
Conventional clinching (CCL)	1420	1.5	Q215	1.5	galvanized steel	[51]
	AA3004	N/A	SAE 1006	N/A	N/A	[52]
	AA3004	N/A	AISI 304	N/A	N/A	[52]
	AA5052	1.0	ASTM A36	1.0	N/A	[53]
	Al5052	2.0	DP780	1.6	N/A	[54]
	AL5052	1.5	ARC05 (EN 10130)	1.5	N/A	[55]
	AA5182-O	0.85	DX51D+Z	1.2	N/A	[56]
	AA6011-T4	1.0	SAE1004	0.7	pre-strained aluminum, galvanized steel	[57]
	EN AW 6014	1.0	HCT-590X+Z	1.5	N/A	[58]
	EN AW 6014	1.0	HCT-590X+Z	1.5	galvanized steel	[59]
	6061	2.0	304	1.5	N/A	[60]
	aluminum	1.5	mild steel	1.5	N/A	[61]
Single-stage shear-clinching	AA6016-T4	2.0	22MnB5	1.5	aluminum-silicon-coated steel	[62]
Multi-stage clinching with pre-hole	AA6016-T4	2.0	22MnB5	1.5	aluminum-silicon-coated steel	[62]
Conventional clinching with auxiliary layer	AL5052	2.0	HC340LA	2.0	AL1060 auxiliary layer (1.5 mm)	[63]
	AL6061	2.0	HC340LA	2.0	AL1060 auxiliary layer (1.5 mm)	[63]
Mechanical clinching and adhesive bonding	A5052-H34	1.5	JSC780	1.2	CEMEDINE EP138 adhesive	[64]
Hybrid clinching–welding	AA5754	1.5	DQSK	0.8	zinc-coated steel	[65]
Friction stir hole clinching (FSHC)	Al6061	N/A	DP980	N/A	N/A	[66]
Electric-assisted mechanical clinching (EAMC)	AA6061-T6	1.0	DP590	1.5	galvanized steel	[67]

N/A: not applicable.

On the basis of the above, the general objective of our research is the continuation of previous work [64], involving the comparison of aluminum/steel joining technologies and optimizing the technological parameters of joining with respect to different criteria (medium term). The direct aim of the research work and this paper provide newer information about the behavior of aluminum/steel RSW and CCL joints under HCF loading conditions (short term). The novelty content of the manuscript is twofold: On the one hand, a relationship was searched for and found between the magnitude of the high-cycle fatigue loading of joints and the failure modes. On the other hand, the load-bearing capacity of the two joining technologies under HCF loading was compared. The comparison provides a practical opportunity to select the joining technology based on the cyclic load range.

2. Materials and Methods

2.1. Investigated Materials

DP600 (SSAB EMEA AB, Borlange, Sweden) steel was chosen as the steel part, and 6082-T6 and 5754-H22 alloys (both from Comhan Aluminium GMBH, Hagen, (NRW), Germany) were chosen as the aluminum part; these materials are frequently used for automotive applications. The DP600 steel has relatively low strength compared to other dual phase (DP) steels and contains hard martensite islands embedded in a ferrite matrix with a dispersed distribution. The 6082-T6 aluminum alloy is heat-treatable and has a higher strength and lower formability. The material contains mainly Si and Mg alloying elements, causing ageing. Furthermore, it is more sensitive for welding, especially for softening in the heat affected zone (HAZ), and hot cracking can easily occur too. The 5754-H22 aluminum alloy has good formability and appropriate associated strength properties. Its main alloying element is Mg, and its strength can be increased by forming, and then softened to a quarter of the hardness. The weldability of the alloy is favorable compared to other aluminum alloys.

For the investigations, both similar (namely, steel/steel and aluminum alloy/aluminum alloy) and hybrid (such as aluminum alloy/steel) RSW and CCL joints were produced. For a better comparison, the base materials have a 1 mm thickness. Tables 3 and 4 show the chemical composition of the steel and aluminum base materials, respectively, and Table 5 summarizes their essential mechanical properties (yield strength ($R_{p0.2}$), tensile strength (R_m), and elongation (A_{50})). The data in all three tables are taken from quality certificates from the base material manufacturers.

Table 3. Chemical composition of the investigated steel, according to the quality certificate of the manufacturer, weight %.

Material Grade	C	Si	Mn	P	S	Nb	V	B
DP600	0.098	0.2	0.81	0.015	0.002	0.014	0.01	0.0002

Table 4. Chemical composition of the investigated aluminum alloys, according to the quality certificates of the manufacturers, weight %.

Material Grade	Cu	Fe	Mn	Cr	Mg	Ti	Si	Zn
6082-T6	0.09	0.46	0.46	0.02	0.7	0.03	0.9	0.08
5754-H22	0.055	0.294	0.358	0.009	2.796	0.016	0.193	0.034

Table 5. Base mechanical properties of the investigated materials, according to the quality certificates of the manufacturers.

Material Grade	$R_{p0.2}$ (MPa)	R_m (MPa)	$R_{p0.2}/R_m$ (–)	A_{50} (%)
DP600	448	669	0.670	18.7
6082-T6	303	348	0.871	15.0
5754-H22	137	220	0.623	22.0

2.2. Resistance Spot Welding (RSW)

A TECNA 8007-type resistance spot welding machine (TECNA S. p. A., Bologna, Italy) was used with a TE550 control module to form the RSW joints. The machine operates with AC, and the frequency of the transformer was 50 Hz. A pneumatic system was used to ensure the welding force. The process was controlled for a constant current, meaning that the welding current remained the same throughout the process, while the voltage varied. The same welding electrode was applied for all material combinations; the electrode material was CuCrZr (AVIVA Metals, La Seyne sur Mer, France) (RWMA (Resistance Welding Manufacturing Alliance), class 2, where Cr = 0.5–1.5%, Zr = 0.02–0.2%, Cu = balance), and furthermore, Figure 1 shows the electrode geometry (adapted from [68]).

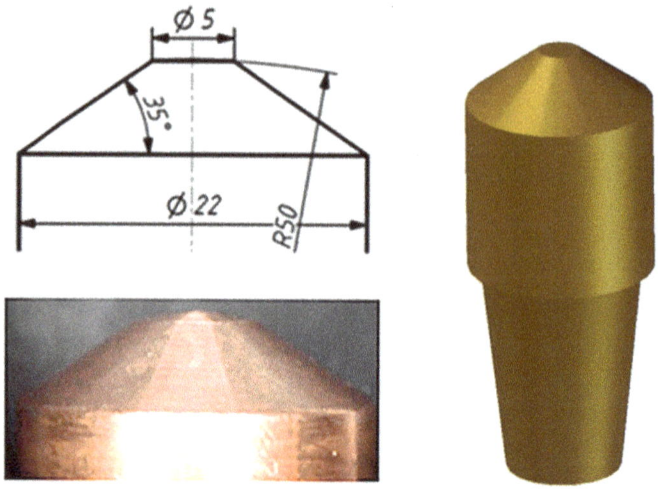

Figure 1. The geometry of the resistance spot welding (RSW) electrode.

The electrode pin diameter was 5 mm (which is recommended for 1 mm thick base materials), but it was not flat, a radius of R = 50 mm was applied. There was no difference between the upper and lower electrodes. Different welding parameters were used for the different base material combinations; the parameter optimization was based on previous experiments [69–71] and possible practice (e.g., [72]). The optimization was aimed at achieving the maximum tensile-shear force by varying the welding current and the welding time. Table 6 shows the selected welding parameters for all material combinations.

Table 6. Welding parameters for the resistance spot welding (RSW) experiments.

Material Combination	Welding Current (kA)	Welding Time (ms)	Welding Force (kN)
DP600/DP600	8.5	320	4.0
6082-T6/6082-T6	23.0	100	2.5
5754-H22/5754-H22	24.0	100	2.5
DP600/6082-T6	15.0	220	2.5
DP600/5754-H22	16.5	220	2.5

The DP600/DP600 base material combination was welded with the lowest welding current and the longest welding time (compared with the others). The aluminum/aluminum RSW joints required the highest welding current and the shortest welding time. In the case of the aluminum/steel RSW joints, almost twice the welding current was applied compared with the steel/steel combination, but the welding time was shorter, because these parameters resulted in a thin intermetallic compound (IMC) layer with good tensile-shear strength. The reason for the large difference in parameters comes from the significant differences in the electrical resistance and thermal conductivity of the steel and the aluminum alloy. Based on our own investigations, Figure 2 shows the differences in the IMC layers in the case of the aluminum/steel RSW joints, illustrated by an optical microscope.

The shape and the geometrical dimensions of the RSW specimens can be seen in Figure 3 (adapted from [68]).

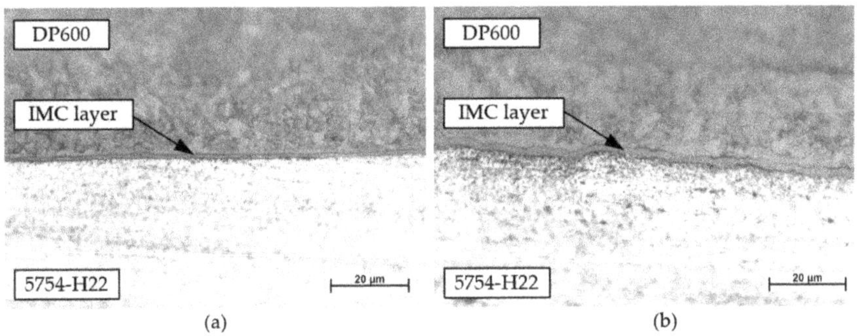

Figure 2. The result of welding parameter optimization in the case of the aluminum/steel joints: (**a**) the thinnest IMC layer (1.1 µm average) with 16.5 kA, 220 ms, and 2.5 kN; (**b**) the thickest IMC layer (2.5 µm average) with 16 kA, 400 ms, and 2.5 kN.

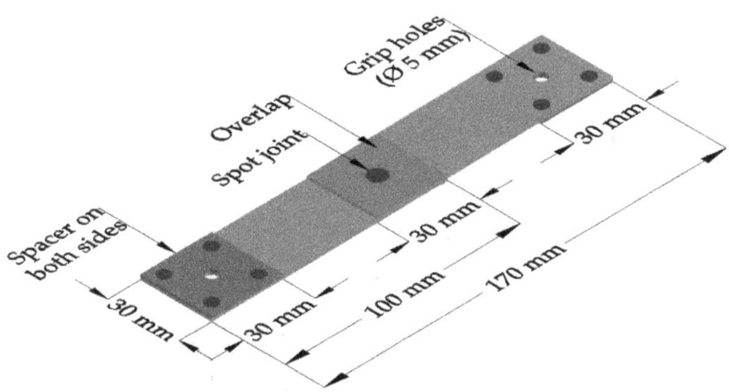

Figure 3. The shape and the geometry of the resistance-spot-welded (RSW) specimens.

2.3. Conventional Clinching (CCL)

The clinching method is one of the most versatile mechanical joining methods, connecting two or more sheets through a one-step process of local plastic deformation. The main elements of the technology are the die (body and insert), the punch, and the blank holder, as shown in Figure 4a. During the conventional clinching process, the sheets are placed on the die and the blank holder moves down, pushing them. After that, the punch is moved downwards to stamp the sheets. The materials can flow in a downward direction, fill the gaps on the die, and form a mechanical interlock between the sheets. The geometrical parameters, such as the neck thickness (t_N), the bottom thickness (t_B), and the undercut (t_C), are the most significant parameters affecting the strength of the clinched joints (see Figure 4b).

The clinched joints were made on an MTS electrohydraulic, computer-controlled, universal materials testing system (MTS Systems, Eden Prairie, MN, USA) with a 250 kN maximum compressive loading capacity. The equipment, together with a TOX tool (TOX® PRESSOTECHNIK GmbH & Co.KG, Weingarten, Germany) mounted on the load frame, is shown in Figure 5.

The shapes of the specimens and the tool, along with the geometrical dimensions, can be seen in Figure 6. The bottom thickness (see Figure 4b) of the investigated clinched joints was t_B = 0.5 mm. The aim of the selection of the (optimal) bottom thickness was to achieve the maximum tensile-shear force, taking into account the favorable clinching force [73,74].

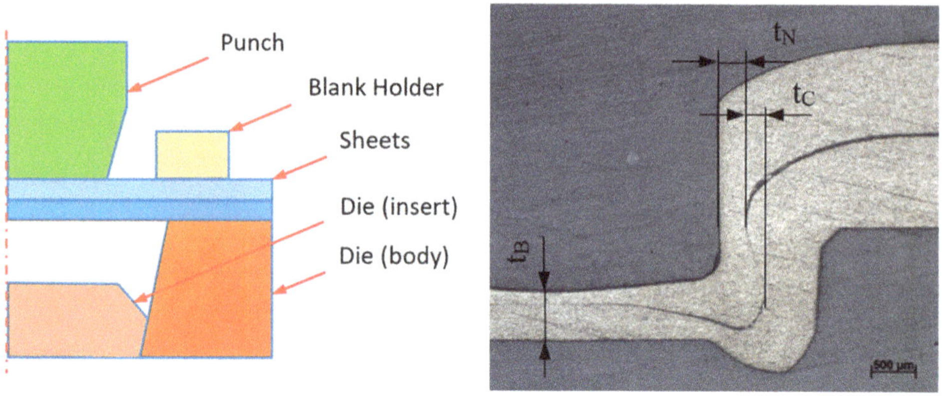

(**a**) (**b**)

Figure 4. Characteristics of the conventional clinching technology (adapted from [73]): (**a**) the main elements of conventional clinching technology, and (**b**) the structure of a clinched joint.

Figure 5. MTS electrohydraulic materials testing system with the clinching tool and the main elements of the tool.

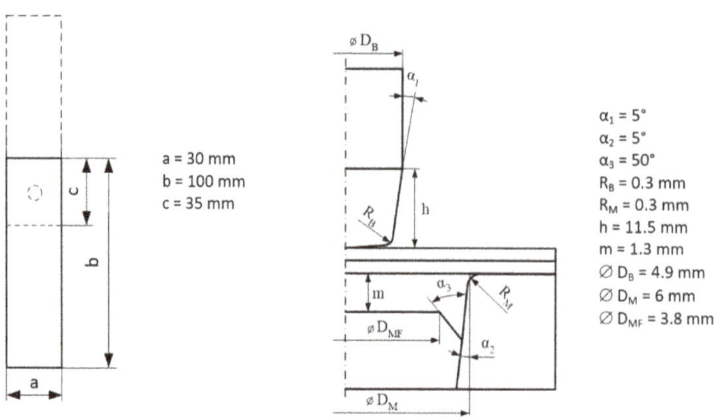

Figure 6. The shape of the specimens and the clinching tool and their geometrical dimensions.

2.4. High-Cycle Fatigue (HCF) Tests

The shape and the geometry as well as the major dimensions of the test specimens of all the base material combinations for the RSW and CCL joints can be seen in Figures 3 and 6, respectively. The two parts of the specimens were cut from 1000 mm × 2000 mm sheets into 100 mm × 30 mm strips with laser cutting. For the RSW specimens an overlap of 30 mm was used and for the CCL specimens an overlap of 35 mm. The investigated joints of both technologies can be found at the geometrical center of the overlapped area. Figures 7 and 8 show the RSW and CCL joints, respectively, for all of the material combinations.

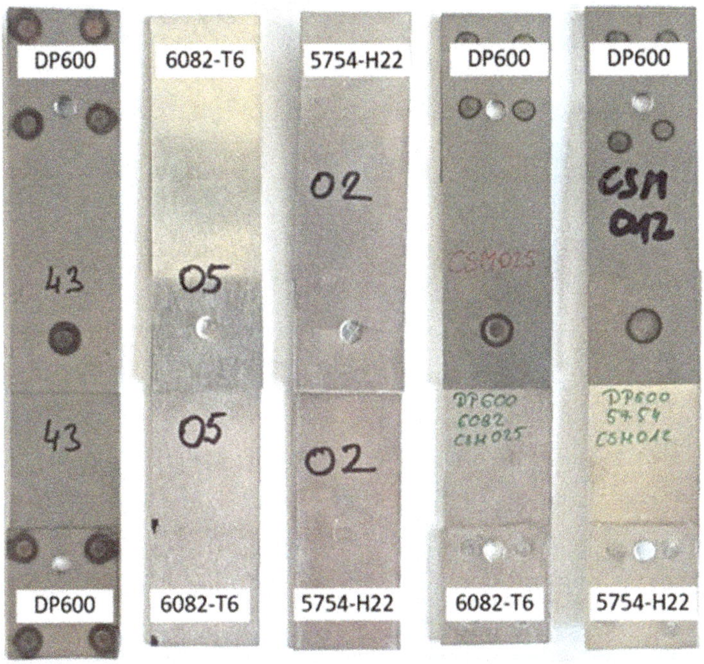

Figure 7. An example of the RSW joints/investigated high-cycle fatigue specimens for all material combinations.

The HCF tests were executed based on the instructions of the relevant standard [75]. The tests were performed using MTS electrohydraulic, universal materials testing equipment with an MTS FlexTest 40 controller (MTS Systems, Eden Prairie, MN, USA). The 5 mm overlapping difference has no effect on the feasibility of the investigations and the comparability of the test results.

A sinusoidal loading wave form was applied, and during the entire test phase the load ratio (in our case F_{min}/F_{max}) was R = 0.1, with a frequency of f = 30 Hz. Several load levels were selected and used during testing of the RSW and CCL joints. Considering that the investigations were evaluated according to [76,77], the load range levels were chosen as described in these sources. All tests were performed at room temperature and in a laboratory environment (controlled air and humidity). Compliance with the testing conditions detailed above ensured that they did not significantly affect the reliability of the results.

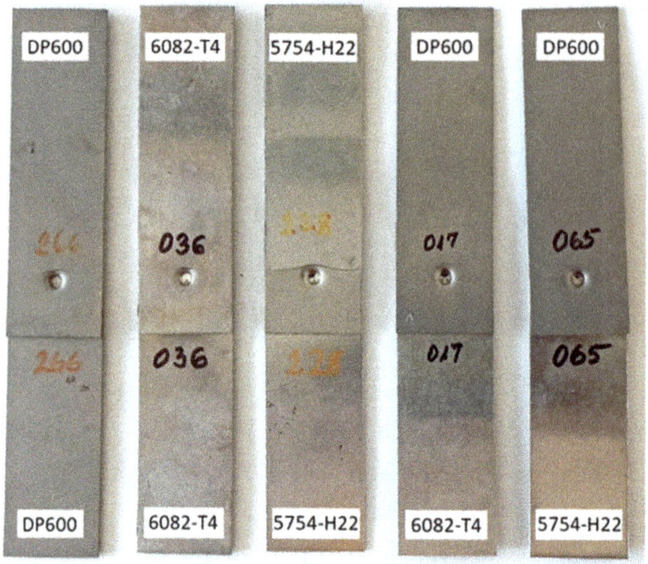

Figure 8. An example of the CCL joints/investigated high-cycle fatigue specimens for all the material combinations.

3. Results and Discussion

3.1. Failure Modes

After performing the high-cycle fatigue tests, the mode of failure was determined by visual inspection (VT) of each specimen, for both technologies, one by one.

For the RSW technology, among the variations presented for static and cyclic failures ([46,78,79] and [79–81], respectively), the following were relevant in our case: pull-out failure; crack initiation in the HAZ and crack growth in the base material; crack initiation in the HAZ and interfacial failure; and interfacial failure. Figure 9 shows an example for each case, for all of the material combinations.

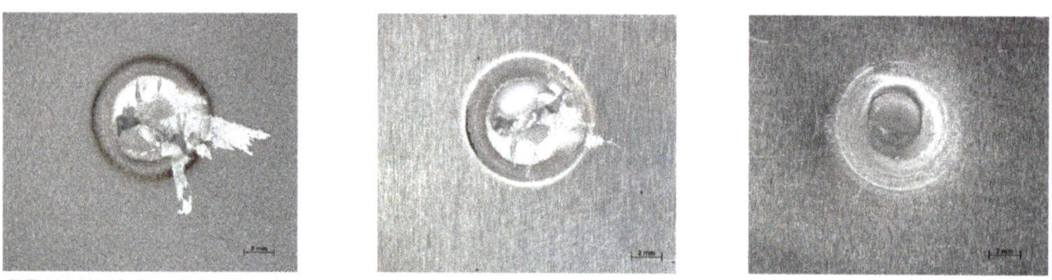

DP600/6082-T6—DP600 side DP600/6082-T6—6082-T6 side DP600/5754-H22—5754-H22 side

(**a**)

Figure 9. *Cont.*

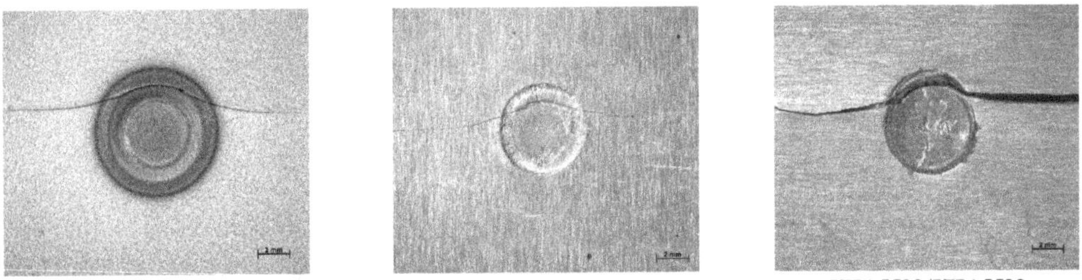

DP600/DP600 6082-T6/6082-T6 5754-H22/5754-H22

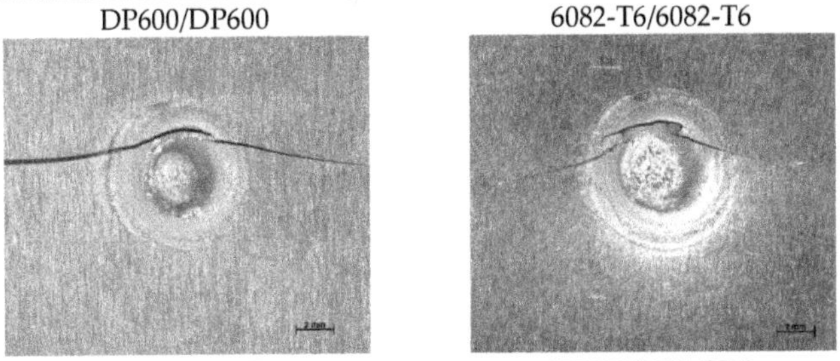

DP600/6082-T6 DP600/5754-H22

(**b**)

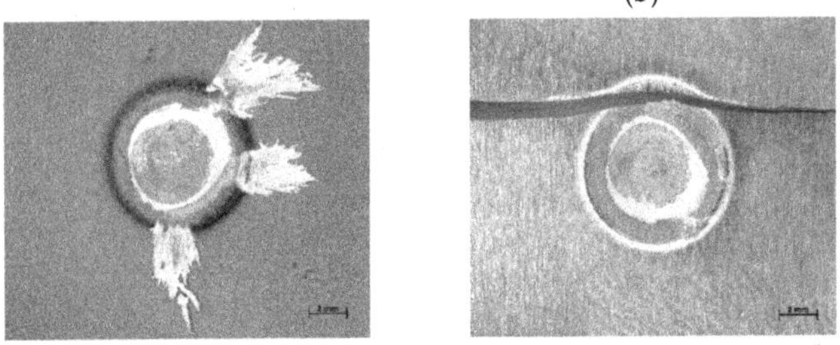

DP600/6082-T6—DP600 side DP600/6082-T6—6082-T6 side

(**c**)

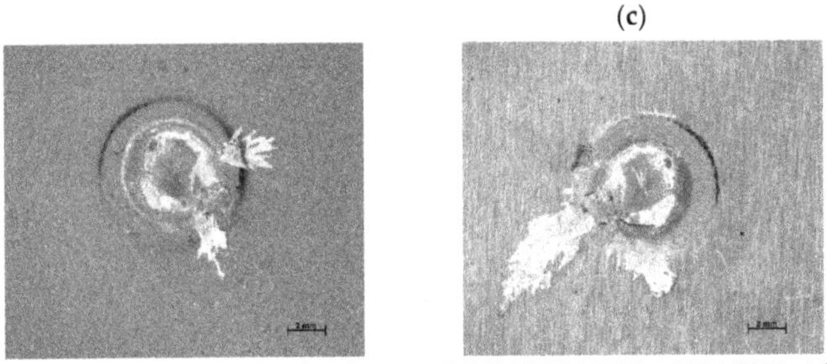

DP600/5754-H22—DP600 side DP600/5754-H22—5754-H22 side

(**d**)

Figure 9. Failure modes during high-cycle fatigue testing of the RSW specimens: (**a**) pull-out failure, (**b**) crack initiation in the HAZ and crack growth in the base material, (**c**) crack initiation in the HAZ and interfacial failure, (**d**) interfacial failure.

For the CCL technology, among the variations presented for static and cyclic failures [82–85] and [86], respectively), the following were relevant in our case: button neck fracture; base material fracture; base material fracture and button neck fracture; and crack initiation in the joint as well as crack growth in the aluminum alloy. Figure 10 shows an example for each case, for all of the material combinations.

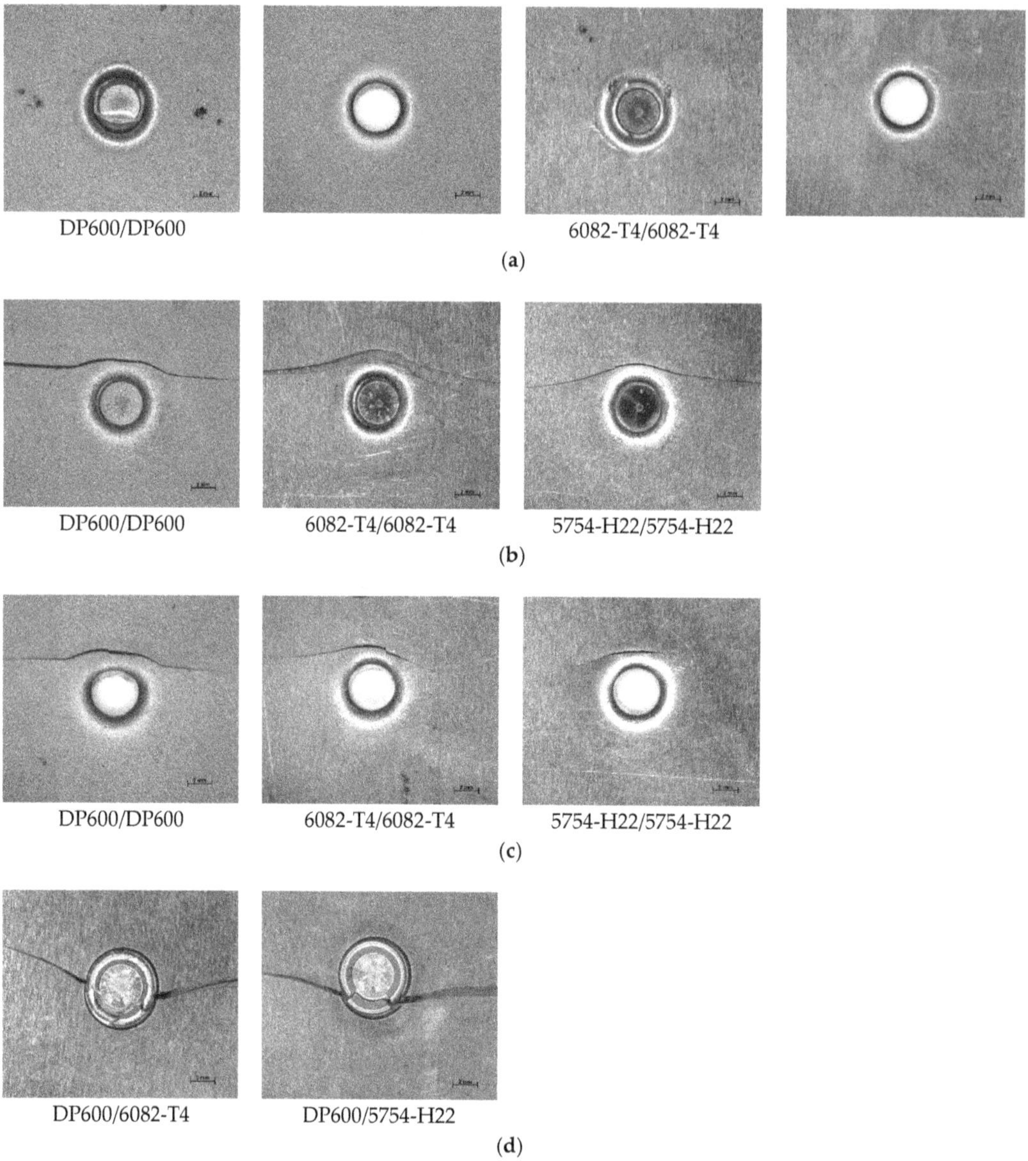

Figure 10. Failure modes during high-cycle fatigue testing of the CCL specimens: (**a**) button neck fracture, (**b**) button neck fracture, (**c**) base material fracture and button neck fracture, (**d**) crack initiation in the joint and crack growth in the aluminum alloy.

3.2. Results of the High-Cycle Fatigue (HCF) Tests

The number-of-cycles-to-failure values were registered at the end of the high-cycle fatigue tests. The results were illustrated by the diagrams plotting load range (ΔL) against the number of cycles to failure (N), applying a logarithmic scale for the number of cycles in the ΔL-N curves.

Straight lines are fitted to the measured ΔL-N points. The straight lines were determined by applying the least squares method (in the lifetime stage) and calculating the mean values (in the endurance limit stage, ΔL_{el}); therefore, these lines are associated with a 50% probability of failure.

Figure 11 shows the ΔL-N curves for the similar and hybrid RSW joints, including the failure modes. The figure clearly demonstrates that the most common failure modes across the entire range were "crack initiation in the HAZ and crack growth in the base material" and "crack initiation in the HAZ and interfacial failure" (Figure 9b,c); the mode "pull-out failure" (Figure 9a) occurred at higher loads and lower cycles, and the mode "interfacial failure" (Figure 9d) occurred at lower loads and higher cycles.

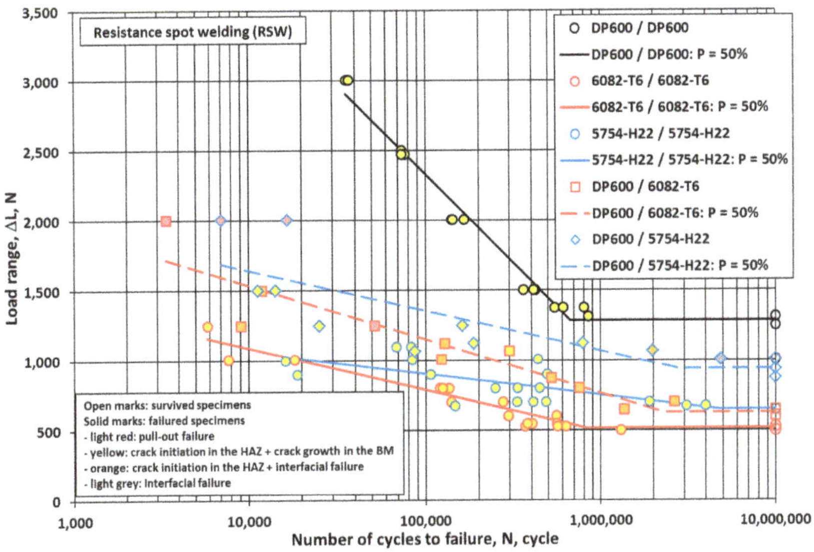

Figure 11. ΔL-N curves for the investigated similar and dissimilar RSW joints including the failure modes.

Figure 12 presents the HCF test results of all the RSW joint combinations, compared with data from the literature from two sources [36,87]. The HCF resistance of the DP600/5754-H22 hybrid joints is significantly better than that of the DP600/6082-T6 hybrid joints, and, furthermore, they are competitive with the results found in the literature. According to data found in the literature [36], the endurance limit is close to that of IF/6022-T4 hybrid and 6022-T4/6022-T4 similar joints, and demonstrates a slight difference compared to our result (DP600/6082-T6, 5754-H22/5754-H22). It should be noted that in source [36] the fatigue limit values were not determined; furthermore, wider and thicker specimens were used. For the DP590/DP590 material combination [36], there were few measuring points, no fatigue limit value was specified, and the tests were carried out on wider and thicker specimens. From this point of view, the difference between the data measured on the DP600 and DP590 materials can be evaluated as large.

Figure 13 shows the ΔL-N curves of similar and hybrid CCL joint combinations, including the failure modes. The figure clearly illustrates that different failure modes occurred for the similar DP600/DP600 and the dissimilar joints, with the typical modes "base material fracture" (Figure 10b) and "button neck fracture" (Figure 10a) occurring for higher loads and lower cycles, and the mode "base material fracture and button neck fracture" (Figure 10c) occurring at lower loads and higher cycles. For the dissimilar joints, all specimens failed in the mode "initiation in the joint and crack growth in the aluminum alloy" (Figure 10d).

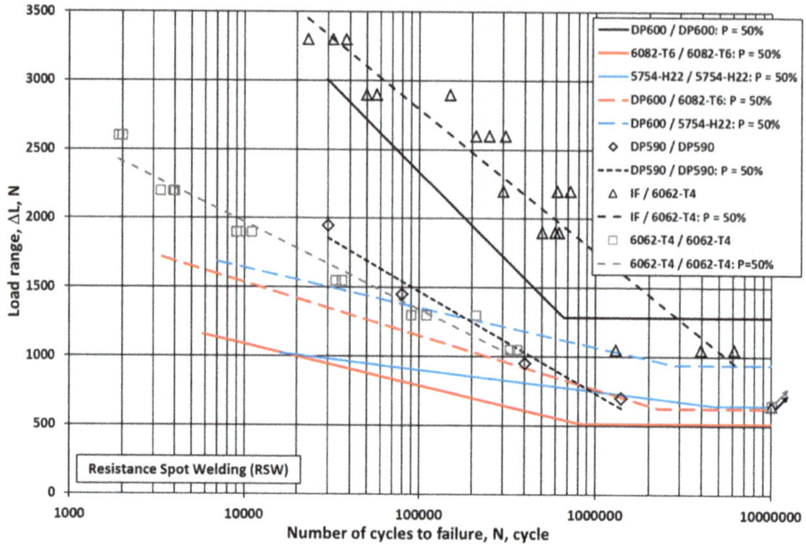

Figure 12. Comparing our results on RSW joints with relevant results that can be found in the literature.

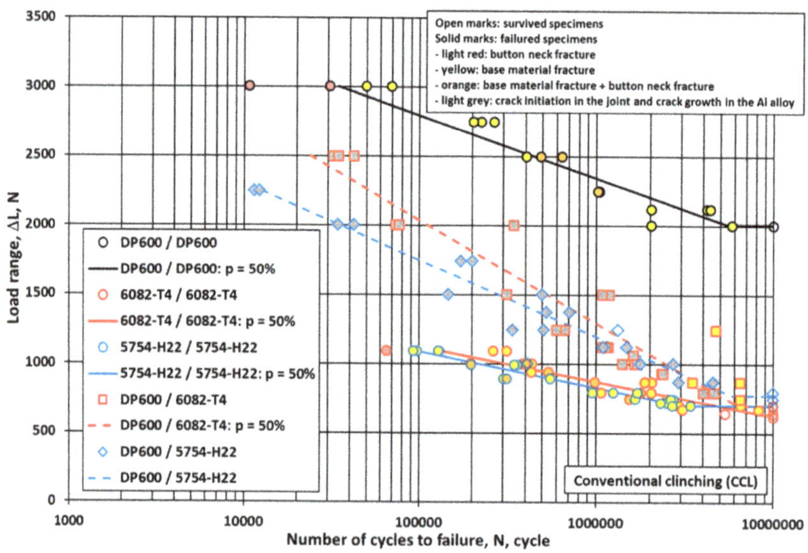

Figure 13. ΔL-N curves for the investigated similar and dissimilar CCL joints including the failure modes.

Figures 14 and 15 show the ΔL-N curves associated with a 50% probability of failure for the similar and dissimilar material combinations. The trend from the figures is that the load-carrying capacity of the conventionally clinched (CCL) joints exceeds that of the resistance-spot-welded (RSW) joints.

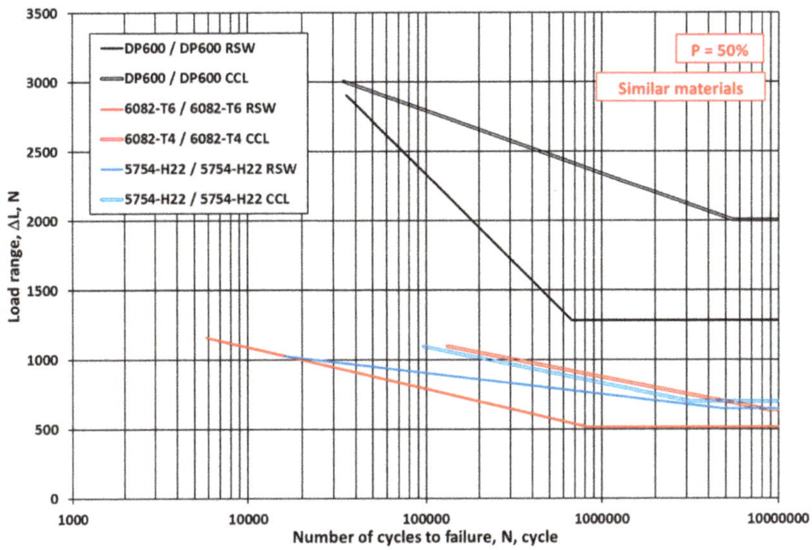

Figure 14. Comparison of the RSW and CCL results on similar joints.

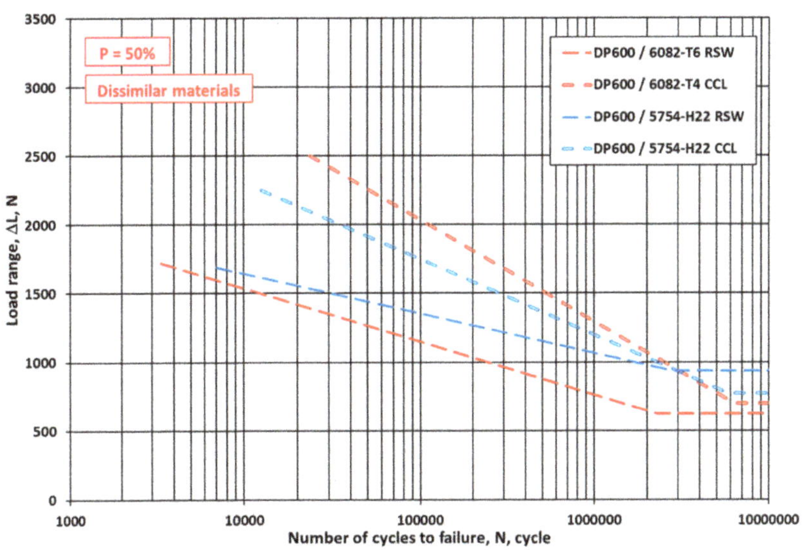

Figure 15. Comparison of the RSW and CCL results on dissimilar joints.

Table 7 summarizes the characteristics of the ΔL-N curves for 50% probability for both joint methods and all material combinations, using the Basquin-type [88] equation in the lifetime section, as follows:

$$\Delta L = A \times \ln(N) + B \quad (1)$$

Among the RSW and the CCL similar joints, the DP600/DP600 combination had the best result, with endurance limit values of 1282.5 N and 2007 N, respectively. Furthermore, the 6082-T6/6082-T6 and the 6082-T4/6082-T4 combinations resulted in the worst values, of 513 N and 636.75 N, respectively. In all cases, results for the dissimilar joints were better than the corresponding results for the similar aluminum joints. The data found in the literature are in accordance with our test results, with the DP600/5754-H22 joints showing a better high-cycle fatigue limit than the DP500/5754-H22 joints [31], where the DP500/5754-H22 joints were performed using an REW process on thicker steel plates with the application of a steel rivet on the aluminum side.

Table 7. Characteristics of the ΔL-N curves for 50% probability based on the Basquin-type equation.

Material Combination	A	B	Correlation Coefficient	ΔL_{el} (N)	Source
			Resistance-spot-welded joints		
DP600/DP600	−552.9	8699	0.985	1285.5	Our previous study [68]
6082-T6/6082-T6	−133.0	2313	0.963	513	Our previous study [68]
5754-H22/5754-H22	−66.4	1668	0.668	648	Our previous study [68]
DP600/6082-T6	−167.9	3083	0.929	625.5	Our previous study [68]
DP600/5754-H22	−125.1	2795	0.770	939	Our previous study [68]
DP590/DP590	−320.4	5160	0.984	N/A	[87]
6022-T4/6022-T4	−274.2	4498	0.981	N/A	[63]
IF/6022-T4	−449.2	7966	0.952	N/A	[63]
DP500/5754-H22	−19.26	2167	N/A	882	[59]
			Conventionally clinched joints		
DP600/DP600	−196.7	5060	0.954	2007	This study
6082-T4/6082-T4	−109.8	2392	0.928	636.75	This study
5754-H22/5754-H22	−113.4	2399	0.974	702	This study
DP600/6082-T4	−322.0	5746	0.945	702	This study
DP600/5754-H22	−240.0	4516	0.975	776.25	This study

N/A: not available.

Two ratios have been defined for the numerical comparison of the endurance limit values of joints, one within and one between the joining technologies, as follows:

$$\Delta L_{el\ ratio(1)} = (\Delta L_{el,i} / \Delta L_{el,DP600/DP600}) \times 100 \quad (2)$$

and

$$\Delta L_{el\ ratio(2)} = (\Delta L_{el\ CCL,i} / \Delta L_{el\ RSW,i}) \times 100 \quad (3)$$

where "i" represents the material combinations. The calculated values of the two ratios are given in Table 8.

Table 8. Numerical comparison of the endurance limit values for the RSW and CCL joints.

Material Combination	ΔL_{el} (N)	$\Delta L_{el\ ratio(1)}$ (%)	$\Delta L_{el\ ratio(2)}$ (%)
	Resistance-spot-welded joints		
DP600/DP600	1285.5	100	N/A
6082-T6/6082-T6	513	39.9	N/A
5754-H22/5754-H22	648	50.4	N/A
DP600/6082-T6	625.5	48.7	N/A
DP600/5754-H22	939	73.0	N/A
DP500/5754-H22 [59]	882	68.6	N/A
	Conventionally clinched joints		
DP600/DP600	2007	100	156.1
6082-T4/6082-T4	636.75	31.7	124.1
5754-H22/5754-H22	702	35.0	108.3
DP600/6082-T4	702	35.0	112.2
DP600/5754-H22	776.25	38.7	82.7

N/A: not applicable.

Considering the steel/steel joints as a base, the load-bearing capacity of the hybrid joints is lower (48.7% and 73.0% for RSW, 35.0% and 38.7% for CCL) and even lower for aluminum/aluminum joints (39.9% and 50.4% for RSW, 31.7% and 35.0% for CCL). The results of our tests are in harmony with the literature data (DP500/5754-H22). With one exception (DP600/5754-H22), the load-bearing capacity of the CCL joints is higher than that of the RSW joints (156.1–108.3%).

3.3. Cost Comparison

Besides the HCF resistance, an overview of the aspects affecting costs can help find the applicable process for joining. Table 9 shows a comparison based on different aspects of the RSW and CCL technologies. The comparisons presented in the table are qualitative and relate only to the two technologies; the qualifying indicators need to be interpreted in relation to the two technologies. Quantitative comparisons require knowledge of global and local data (e.g., product characteristics, energy, wages, and overheads) that are beyond the scope and purpose of this article.

Table 9. Comparison of the RSW and CCL technologies based on different aspects.

Aspects	RSW	CCL
Workpiece preparation	More sensitivity, *more expensive*	Less sensitivity, *cheaper*
Electrical network	Serious, *more expensive*	Regular, *cheaper*
Manufacturing equipment	Complex machine, *more expensive*	Simple machine, *cheaper*
Manpower	Well-educated operator and technologist required, *more expensive*	Basic qualifications for operator and technologist acceptable, *cheaper*
Tool refurbishment possibility	Possible several times, *characteristically cheaper*	Not possible, replacement necessary, *characteristically more expensive*
Tool lifetime	Electrode tip can degrade, especially in the case of aluminum welding [89,90], *more expensive*	Higher than an RSW electrode [91], especially in the case of aluminum, *cheaper*
Maintenance demand of the manufacturing process	Complex machine, several parts can fail, frequent maintenance required, *more expensive*	Simple machine, less parts can fail, higher reliability of equipment, *cheaper*
Energy consumption	Higher, *more expensive*	Lower, *cheaper*

In addition to the aspects presented in Table 9, other criteria may also be important. The loading of joints and the quality requirements basically determine the applicable process. If the joint properties of clinched joints meet the quality requirements, CCL can be cheaper than RSW.

4. Conclusions

Based on the investigations performed and evaluated, and with further regard to their results, the following conclusions can be drawn:

- The investigated joining technologies issued comparable high-cycle fatigue (HCF) test results for the similar and hybrid resistance-spot-welded (RSW) and conventionally clinched (CCL) joints made from DP600, 5754-H22, and 6082-T6 base materials, both within and between the two technologies.
- For both the joining technologies used, the typical failure modes could be identified in the high-cycle fatigue investigations. In the case of RSW, the most common failure

modes across the entire range were "crack initiation in the HAZ and crack growth in the base material" and "crack initiation in the HAZ and interfacial failure"; furthermore, the mode "pull-out failure" occurred at higher loads and lower cycles, and the mode "interfacial failure" occurred at lower loads and higher cycles. In the case of CCL, different failure modes occurred for the similar DP600/DP600 and the dissimilar joints, with the typical modes "base material fracture" and "button neck fracture" occurring at higher loads and lower cycles, and the mode "base material fracture and button neck fracture" occurring at lower loads and higher cycles. For the dissimilar joints, all specimens failed in the mode "initiation in the joint and crack growth in the aluminum alloy".

- Based on a Basquin-type equation, applying the least squares method in the lifetime stage and further calculating the mean values in the endurance limit stage, high-cycle fatigue limit curves (ΔL-N) can be determined corresponding to a 50% probability of failure. In the case of both the similar and the hybrid joints, the fatigue limit curves demonstrated that the load-bearing capacity of the CCL joints exceeds that of the RSW joints. Compared to the steel/steel joints, the load-bearing capacity of the hybrid joints is lower (48.7% and 73.0% for RSW, 35.0% and 38.7% for CCL), and it is even lower for the aluminum/aluminum joints (39.9% and 50.4% for RSW, 31.7% and 35.0% for CCL). The load-bearing capacity of the CCL joints is higher than that of the RSW joints (DP600/DP600: 156.1%, 6082-T4/6082-T4: 124.1%, 5754-H22/5754-H22: 108.3%, and DP600/6082-T4: 112.2%) except for in the case of one joint (DP600/5754-H22: 82.7%).
- Since there is a difference between the two technologies in terms of the load-bearing capacity of the joints under high-cycle fatigue loading conditions, their application can be chosen depending on the real loading of a specified structural element.
- The investigations that have been started should be systematically continued. In order to compare the behavior of hybrid joints, further investigations are needed: joints made with other technologies (e.g., friction stir welding) should be performed; a fatigue crack growth (FCG) test should be prepared, executed, and evaluated on similar and hybrid joints made using different joining technologies; and the two different technologies, as well as the HCF and FCG behaviors, should be compared to specify the optimal conditions and areas for application. Where it is possible, fracture mechanical tests and assessment methods should be prioritized.

Author Contributions: Conceptualization, Á.M. and J.L.; methodology, Á.M., J.L. and P.Z.K.; investigation, J.L.; writing—original draft preparation, J.L.; writing—review and editing, Á.M. and P.Z.K.; visualization, J.L., Á.M. and P.Z.K.; supervision, J.L. and Á.M.; funding acquisition, J.L. All authors have read and agreed to the published version of the manuscript.

Funding: This research work was supported by the European Union and the Hungarian State and co-financed by European Structural and Investment Funds within the framework of the GINOP-2.3.4-15-2016-00004 project, with the aim of promoting cooperation between higher education and industry.

Data Availability Statement: The original contributions presented in the study are included in the article, further inquiries can be directed to the corresponding author.

Acknowledgments: Special thanks to our colleague László Szentpéteri at the Institute of Materials Science and Technology, within the Faculty of Mechanical Engineering and Informatics at the University of Miskolc, for the execution of the long-term high-cycle fatigue tests.

Conflicts of Interest: The authors declare no conflicts of interest.

References

1. Meschut, G.; Hahn, O.; Janzen, V.; Olferman, T. Innovative joining technologies for multi-material structures. *Weld World* **2014**, *58*, 65–75. [CrossRef]
2. Buffa, G.; Fratini, L.; La Commare, U.; Römisch, D.; Wiesenmayer, S.; Wituschek, S.; Merklein, M. Joining by forming technologies: Current solutions and future trends. *Int. J. Mater. Form.* **2022**, *15*, 27. [CrossRef]
3. Merklein, M.; Jäckisch, M.; Kuball, C.M.; Römisch, D.; Wiesenmayer, S.; Wituschek, S. Mechanical joining of high-strength multi-material systems—Trends and innovations. *Mech. Ind.* **2023**, *24*, 16. [CrossRef]

4. Hu, A.; Janczak-Rusch, J.; Sano, T. Joining Technology Innovations at the Macro, Micro, and Nano Levels. *Appl. Sci.* **2019**, *9*, 3568. [CrossRef]
5. Böllinghaus, T.; Byrne, G.; Cherpakov, B.I.; Chlebus, E.; Cross, C.E.; Denkena, B.; Dilthey, U.; Hatsuzawa, T.; Herfurth, K.; Herold, H.; et al. Manufacturing Engineering. In *Springer Handbook of Mechanical Engineering*, 1st ed.; Grote, K.-H., Antonsson, E.K., Eds.; Springer: Berlin, Germany, 2008; pp. 523–785.
6. Martinsen, K.; Hu, S.J.; Carlson, B.E. Joining of dissimilar materials. *CIRP Ann.* **2015**, *64*, 679–699. [CrossRef]
7. Meschut, G.; Merklein, M.; Brosius, A.; Drummer, D.; Fratini, L.; Füssel, U.; Gude, M.; Homberg, W.; Martins, P.A.F.; Bobbert, M.; et al. Review on mechanical joining by plastic deformation. *J. Adv. Join. Process* **2022**, *5*, 100113. [CrossRef]
8. He, X. Clinching for sheet materials. *Sci. Technol. Adv. Mater.* **2017**, *18*, 381–405. [CrossRef]
9. Gullino, A.; Matteis, P.; D'Aiuto, F. Review of aluminum-to-steel welding technologies for car-body applications. *Metals* **2019**, *9*, 315. [CrossRef]
10. EAA Aluminium Automotive Manual—Joining. 11. Joining Dissimilar Materials. European Aluminium Association. 2015. Available online: https://european-aluminium.eu/wp-content/uploads/2022/11/11-joining-dissimilar-materials_2015.pdf (accessed on 10 October 2024).
11. Spisák, E.; Kascák, L. Joining car body steel sheets using the clinching method. *Acta Mech. Slovaca* **2011**, *15*, 28–35. [CrossRef]
12. Mucha, J. Clinching technology in the automotive industry. *Arch. Automot. Eng.* **2017**, *76*, 75–94. [CrossRef]
13. Kascák, L.; Spisák, E.; Majerníková, J. Joining three car body steel sheets by clinching method. *Open Eng.* **2016**, *6*, 566–573. [CrossRef]
14. Yang, Y.; Luo, Z.; Zhang, Y.; Su, J. Dissimilar welding of aluminium to steel: A review. *J. Manuf. Process* **2024**, *110*, 376–397. [CrossRef]
15. Mehta, K.P. A review on friction-based joining of dissimilar aluminium-steel joints. *J. Mater. Res.* **2019**, *34*, 78–96. [CrossRef]
16. Zvorykina, A.; Sherepenko, O.; Jüttner, S. Novel projection welding technology for joining of steel-aluminum hybrid components–part 1: Technology and its potential for industrial use. *Weld World* **2020**, *64*, 317–326. [CrossRef]
17. Zhang, C.; Li, H.; Liu, Q.; Huang, C.; Zhou, K. Ultrasonic welding of aluminum to steel: A review. *Metals* **2023**, *13*, 29. [CrossRef]
18. Haddadi, F. Microstructure reaction control of dissimilar automotive aluminium to galvanized steel sheets ultrasonic spot welding. *Mater. Sci. Eng. A* **2016**, *678*, 72–84. [CrossRef]
19. Zhao, D.; Ren, D.; Zhao, K.; Pan, S.; Guo, X. Effect of welding parameters on tensile strength of ultrasonic spot welded joints of aluminum to steel—By experimentation and artificial neural network. *J. Manuf. Process.* **2017**, *30*, 63–74. [CrossRef]
20. Patel, V.K.; Bhole, S.D.; Chen, D.L. Ultrasonic spot welding of aluminum to high-strength low-alloy steel: Microstructure, tensile and fatigue properties. *Metall. Mater. Trans. A* **2014**, *45*, 2055–2066. [CrossRef]
21. Sizhe, N.; Ming, L.; Yunwu, M.; Yongbing, L. Study on the microstructure and mechanical performance for integrated resistance element welded aluminum alloy/press hardened steel joints. *Mater. Sci. Eng. A* **2021**, *800*, 140329. [CrossRef]
22. Sizhe, N.; Ming, L.; Zixuan, C.; Zelong, W.; Yunwu, M.; Shanqing, H.; Haiyang, L.; Yongbing, L. Resistance rivet welding of multilayer dissimilar materials. *J. Mater. Res. Technol.* **2023**, *27*, 6424–6437. [CrossRef]
23. Guotao, Z.; Hang, Z.; Xianghe, X.; Guohua, Q.; Yongbing, L.; Zhongqin, L. Metallic bump assisted resistance spot welding (MBaRSW) of AA6061-T6 and Bare DP590: Part II-joining mechanism and joint property. *J. Manuf. Process.* **2019**, *44*, 19–27. [CrossRef]
24. Farid, H.; Fadi, A.F. Microstructural and mechanical performance of aluminium to steel high power ultrasonic spot welding. *J. Mater. Process. Technol.* **2015**, *225*, 262–274. [CrossRef]
25. Ying, L.; Ellis, M.; Hyeyun, S.; Menachem, K.; Wei, Z. Dissimilar metal joining of aluminum to steel by ultrasonic plus resistance spot welding—Microstructure and mechanical properties. *Mater. Des.* **2019**, *165*, 107585. [CrossRef]
26. Baskoro, A.S.; Muzakki, H.; Kiswanto, G.; Winarto, W. Effect of interlayer in dissimilar metal of stainless steel SS 301 and aluminum alloy AA 1100 using micro resistance spot welding. *AIP Conf. Proc.* **2018**, *1983*, 040014. [CrossRef]
27. Shan, S.; Shujun, C.; Yu, M.; Jun, X.; Anupam, V.; Glenn, D. Joining Aluminium Alloy 5A06 to Stainless Steel 321 by Vaporizing Foil Actuators Welding with an Interlayer. *Metals* **2019**, *9*, 43. [CrossRef]
28. Spisák, E.; Kascák, L. Mechanical Joining of Steel Sheets in Automotive Industry. *Acta Mech. Slovaca* **2014**, *18*, 6–13. [CrossRef]
29. Ranfeng, Q.; Chihiro, I.; Shinobu, S. Interfacial microstructure and strength of steel/aluminum alloy joints welded by resistance spot welding with cover plate. *J. Mater. Process. Technol.* **2009**, *209*, 4186–4193. [CrossRef]
30. Jianbin, C.; Xinjian, Y.; Zhan, H.; Ting, L.; Kanglong, W.; Ci, L. Improvement of resistance-spot-welded joints for DP 600 steel and A5052 aluminum alloy with Zn slice interlayer. *J. Manuf. Process.* **2017**, *30*, 396–405. [CrossRef]
31. Aleksija, Đ.; Dragan, M.; Zijah, B.; Damjan, K.; Miodrag, M.; Biljana, M.; Vladislav, K. Microstructure and Fatigue Properties of Resistance Element Welded Joints of DP500 Steel and AW 5754 H22 Aluminum Alloy. *Crystals* **2022**, *12*, 258. [CrossRef]
32. Sun, X.; Stephens, E.V.; Khaleel, M.A.; Shao, H.; Kimchi, M. Resistance Spot Welding of Aluminum Alloy to Steel with Transition Material From Process to Performance Part I: Experimental Study. *Weld. J.* **2004**, *2*, 188–195.
33. Weihua, Z.; Daqian, S.; Lijun, H.; Yongqiang, L. Optimised design of electrode morphology for novel dissimilar resistance spot welding of aluminium alloy and galvanised high strength steel. *Mater. Des.* **2015**, *85*, 461–470. [CrossRef]

34. Weihua, Z.; Daqian, S.; Lijun, H.; Dongyang, L. Interfacial microstructure and mechanical property of resistance spot welded joint of high strength steel and aluminium alloy with 4047 AlSi12 interlayer. *Mater. Des.* **2013**, *57*, 186–194. [CrossRef]
35. Rendong, L.; Yanjun, W.; Shuangjian, C.; Dong, L.; Xing, X.; Shanglu, Y. Improving the performance of steel aluminum resistance spot welding joints based on steel coating design. *Mater. Lett.* **2023**, *353*, 135291. [CrossRef]
36. Jidong, K.; Harish, M.R.; David, R.S.; Blair, E.C. Tensile and Fatigue Behaviour of AA6022-T4 to IF Steel Resistance Spot Weld. *Procedia Struct. Integr.* **2017**, *5*, 1425–1432. [CrossRef]
37. Harish, M.R.; Jidong, K.; Liting, S.; David, R.S.; Blair, E.C. Effect of specimen configuration on fatigue properties of dissimilar aluminum to steel resistance spot welds. *Int. J. Fatigue* **2018**, *116*, 13–21. [CrossRef]
38. Nannan, C.; Hui-Ping, W.; Blair, E.C.; David, R.S.; Min, W. Fracture mechanisms of Al/steel resistance spot welds in lap shear test. *J. Mater. Process. Technol.* **2017**, *243*, 347–354. [CrossRef]
39. Zixuan, W.; Hui-Ping, W.; Nannan, C.; Min, W.; Blair, E.C. Characterization of intermetallic compound at the interfaces of Al-steel resistance spot welds. *J. Mater. Process. Technol.* **2017**, *242*, 12–23. [CrossRef]
40. Liting, S.; Jidong, K.; Mark, G.; Xu, C.; Amberlee, S.H.; Blair, E.C. Fatigue life assessment of Al-steel resistance spot welds using the maximum principal strain approach considering material inhomogeneity. *Int. J. Fatigue* **2020**, *140*, 105851. [CrossRef]
41. Liting, S.; Jia, X.; Jidong, K.; Amberlee, S.H.; Hassan, G.-A.; Blair, E.C. Fatigue behavior of three-sheet aluminum-steel dissimilar resistance spot welds for automotive applications. *Procedia Struct. Integr.* **2023**, *51*, 102–108. [CrossRef]
42. Al-Filfily, A.A.; Al-Adili, A.S.; Sar, M.H. Strength of resistance spot welding of aluminum alloy AA6061 to carbon steel using different filler materials. *IOP Conf. Ser. Mater. Sci. Eng.* **2020**, *881*, 012067. [CrossRef]
43. Daqian, S.; Yueying, Z.; Yanjun, L.; Xiaoyan, G.; Hongmei, L. Microstructures and mechanical properties of resistance spot welded joints of 16Mn steel and 6063-T6 aluminum alloy with different electrodes. *Mater. Des.* **2016**, *109*, 596–608. [CrossRef]
44. Seungmin, S.; Dae-Jin, P.; Jiyoung, Y.; Sehun, R. Resistance Spot Welding of Aluminum Alloy and Carbon Steel with Spooling Process Tapes. *Metals* **2019**, *9*, 410. [CrossRef]
45. Baskoro, A.S.; Muzakki, H.; Kiswanto, G. Mechanical Properties and Microstructures on Dissimilar Metal Joints of Stainless Steel 301 and Aluminum Alloy 1100 by Micro-Resistance Spot Welding. *Trans. Indian Inst. Met.* **2019**, *72*, 487–500. [CrossRef]
46. Karim, M.A.; Manladan, S.M.; Afroz, H.M.M.; Jin, W.; Krishna, T.; Ji, C.; Kim, D.B.; Park, Y.-D. Critical effect of heat input on joint quality in resistance element welding of Al and steel. *J. Manuf. Process.* **2023**, *95*, 91–104. [CrossRef]
47. Günter, H.; Meschut, G. Joining of ultra-high-strength steels using resistance element welding on conventional resistance spot welding guns. *Weld World* **2021**, *65*, 1899–1914. [CrossRef]
48. Mirza, F.A.; Macwan, A.; Bhole, S.D.; Chen, D.L.; Chen, X.G. Effect of welding energy on microstructure and strength of ultrasonic spot welded dissimilar joints of aluminum to steel sheets. *Mater. Sci. Eng. A* **2016**, *668*, 73–85. [CrossRef]
49. Dash, S.S.; Biswas, S.; Peng, H.; Jiang, X.Q.; Li, D.Y.; Chen, D.L. Deformation behavior of dissimilar ultrasonic spot-welded joints of a clad 7075 aluminum alloy to galvanized highstrength low-alloy steel. *Mater. Sci. Eng. A* **2024**, *894*, 146179. [CrossRef]
50. Kapil, A.; Vivek, A.; Daehn, G. Dissimilar impact spot welding of high-thickness aluminum-steel sheets using vaporizing foil actuators. *Manuf. Lett.* **2024**, *41*, 38–45. [CrossRef]
51. Chu, M.; He, X.; Zhang, J.; Lei, L. Clinching of similar and dissimilar sheet materials of galvanized steel, aluminium alloy and titanium alloy. *Mater Trans.* **2018**, *59*, 694–697. [CrossRef]
52. Mozetic, H.; da Rocha, R.P.; Riffel, M.H.; Schaeffer, L. Application of the process of joining sheet by mechanical cold forming (clinching) using dissimilar materials. *Obs. De La Econ. Latinoam.* **2023**, *21*, 27717–27734. [CrossRef]
53. Ismail, M.; Buang, A. Precision joining of steel-aluminum hybrid structure by clinching process. *J. Adv. Manuf. Technol.* **2018**, *12*, 25–36. Available online: https://jamt.utem.edu.my/jamt/article/view/2848 (accessed on 10 October 2024).
54. Lee, C.J.; Kim, J.Y.; Lee, S.K.; Ko, D.C.; Kim, D.M. Parametric study on mechanical clinching process for joining aluminum alloy and high-strength steel sheets. *J. Mech. Sci. Technol.* **2010**, *24*, 123–126. [CrossRef]
55. Tenorio, M.B.; Lajarin, S.F.; Gipiela, M.L.; Marcondes, P.V.P. The influence of tool geometry and process parameters on joined sheets by clinching. *J. Braz. Soc. Mech. Sci. Eng.* **2019**, *41*, 67. [CrossRef]
56. Ge, Y.; Xia, Y. Dynamic behavior of self-piercing riveted and mechanical clinched joints of dissimilar materials: An experimental comparative investigation. *Adv. Mater. Sci. Eng.* **2019**, *2019*, 6463576. [CrossRef]
57. Jiang, T.; Liu, Z.X.; Wang, P.C. Quality inspection of clinched joints of steel and aluminum. *Int. J. Adv. Manuf. Technol.* **2015**, *76*, 1393–1402. [CrossRef]
58. Ewenz, L.; Kuczyk, M.; Zimmermann, M. Approach to transferring force-based fatigue curves into stress-related fatigue curves for clinch joints. *Mater. Res. Proc.* **2023**, *25*, 141–146. [CrossRef]
59. Harzheim, S.; Ewenz, L.; Zimmermann, M.; Wallmersperger, T. Corrosion phenomena and fatigue behavior of clinched joints: Numerical and experimental investigations. *J. Adv. Join. Process* **2022**, *6*, 100130. [CrossRef]
60. Yang, C.; Yao, J.; Niu, Y.; Wang, R.; Kang, J. Fatigue life analysis of steel-aluminium non-rivet connection. *J. Plast. Eng.* **2021**, *28*, 154–162. [CrossRef]
61. Kamble, P.; Mahale, R. Simulation and parametric study of clinched joint. *Int. Res. J. Eng. Technol.* **2016**, *3*, 2730–2734. Available online: https://www.irjet.net/archives/V3/i5/IRJET-V3I5571.pdf (accessed on 10 October 2024).
62. Hörhold, R.; Müller, M.; Merklein, M.; Meschut, G. Mechanical properties of an innovative shear-clinching technology for ultra-high-strength steel and aluminium in lightweight car body structures. *Weld World* **2016**, *60*, 613–620. [CrossRef]

63. Gao, X.; Chen, C. Transformation of the failure state for the clinched joints of dissimilar materials using an auxiliary layer. *Eng. Fail Anal.* **2023**, *146*, 107106. [CrossRef]
64. Ma, Y.; Akita, R.; Abe, Y.; Geng, P.; Luo, P.; Tsutsumi, S.; Ma, N. Mechanical performance evaluation of multi-point clinch-adhesive joints of aluminum alloy A5052-H34 and high-strength steel JSC780. *Automot. Innov.* **2023**, *6*, 340–351. [CrossRef]
65. Zhang, Y.; Wang, C.; Shan, H.; Li, Y.; Luo, Z. High-toughness joining of aluminum alloy 5754 and DQSK steel using hybrid clinching-welding process. *J. Mater. Process Technol.* **2018**, *259*, 33–44. [CrossRef]
66. Gao, L.H.; Kang, G.S.; Lee, C.J.; Kim, B.M.; Ko, D.C. Application of Friction Stir Hole Clinching to Joining of Dissimilar Materials. International Symposium on Green Manufacturing and Applications, 2017, PP152. Available online: https://www.presm.org/data/Poster%20Session/PP152.pdf (accessed on 10 October 2024).
67. Barimani-Varandi, A.; Aghchai, A.J. Electrically-assisted mechanical clinching of AA6061-T6 aluminum to galvanized DP590 steel: Effect of geometrical features on material flow and mechanical strength. *Mech. Ind.* **2020**, *21*, 529. [CrossRef]
68. Meilinger, Á.; Cserépi, M.F.; Lukács, J. Behaviour of aluminium/steel hybrid RSW joints under high cycle fatigue loading. *Weld World* **2024**, *68*, 427–440. [CrossRef]
69. Gáspár, M.; Tervo, H.; Kaijalainen, A.; Dobosy, Á.; Török, I. The Effect of Solution Annealing and Ageing During the RSW of 6082 Aluminium Alloy. In *Vehicle and Automotive Engineering 2. VAE 2018*; Lecture Notes in Mechanical Engineering; Jármai, K., Bolló, B., Eds.; Springer: Cham, Switzerland, 2018; pp. 694–708. [CrossRef]
70. Meilinger, Á.; Prém, L.; Abd Al Al, S.; Gáspár, M. Comparion of RSW technologies on DP steels with modified instrumented Charpy impact test. *Weld World* **2023**, *67*, 1911–1922. [CrossRef]
71. Alden, A.A.A.S.; Gáspár, M.; Meilinger, Á. Properties of hybrid aluminium-steel joints made by resistance spot welding. *Defect Diffus. Forum* **2022**, *416*, 131–138. [CrossRef]
72. Abd, A.A.S.A.; Meilinger, Á.; Gáspár, M.; Lukács, J. High Cycle Fatigue Testing of Lap Shear RSW Joints from Martensitic MS1400 Steel Sheets. *Mater. Sci. Forum* **2023**, *1095*, 139–151. [CrossRef]
73. Kovács, P.Z.; Lukács, J. Mechanical Investigations of EN AW 5754 Aluminium Alloy Clinched Joints. In *Research Results at the Institute of Materials Science and Technology, Faculty of Mechanical Engineering and Informatics, University of Miskolc—Yearbook 2022*, 1st ed.; Koncsik, Z., Lukács, J., Eds.; University of Miskolc: Miskolc, Hungary, 2022; pp. 17–24.
74. Kovács, P.Z.; Tisza, M. Investigation and finite element modelling of technological parameters of clinched joints. *Anyagmérnöki Tudományok* **2016**, *39/1*, 7–18.
75. *ISO 14324*; Resistance Spot Welding—Destructive Tests of Welds—Method for the Fatigue Testing of Spot Welded Joints. International Organization for Standardization: Geneva, Switzerland, 2013.
76. *JSME S 002*; Standard Method of Statistical Fatigue Testing. The Japan Society of Mechanical Engineers: Tokyo, Japan, 1981.
77. Nakazawa, H.; Kodama, S. Statistical S-N Testing Method with 14 Specimens: JSME Standard Method for Determination of S-N Curves. In *Statistical Research on Fatigue and Fracture. Current Japanese Materials Research*, 1st ed.; Tanaka, T., Nishijima, S., Ichikawa, M., Eds.; Elsevier Applied Sciences: Essex, UK; The Society of Materials Science, Japan: Kyoto, Japan, 1987; Volume 2, pp. 59–69.
78. Qian, C.; Ghassemi-Armaki, H.; Shi, L.; Kang, J.; Haselhuhn, A.S.; Carlson, B.E. Competing fracture modes in Al-steel resistance spot welded structures: Experimental evaluation and numerical prediction. *Int. J. Impact. Eng.* **2024**, *185*, 104838. [CrossRef]
79. Aydin, K.; Hıdıroğlu, M.; Kahraman, N. Enhancing weld strength in high-strength steels: The role of regional preheating in RSW. *Mater Test.* **2024**, *66*, 328–346. [CrossRef]
80. Khaleel, H.H.; Mahmood, I.A.; Khoshnaw, F. Fatigue and impact properties of single and double resistance spot welding for high-strength steel used in automotive applications. *J. Braz. Soc. Mech. Sci. Eng.* **2023**, *45*, 155. [CrossRef]
81. Yagita, R.; Abe, Y. Effects of sheared edge and overlap length on reduction in tensile fatigue limit before and after hydrogen embrittlement of resistance spot-welded ultra-high-strength steel sheets. *Metals* **2023**, *13*, 2002. [CrossRef]
82. Chen, C.; Zhao, S.; Han, X.; Zhao, X.; Ishida, T. Experimental investigation on the joining of aluminum alloy sheets using improved clinching process. *Materials* **2017**, *10*, 887. [CrossRef] [PubMed]
83. Lei, L.; He, X.; Yu, T.; Xing, B. Failure modes of mechanical clinching in metal sheet materials. *Thin-Walled Struct* **2019**, *144*, 106281. [CrossRef]
84. He, X.; Zhang, Y.; Xing, B.; Gu, F.; Ball, A. Mechanical properties of extensible die clinched joints in titanium sheet materials. *Mater Des.* **2015**, *71*, 26–35. [CrossRef]
85. Dzupon, M.; Kascák, L.; Cmorej, D.; Ciripová, L.; Mucha, J.; Spisák, E. Clinching of high-strength steel sheets with local preheating. *Appl. Sci.* **2023**, *13*, 7790. [CrossRef]
86. Liu, F.; Chen, W.; Deng, C.; Guo, J.; Zhang, X.; Men, Y.; Dong, L. Research advances in fatigue behaviour of clinched joints. *Int. J. Adv. Manuf. Technol.* **2023**, *127*, 1–21. [CrossRef]
87. Yu, Z.; Ma, N.; Murakawa, H.; Watanabe, G.; Liu, M.; Ma, Y. Prediction of the fatigue curve of high-strength steel resistance spot welding joints by finite element analysis and machine learning. *Int. J. Adv. Manuf. Technol.* **2023**, *128*, 2763–2779. [CrossRef]
88. Basquin, O.H. The exponential law of endurance tests. In Proceedings of the Thirteenth Annual Meeting (X), Atlantic City, NJ, USA, 28 June–2 July 1910; American Society for Testing and Materials: West Conshohocken, PA, USA, 1910; pp. 625–630.
89. Peng, J.; Fukumoto, S.; Brown, L.; Zhou, N. Image analysis of electrode degradation in resistance spot welding of aluminium. *Sci. Technol. Weld. Join.* **2004**, *9*, 331–336. [CrossRef]

90. Zhong, L.; Yu, W.; Jia, L.; Han, L. Study on Influence of Electrode Type on Weld ability of TL091 Al. Alloy in Medium Frequency RSW. *J. Phys. Conf. Ser.* **2023**, *2706*, 012028. [CrossRef]
91. Varis, J. Economics of clinched joint compared to riveted joint and example of applying calculations to a volume product. *J. Mater. Process. Technol.* **2006**, *172*, 130–138. [CrossRef]

metals

MDPI

Review

Fatigue Behaviour of Mechanical Joints: A Review

Animesh Kumar Basak [1,2,*], Dharamvir Singh Bajwa [3] and Alokesh Pramanik [3,*]

[1] Adelaide Microscopy, The University of Adelaide, Adelaide, SA 5005, Australia
[2] Centre for Research Impact & Outcome, Chitkara University Institute of Engineering and Technology, Chitkara University, Rajpura 140401, Punjab, India
[3] School of Civil and Mechanical Engineering, Curtin University, Bentley, WA 6102, Australia; dharamvir.bajwa@postgraduate.curtin.edu.au
* Correspondence: animesh.basak@adelaide.edu.au (A.K.B.); alokesh.pramanik@curtin.edu.au (A.P.)

Abstract: Mechanical joints, regardless of materials, are useful when joining multiple components, though there are certain limits when applying them in engineering applications such as fatigue loading. The purpose of this research is to provide a comprehensive review of the trend of fatigue properties of common non-thermal mechanical connections such as adhesive, bolted, clinched and riveted joints. Towards that, a narrative approach was taken. In modern engineering applications, most of the joints contain both metallic and non-metallic components. The relevant experimental studies have proven many factors that can affect each type of joint and how they can be implemented in real-time appliances. For instance, the fatigue behaviour of adhesive joints is affected by the bond length, thickness and the use of different materials. Increasing the bond length can enhance its fatigue resistance up to a certain length, whilst increasing the thickness of laminate or adhesive decreases the fatigue life unless the surface roughness increases. On the other hand, different laminate materials can affect the fatigue performance depending on their mechanical properties. These findings will allow readers to have an overall concept of the fatigue behaviour of mechanical joints and the influence of various internal and external parameters on that.

Keywords: fatigue; mechanical joints; loading; failure; endurance limit

1. Introduction

Mechanical joints have always been a vital aspect of the construction industry. Mechanical joints are defined as using two or more components such as mechanical fasteners [1]. They are used to transfer load, which involves compression between connected objects whilst relative motion occurs simultaneously. Mechanical joints depend on compressive stresses between compartments while reallocating the tensile stress away from the system [1]. The most common mechanical joints are adhesive joints, bolted joints, riveted joints and clinching joints. Some of these joints have similar joining methods such as bolted and adhesive joints, where two components are joined together using bolts or adhesives and heat is not applied. In many cases, some of these methods are combined to generate hybrid joints [2–4]. Mechanical joints are used in various industries such as automotive, aerospace, fabrication processes and other construction appliances [5]. The most common materials used in mechanical joints are metals, ceramics, plastics (thermoplastics) and adhesives [6,7]. Research shows that mechanical joints with fasteners through the use of drilled holes are more inexpensive compared to welding and other joining methods [8,9]. Mechanical joints are also critical due to their ease of assembly and disassembly, which is one of the reasons why bolted joints are being used in aerospace industries, whether

 https://doi.org/10.3390/met15010025

constructed as a single- or double-lap joint. Moreover, some mechanical joints do not require sufficient surface preparation such as fasteners used in bolted joints, which allows quick assembly and disassembly without harming objects [8,10]. This also ensures easy inspection of mechanical joints as it helps to detect any signs of wear, corrosion [11] or degradation of the joints [12].

Mechanical joints are usually used in many industries despite the constraints involved within different joints. The restrictions of these joints are not the main concern when implementing mechanical joints in engineering appliances but rather the way they behave under fatigue. By understanding the concepts of fatigue parameters and underlying mechanisms, solutions can be formed that will mitigate fatigue failures experienced within mechanical joints. The purpose of this review is to provide a comprehensive understanding of the fatigue behaviour of mechanical joints based on the latest publications that conducted experimental analysis. The fatigue characteristics of four common mechanical joints, namely adhesive, bolted, clinched and riveted joints, were investigated since they are used in most engineering applications. Various input parameters such as cyclic loading conditions, materials used, geometric parameters, and environmental conditions were taken into consideration, along with their interplay. The fatigue life [13] S-N curves and the fatigue crack nucleation and propagation of the joints were analysed to understand the fatigue behaviour of the joints.

Fatigue behaviour is an important aspect in many industries, particularly aerospace and automotive, where parts and components are subjected to repetitive and fluctuating loads. This causes premature failure of components that leads to catastrophic failure [14–16]. Though there are extensive reports that are available in the literature, there are inconsistencies regarding the performance of the mechanical joints under fatigue, which give rise to this review. This present review will inform the reader regarding the influence of various internal and external parameters that affect the fatigue of mechanical joints, starting with the basic concept of fatigue. Moreover, this review will identify the current trends and suggest future research directions. Towards that, a narrative approach was taken where the research outcomes of the last 15 years were analysed and 'Preferred Reporting Items for Systematic reviews and Meta-Analyses (PRISMA)' [17,18] were also considered. However, instead of meta-analysis, the 'narrative approach' was considered based on keyword (fatigue; mechanical joints; loading; failure; endurance limit) search of the major scientific databases such as Web of Science, ScienceDirect, Google Scholar and so on.

Analysing these fatigue parameters would help engineers/researchers to understand how different joints behave under such conditions and how some changes in parameters would affect the overall fatigue behaviour of the joint. Moreover, this work will provide an overview of experimental studies conducted over the past 15 years which will highlight the latest research and advancements in this field, together with highlighting the inconsistencies and gaps for further research directions.

2. Fatigue of Mechanical Joints

Understanding the fatigue mechanisms is important to evaluate the fatigue performance of these joints when they undergo cyclic loading. Cyclic loading occurs when there is a change in the range of stress, which causes damage to the component usually below its prescribed strength values [19]. The main aspects that describe the fatigue are fatigue life, crack propagation/nucleation, endurance limit and fatigue strength. The repeated cyclic loading and unloading of a component is known as the cyclic stress, which initiates crack development and propagation [20]. This may be influenced by the existing residual stresses in the components that resulted during manufacturing processes [21,22]. The crystalline structure and the grain size of the materials also affect the fatigue life and dictate fatigue crack development [23,24]. According to Wang et al. [23], minimising the grain size of involved materials can enhance the fatigue strength, which reduces the rate of the crack nucleation process

The fatigue life in mechanical joints is usually influenced by the number of cycles experienced within the joint over time and high magnitudes of stress were needed for crack initiation compared to crack propagation [25]. However, for joints that experienced low load/high cycle fatigue, the crack initiation period dictates the fatigue life of the joint, which means that the majority of cyclic loads influence crack initiation [26]. This may not always be the case as some joints acquire defects during fabrication, which causes the crack initiation that is influenced by the surface imperfections such as the fillet, edge, weld toe or other geometric discontinuities [27]. Ideally, joints that consist of higher fatigue life can resist failure while experiencing longer cyclic loads.

When a mechanical joint undergoes cyclic loading, fracture can result in the joint, which is caused by crack propagation and nucleation and is known as fatigue fracture. Microscopic cracks will occur in such areas on the joint where stress concentration is the highest [28]. The fracture will be further intensified, which causes the fracture to happen instantly. Some mechanical joints that can resist such loads being applied despite considerable crack sizes are said to be damage tolerant. The fatigue crack behaviour of mechanical joints can be calculated using the fatigue crack law and is used to determine the fatigue crack process of different mechanical joints [28]:

$$\frac{da}{dN} = f(\sigma, a) \tag{1}$$

where σ—applied stress or stress amplitude, a—crack length, N—lifetime cycles. The lifetime cycles of the joint can be determined through the integration of the initial to the final crack length, as follows [28]:

$$N_f = \int_{a_i}^{a_f} \frac{da}{f(\sigma, a)} \tag{2}$$

The number of cycles is recorded and plotted in an S-N curve, which is used to compare and predict the joint's fatigue life. This allows us to determine the time it takes for a joint to experience failure. According to Zerbst et al. [29], the average initial and final crack lengths are usually 1 mm and 10 mm, respectively, for materials under fatigue as shown in Figure 1. However, this would vary depending on the joint material chosen, the size of joints, adhesive type etc.

There are usually two different approaches to describing fatigue, namely, the S-N curve method and the fatigue crack growth approach [30,31]. In the S-N curve method [32], crack nucleation typically occurs during 80 to 90% of the loading cycles, with crack propagation governed by Equation (1). However, this takes place only afterwards. On the other hand, the fatigue crack growth approach helps to understand the growth mechanism and nature of the crack from the beginning and how it propagates [33].

The endurance limit, $S\prime_e$, is known for a material that can withstand loads below the stress level for an infinite number of loading cycles without failure [34]. Some materials do not have an endurance limit. In that case, it is vital to design a mechanical joint that has high resistance to fatigue for materials to prevent failure [35]. In terms of calculating the crack nucleation/propagation, the stress and strain factors need to be considered [36]. The endurance limit is usually 10^6 cycles for most materials. However, this is not always the case as there are factors that affect the endurance limit such as the geometrical structure, notch sensitivity, surface finish, loading type, etc., as mentioned earlier [36]. The true endurance limit can only be attained when there are no microstructural anomalies that involve second-phase particles, inclusions or the existence of a number of phases [37].

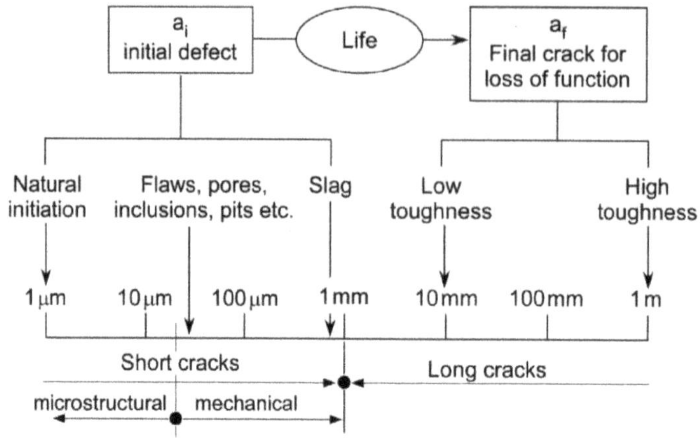

Figure 1. Average crack lengths of components under frequent cyclic loading. Reprinted with permission from Ref. [29]. Copyright 2024 Elsevier.

An S-N curve plots the fatigue-life data for different cyclic loading, which usually ranges the fatigue life, N, between 10^3 and 10^6 [36]. The curve shows a log-log plot which plots the stress amplitude against the number of cycles to failure [37]. The curve is determined through experimental data when the component undergoes uniaxial, fully reversed cyclic loading where the loading can either be bending or tension. S-N curves can be useful to identify the number of load cycles to failure and not the amount of damage a material experiences from fatigue before failure. The curves can also be used to determine which part of the material experienced fatigue crack initiation, growth and failure at a specific number of cycles.

As mentioned earlier, fatigue strength is important to determine during fatigue testing as this would estimate how well a component can withstand repeated high-cyclic loading without failure. In real-life applications using mechanical joints, they undergo high fluctuation in loads, which is why mechanical joints must have reasonable fatigue strength which ensures buildings can endure for longer periods. Fatigue strength is known as the stress value where failure in fatigue could occur due to an infinite number of cyclic loading cycles [26]. In terms of mechanical joints, it depicts how well it can tolerate high stresses over its service life. Out of all the factors that affect the fatigue strength of components, material selection is crucial before making a joint since some have different fatigue strength values. For example, ferrous and titanium alloys have lower fatigue strength than the fatigue life that these materials can obtain without experiencing failure [26]. The usual number of cycles that are tested for fatigue strength is between 10^7 and 10^9 [38]. It is found that at least 90% of failures are due to fatigue failures in components that contain metallic materials. To prevent this, it is recommended to carry out fatigue testing to ensure that the component is still in a good condition to resist fatigue.

2.1. Adhesive Joints

Adhesive joints are created by joining two or more components together using adhesive material [39,40]. The strength of the bond between the surfaces is dependent upon the adhesive material. There are several types of adhesive joints which are single/double-lap joints, scarf joints and stepped-lap joints, as shown in Figure 2.

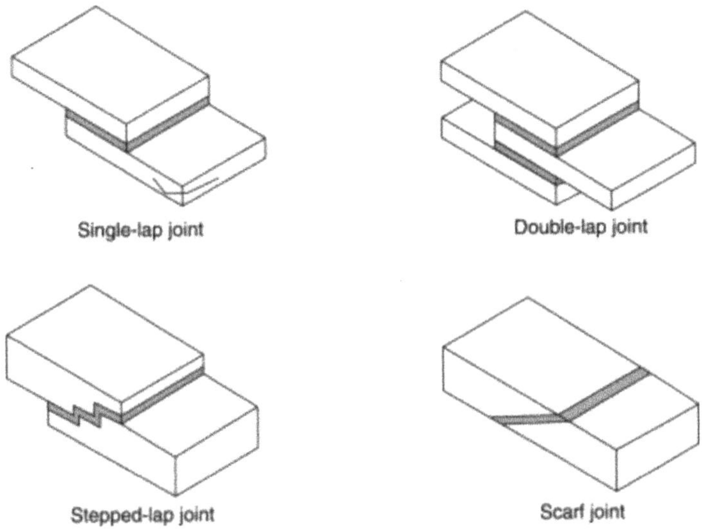

Figure 2. Common types of adhesive joints. Reprinted with permission from Ref. [41]. Copyright 2024 Elsevier.

The typical materials used in adhesive joints include magnesium and titanium alloys, as well as other common materials such as metals, plastics, ceramics, etc. [42]. Surface preparation is vital as contaminants such as dust and oil must be cleared to ensure strong adhesion is achieved between the adherents joined together. There have also been instances where dissimilar materials have joined together in adhesive joints such as carbon fibre-reinforced polymers (CFRPs) [43] and fibre metal laminates (FMLs) [42], which can be used in applications such as aircraft, automotive and wind turbines industry.

It is also vital to select the right materials for the adhesives that give an extra edge to resist the fatigue and blows the joint may encounter. Epoxy is one of the most common materials used in adhesive joints, mainly in plastics and ceramics. It has high strength, durability, and excellent heat resistance, as well as chemical and environmental aspects [44]. Another type of adhesive used is called cyanoacrylate, which also forms a strong bond between similar adherend materials used for epoxy adhesives. They have fast curing times and a fast solidifying process and are commonly used with rubber, metals, ceramics and other applications [44]. Acrylics are another adhesive material that can be used in replacement of the other two and have similar benefits to epoxy adhesives, which are mainly used in automotive vehicles and buildings [44]. Despite having excellent mechanical properties, adhesive joints are prone to failure through fatigue since other factors affect the fatigue characteristics of adhesive joints, which will be discussed in the following sections.

2.1.1. Bond Length

Studies have shown that altering the bond length of the joint allows the joint to either be more resistant to fatigue or susceptible, which can potentially lead to failure. Altering the bond length can affect the peel stresses of the joint, as well as the bending moments, which can disrupt the start of the damage and impact the fatigue life of the joint [45]. A typical evolution of fatigue life concerning different bond lengths (20 and 40 mm) between two carbon-fibre fabric polymer (CFRP) adherents joined together using epoxy adhesive is shown in Figure 3. As can be seen, the increase in bond length enhances the fatigue strength. The fatigue life increased by a factor of 3 for the 40 mm bond length compared to the 20 mm, where the thickness of adhesive was 0.15 mm. It can be said that altering the bond length significantly influences the fatigue strength compared to modifying the ply orientation or changing the fillet geometry of the joint.

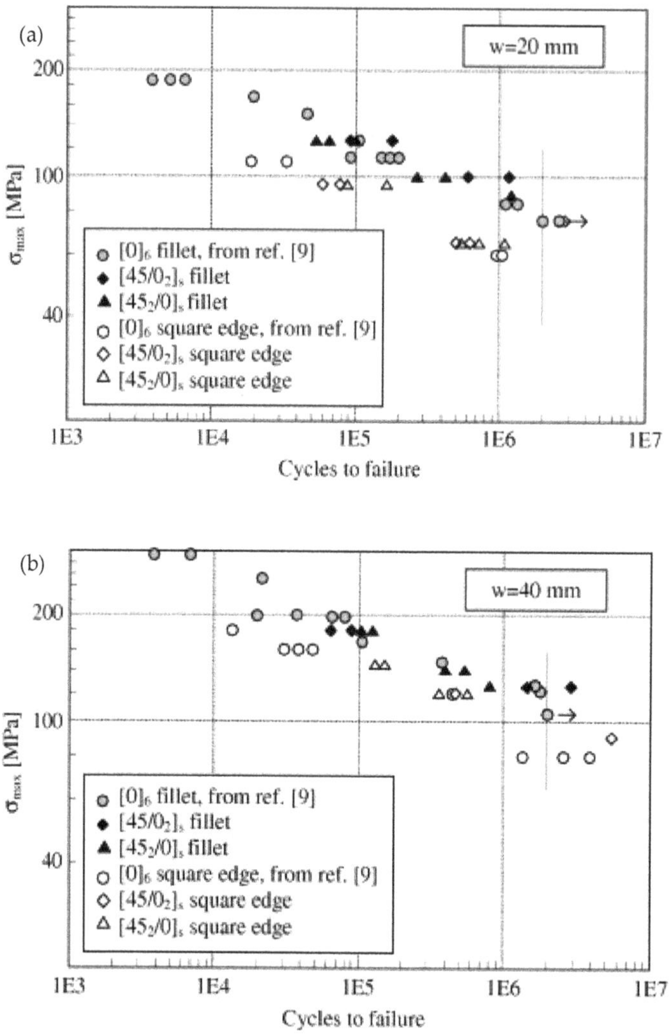

Figure 3. The tensile fatigue life of CFRP adhesive (epoxy adhesive 9323 B/A) joint with different bond lengths: (**a**) 20 mm and (**b**) 40 mm. Reprinted with permission from Ref. [45]. Copyright 2024 Elsevier.

Increasing the bond length also increased the propagation phase and fatigue life to crack initiation. This was because shorter bond lengths consist of higher stress intensity values resulting in shorter fatigue life. This indicates that the area of the adhesive layer at the interface reduces, though the applied load remains constant. The same load on the reduced area causes higher stress. Similar results were also found in another investigation conducted by Mora et al. [46], where they analysed 6061-T6 aluminium alloy bonded to a dual-phase steel adherend using three different types of epoxy adhesive with different mechanical properties.

An investigation conducted by Moreira et al. [47] studied the effects of single-strap adhesive joints using araldite epoxy adhesive bonded with CFRP through high-cycle fatigue loading. The effect of bond length on fatigue life is shown in Figure 4. The fatigue life increases when the bond length alters from 5 mm to 15 mm. However, when the bond length increased from 15 mm to 20 mm, there was slight to no change in the fatigue life. Similar results were also found in another investigation conducted by Moura et al. [48]. Hence, Moreira et al. [47] concluded that increasing the bond length did not necessarily improve the fatigue life of the joints. However, altering the bond thickness and the adherend thickness may show different effects on the fatigue strength and life if the bond length was kept the same. The fracture mode was mainly cohesive failure [49] with the presence of micro-defects [47].

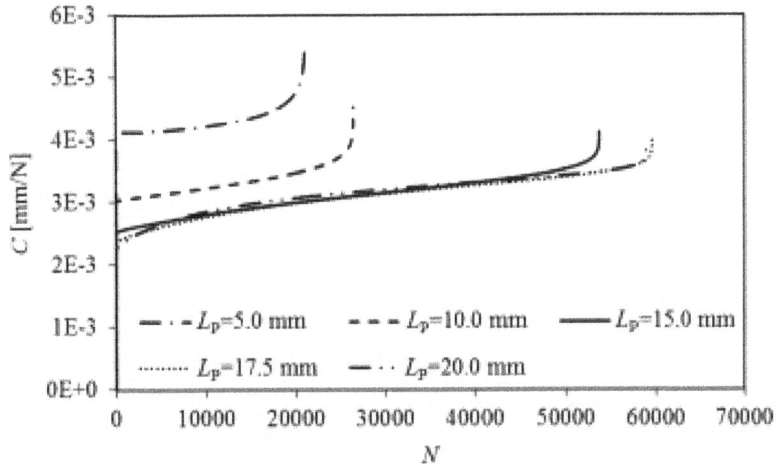

Figure 4. Effects of bond length on fatigue life. Reprinted with permission from Ref. [47]. Copyright 2024 Elsevier.

2.1.2. Bond Line and Laminate Thickness

A study conducted by Tang et al. [50] investigated the effects of the bond-line thickness on single-lap adhesive joints. They also proposed appropriate stress intensity factors to give an accurate fatigue life prediction and estimate the fatigue crack propagation using the Paris law [51]. The evolution of fatigue life on 2.5 mm and 5.5 mm bond-line thicknesses is shown in Figure 5 on glass fibre-reinforced polymer (GFRP) with epoxy adhesive. It is evident that the 5.5 mm bond-line thickness depicts a lower fatigue life than the 2.5 mm.

The failure of the above-mentioned joint is shown in Figure 6, where a mixed failure fracture is evident. Each of the joints experienced fracture nucleation between the adhesive and the adherend, especially in the corner between the two sides (Figure 6c). As the number of cycles increased, an interlayer failure caused an initiation of crack propagation on the side of the joint. The crack continues to propagate between the adhesive and the bond before the separation occurs, which results in final fracture failure. The last stage was where the joint split into two components due to interfacial and interlayer propagation [50]. This was confirmed by an investigation conducted by Banea et al. [52] on single-lap joints with silicone adhesives with bond-line thicknesses of 0.5 mm and 1 mm.

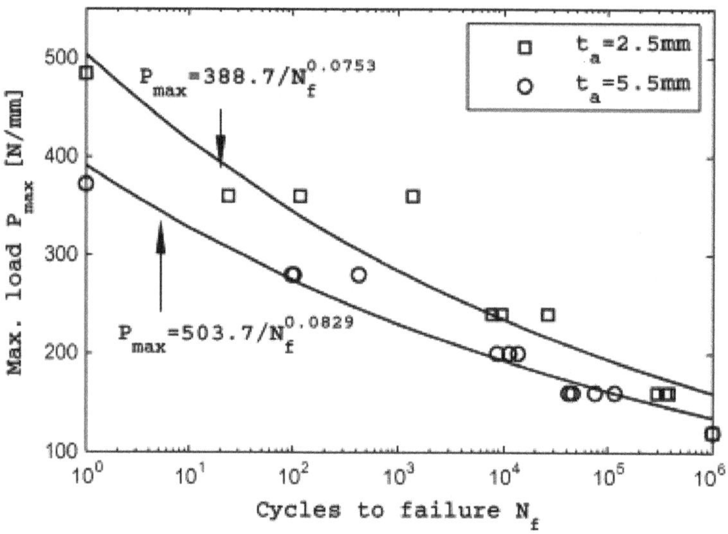

Figure 5. Tensile fatigue life for the bond-line thicknesses of 2.5 mm and 5.5 mm between CFRP and aluminium. Reprinted with permission from Ref. [50]. Copyright 2024 Elsevier.

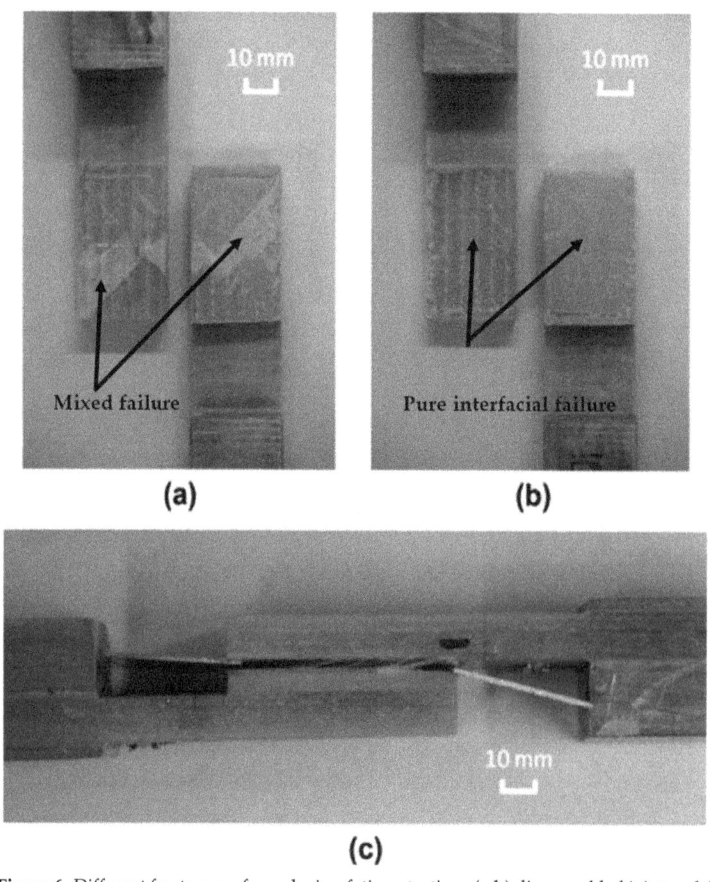

Figure 6. Different fracture surfaces during fatigue testing: (**a**,**b**) disassembled joint and (**c**) assembled joint with fracture. Reprinted with permission from Ref. [50]. Copyright 2024 Elsevier.

The higher the thickness of the adhesive, the more out-of-plane deflection and bending stress in the joint increase, which contributes to lower strength at higher adhesive thicknesses. It is also reported that the singularity and stress concentration at the adherend–adhesive interface intensify as bond-line thickness increases [53]. The variation in bond-line thickness significantly affects joint strength as the inclination of the thinner bonded area induces stress concentration. It was reported that a ±0.254 mm variation in the 0.254 mm thick bond line is far worse than that of a 2.5 mm thick bond line [54].

The effects of surface roughness on the fatigue life of adhesive joints were studied by Boutar et al. [55], as shown in Figure 7. Effects of four different surface roughness levels, from 0.6 μm to 1.5 μm with different bond-line thicknesses between 0.3 mm and 2 mm on aluminium–copper alloy adherents were investigated. It should be noted that in cohesive failure, a layer of adhesive remains on both surfaces, which means that the crack propagates through the adhesive layer. In case of interfacial failure, cracks occur at the interface between adhesive and joint parts. Practical joint failures have both cohesive and interfacial failure, which are known as mixed failure [56,57].

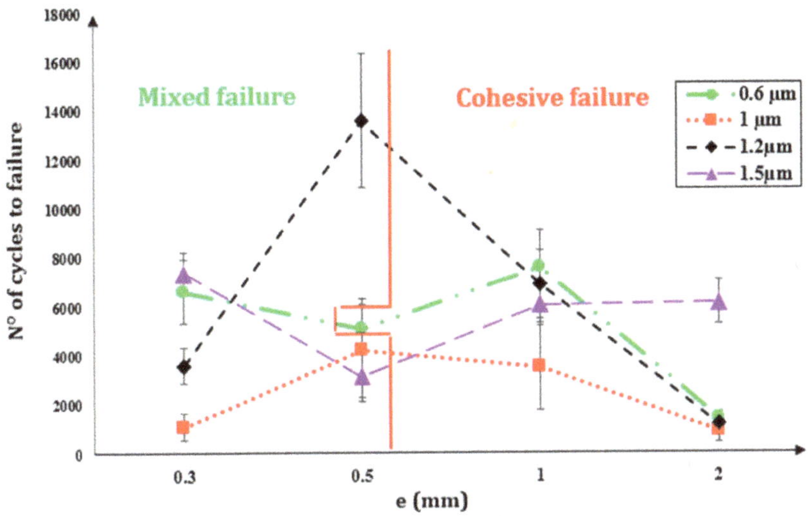

Figure 7. Effects of surface roughness and adhesive thicknesses on fatigue life. Reprinted with permission from Ref. [55]. Copyright 2024 Elsevier.

For a surface roughness of 0.6 μm, the maximum fatigue life was recorded when the bond-line thickness was at 1 mm but decreased later between 1 and 2 mm. The fatigue lifetime increased from 0.3 mm to 0.5 mm bond-line thicknesses for the surface roughness of 1 μm and decreased towards the end for the bond-line thickness of 0.5 mm to 2 mm. Furthermore, when tested against the 1.2 μm surface roughness, the fatigue life increased drastically at a bond-line thickness of 0.5 mm, and when the surface roughness was tested at 1.5 μm, the highest fatigue lifetime was recorded at a bond-line thickness of 0.3 mm and decreased at 0.5 mm. It can also be stated that different surface roughness values experienced either a cohesive or mixed failure mode within the adhesive joint. A scanning electron microscopy (SEM) analysis of the failed surface revealed the presence of 'air bubbles', which were responsible for crack initiation as shown in Figure 8. It can be stated that for 1 mm or higher adhesive thickness, the crack propagates depending on the size of the air bubbles in the adhesive and cracks within the joint. Figure 8 also indicates that the size of the bubbles and the number of bubbles are reduced with the increase in surface roughness due to the variation in wettability of the surface.

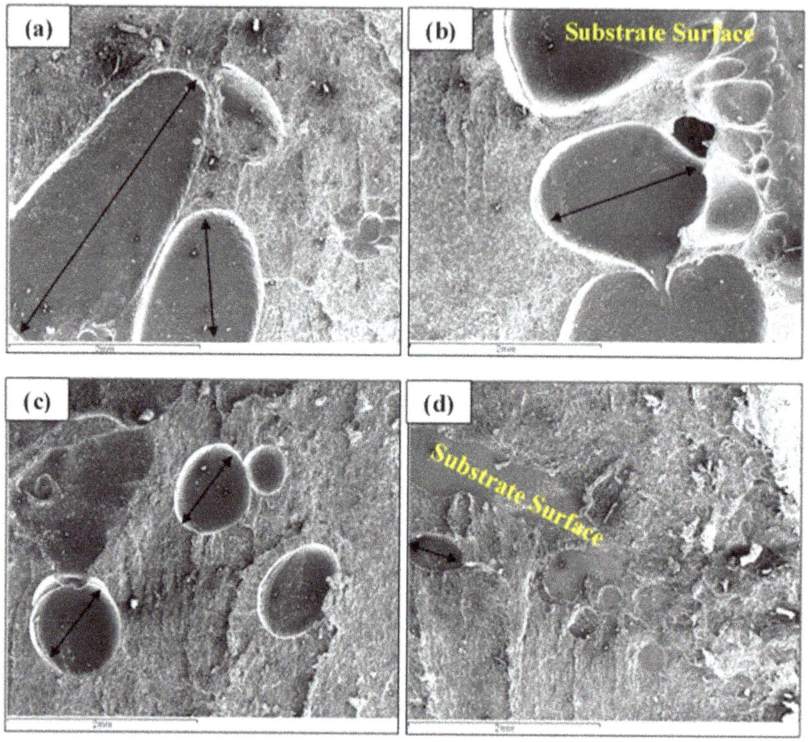

Figure 8. The fracture surface of 0.3 mm bond-line thickness for different surface roughness values: (**a**) 0.6 μm, (**b**) 1 μm, (**c**) 1.2 μm and (**d**) 1.5 μm. Reprinted with permission from Ref. [55]. Copyright 2024 Elsevier.

Here, it was found that the bond-line thickness had a very low porosity concentration, which prevented the cracks to be controlled along the paths. The surface roughness influences the stress concentration in the interface, which contributes to the fatigue crack growth. It was concluded that when the surface roughness decreased despite the bond line thickness, the wettability on the surface increased, which leads to a decrease in the fatigue strength of the joint, which can be seen when the surface roughness was 0.6 μm at a bond-line thickness of 0.5 mm. On the other hand, higher surface roughness showed reduced wettability, which led to an increase in fatigue lifetime.

Akpinar et al. [58] examined how altering the adherend thickness can enhance the fatigue performance of some joints. The adherence of Araldite-2015 adhesives of 2 to 6 mm thickness to aluminium alloy AA2024-T3 on single-lap joints was investigated for that. Surface treatment was implemented before fatigue testing of the adherents due to the prevention of external factors that may potentially affect the failure modes of the joints. The S-N curves produced were based on the eight loading levels as shown in Figure 9. It was reported that there was the presence of an adhesive layer between the upper and bottom adherend materials during the fatigue loading. It was said to have cohesion failure mode since there was an equal amount of the adhesive layer on both surfaces of the laminate [59,60]. After the fatigue testing, they also discovered the absence of any plastic deformation for all the thicknesses except the single-lap joint with the 2 mm adherend thickness during maximum fatigue loading. Overall, by examining the fracture characteristics of these joints, it can be stated that the fatigue and fracture strength do not correlate with each other well due to some deformation occurring in the laminate and being affected by the geometrical parameters. A similar observation was also reported by other researchers [61–63] on single-lap adhesive joints with aluminium 2024-T3 laminates. When the joints were tested until 10^6 lifecycles, they also found the fatigue strength increased as the adherend thickness was increased.

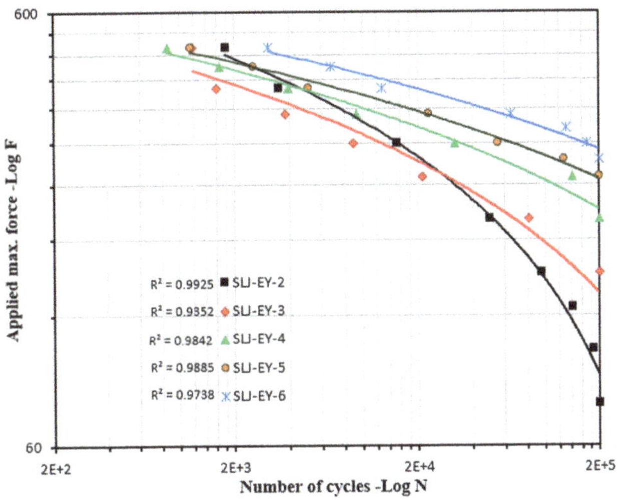

Figure 9. The fatigue life of different adherend thicknesses, where SLJ indicates single lap joint, EY indicates fatigue loading and the numerical values 2, 3, 4, 5 and 6 indicate adherend thicknesses in mm. Reprinted with permission from Ref. [58]. Copyright 2024 Elsevier.

2.1.3. Dissimilar Adherent Materials

While the thickness of adherends is useful to enhance fatigue performance, the type of materials used in the adhesive joints must also be taken into consideration. This is because some materials have different mechanical properties (such as stiffness, surface energy or thermal expansion) that can sustain such high fatigue loadings and prevent early failure modes. Zhang et al. [64] studied the adhesive single-lap joints between CFRP and aluminium alloy bonded with Araldite adhesives. They reported that (Figure 10), similar joints (CFRP-CFRP, denoted as J-CC) can withstand high fatigue loads with a lower fatigue life without failure compared to dissimilar joints. This suggests that while similar joints were suitable for high loads, they cannot sustain repeated cycles as they will fail quicker due to low fatigue life compared with CFRP–aluminium and aluminium–aluminium joints. Thus, CFRP–aluminium and aluminium–aluminium joints are suitable for long-lasting fatigue performance under cyclic loads as they both endure similar fatigue lives [65–67].

Machado et al. [68] concluded that joints between similar materials (CFRP–CFRP) failed through substrate delamination while dissimilar joints (CFRP–aluminium) gave lower fatigue strength and failed at the aluminium substrate. Mariam [69] noted that dissimilar adhesive joints performed superiorly to the similar joints of two materials. Similar joints having flexible substrates worsen the fatigue performance. Fatigue strength of dissimilar joints is related to adhesive ductility [70,71].

Stiffness degradation is an important factor for the fatigue performance of these joints and had three stages [64]. The first stage is the initiation of the fatigue process and usually has minimal to zero damage to the joint. The second stage is when the stiffness decreases as the number of cycles increases and the third stage is towards the end of the fatigue process where the stiffness decreases drastically until failure is present [64]. Single-lap joints with similar adherents have better stiffness degradation since the degradation occurs more gradually. Similar findings were also reported by other researchers such as Mariam et al. [72] on dissimilar adherents of single-lap adhesive joints using aluminium alloy (AA7075) and GFRP bonded with Araldite adhesives.

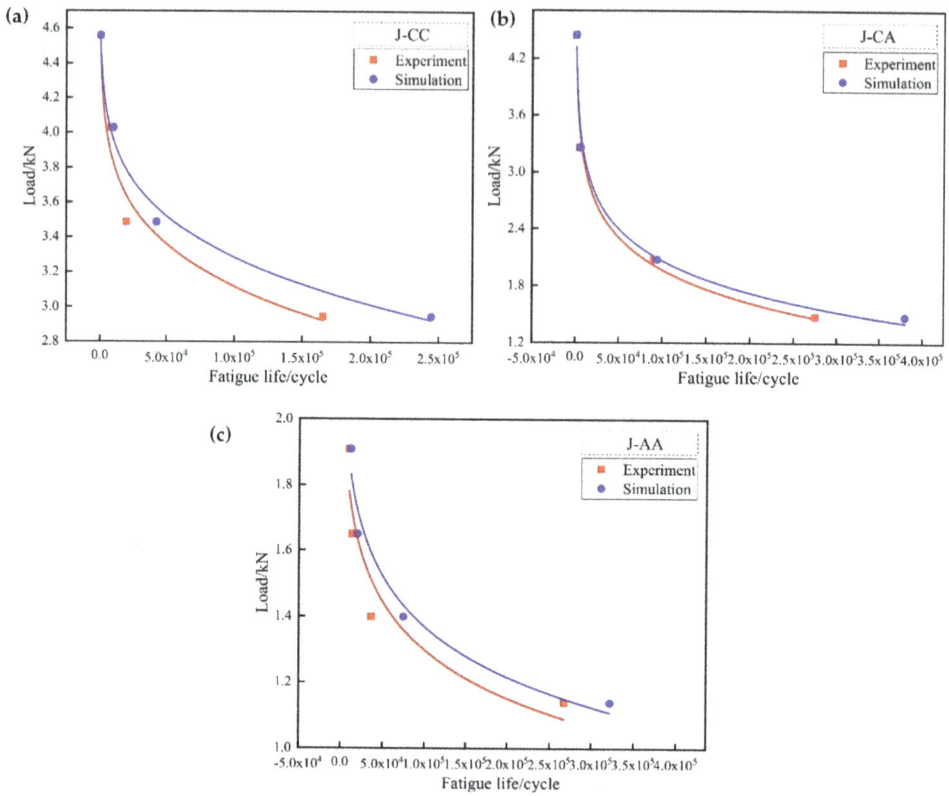

Figure 10. The fatigue life of three different types of single-lap joints: (**a**) CFRP-CFRP, (**b**) CFRP–aluminium, and (**c**) aluminium–aluminium. Reprinted with permission from Ref. [64]. Copyright 2024 Elsevier.

2.2. *Bolted Joints*

Bolted joints have a similar technique when joining two components together; however, they use bolts and screws instead of adhesives [73]. Nuts are also implemented in the joint to ensure tightness of the component. The bolts are inserted through a hole which is tightened, causing tension forces between the adherents, which locks both the laminates together. Examples of bolted joints include single- and double-lap bolted joints as shown in Figure 11.

Bolted joints are mainly used in common applications such as automobiles, machinery, building construction (mainly in beams), aerospace, nuclear and many more [74]. Most engineers and scientists use bolted joints due to their inexpensiveness and capability to easily assemble and disassemble parts for inspection. However, these joints also have some constraints since they can develop high stress concentrations due to the location of the holes which can potentially lead to failure especially when the joint undergoes repeated cycles of stress [75]. There are many factors that affect the fatigue behaviour of bolted joints and the most common ones are discussed below.

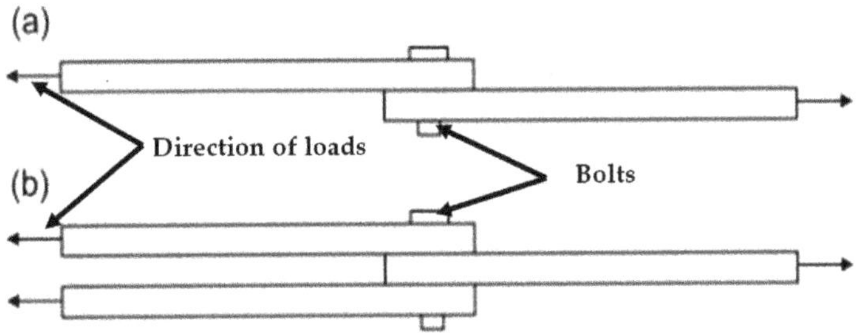

Figure 11. (**a**) Single- and (**b**) double-lap bolted joints. Reprinted with permission from Ref. [73]. Copyright 2024 Elsevier.

2.2.1. Cold Extrusion/Expansion of Bolted Joints

Cold extrusion is a method to enhance the fatigue life of bolted joints and prevent fatigue failures at an early stage and has been used in many engineering applications, especially in ships and aerospace. It is placed in a die which forms the structure of a desired shape and in terms of bolted joints; the die is fitted in the hole to restructure the hole so it can counteract and increase the overall property of the hole to prevent fatigue cracks in the joint [76]. During cold expansion, a mandrel is forced through a hole to expand it, which increases the compressive residual stress (CRS) around the hole and prevents crack initiation of the joint as it undergoes high-cyclic loads [77,78]. A study conducted by Wan et al. [79] explored the fatigue characteristics of single-lap bolted joints using 6.2 mm thick 42CrMo4 high-strength steel while two M12 flat-head bolts made out of Ti-6Al-4V titanium alloy were clamped to the steel laminates. Their main observation on that was the change in the preload of bolts affected the fatigue life. It was also stated that increased extrusion amounts led to higher fatigue life. They also noticed that during fatigue testing, the specimen had fatigue cracks near the threaded hole and propagated more towards that region, whereas specimens that experienced the cold-extrusion process had their crack propagating towards the edge of the threaded hole. The cold-extrusion process caused an alteration in the stress distribution of the plate, which prevents the threaded hole from cracking. The investigation of the failed specimens showed that there were three different sections of crack propagation with multiple slippages in the surface fracture as shown in Figure 12. These findings indicate how different amounts of cold-extrusion or varying preloads can influence the fatigue life and crack propagation of the joints; however, the cold-expansion process may slightly defer these processes.

Chakherlou et al. [80] investigated the cold expanded holes of aluminium alloy (2024-T3) double-shear bolted joints using M6 bolts. Two 4.5 mm thick aluminium alloy (7075-T6) laminates were joined together, while the 2024-T3 component was the middle laminate as part of the whole bolted joint. They investigated five specimens where two had holes created by cold expansion (CE), another two specimens had interference fit (IF) and the last one had a normal hole drilled in the laminates. An interference fit was another method to reduce fatigue and increase its resistance in bolted joints. The S-N curves of their investigation were shown in Figure 13. Similar to the findings of Wan et al. [79], Chakherlou et al. also reported that the cold-expansion specimens had a more effective fatigue life compared to the plain hole, but the interference fit had a slightly higher fatigue life compared to the CE joints. It can be stated that the IF and CE specimens had minimal to no difference in terms of fatigue life resistance.

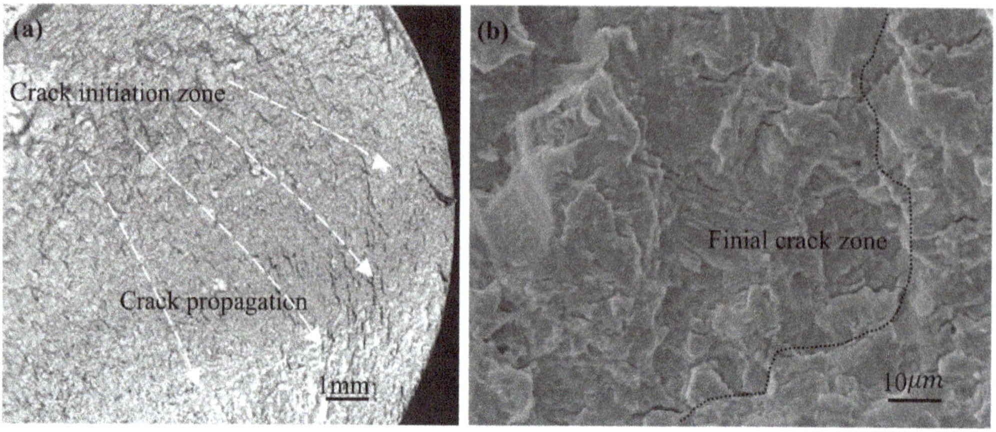

Figure 12. Fatigue crack zones for cold extrusion/expansion specimens: (**a**) overall view and (**b**) zoomed view of the crack (at a typical location). Reprinted with permission from Ref. [79]. Copyright 2024 Elsevier.

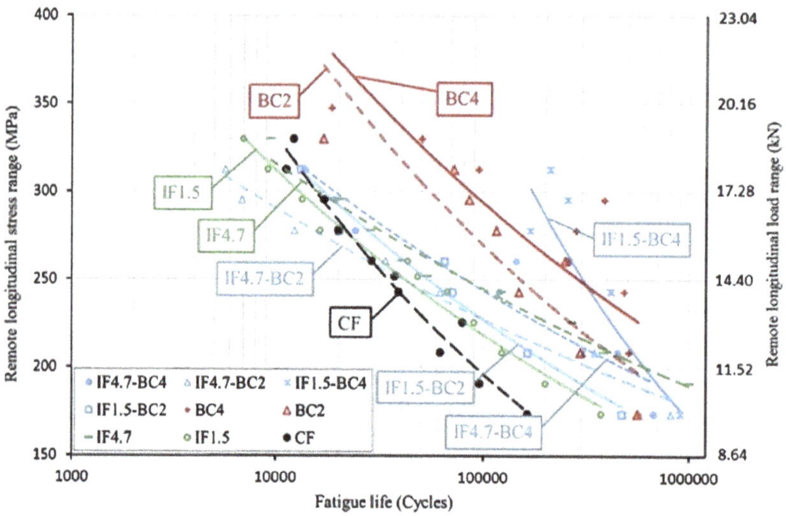

Figure 13. The fatigue life of interference fit (IF), cold-expansion (CE) and plain hole (PH) double-lap bolted joints with 1.5% and 4.7% preload, respectively. Reprinted with permission from Ref. [80]. Copyright 2024 Elsevier.

In the above discussion, interference fit (IF) in joints indicates when the minimum diameter of the bolts was bigger than the maximum diameter of the hole. In this case, the mating surfaces are embedded with compressive residual stress and experience deformation after assembly. In cold expansion (CE), a tapered mandrel having a diameter bigger than that of the hole is pulled through to induce compressive hoop stress in the internal surface of the hole. The compressive stress in the internal surfaces of the holes deter crack initiation and propagation in the bolted joint. No stress is introduced in the plain holes (PHs) except those generated from drilling or other fabrication processes.

Regarding the crack morphology, it was found that the holes experienced crack initiation at the entrance of the plate in mid-plane. As the load levels gradually increased, the cracks propagated from the edge of the plate to the hole's edge, which was due to the increase in the longitudinal distribution [80]. The increase in expansion and interference levels did not make a difference in the fatigue life since at higher cyclic loads, the level of stress distribution around the hole was kept constant. Thus, the improved fatigue life for cold expansion was due to the geometrical parameters that affected the CRS and pre-stress of the components. The plastic deformation of the components was enhanced where the grains had become more isotropic, which delayed the fatigue crack growth, hence leading to higher fatigue lives than those of plain holes [81].

2.2.2. Bolted Spherical Joints

Bolted spherical joints (BSJs) have recently been studied to investigate their fatigue properties. BSJs consist of multiple components such as seals, cone heads, screws and other connections that are joined to a spherical shape bolt as shown in Figure 14. These joints have started to be used in many industries due to their compatibility in many engineering applications and their beneficial mechanical properties [82]. However, they are susceptible to high fatigue loading and experience fatigue failures due to the high-strength bolts used in the BSJs, which increases the stress concentration leading to onset fatigue crack propagation.

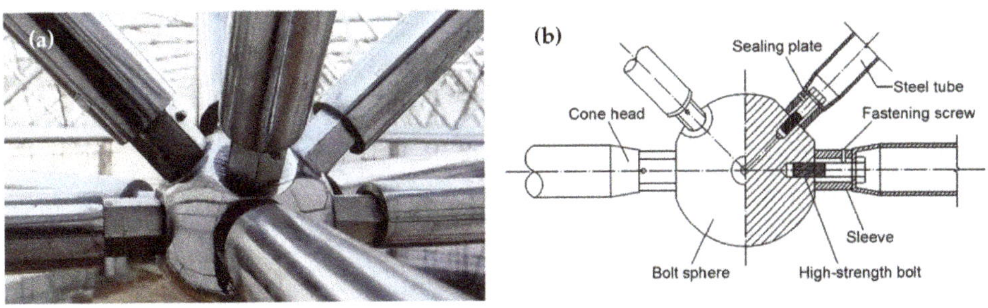

Figure 14. Spherical bolted joint: (**a**) real-life example and (**b**) schematic diagram. Reprinted with permission from Ref. [83]. Copyright 2024 Elsevier.

A study conducted by Qiu et al. [84] investigated the use of M30 high-strength bolts in BSJs and the effect it had on fatigue performance. The bolts were manufactured using material steel 40Cr alloy and the sphere bolts were made from steel, where the outside diameter was 200 mm. The sphere was manufactured in a way where four threads of 30 mm diameter could be used to fit the M30 bolts. The experimental and simulation (regression analysis) S-N curves are shown in Figure 15 and the BSJs displayed better fatigue life for lower stress ranges. Furthermore, the fact that the BSJs have better fatigue performance compared to codes indicates that they have better safety considerations and are more reliable.

The fatigue failure of one high-strength bolt occurred in the first engaged thread between the bolt and sphere as shown in Figure 14b. Figures 16 and 17 correspond to the failed bolt. The fracture surface indicated that the crack initiation occurred at the bolt thread, which further propagates, causing slippage cracks between the threads and the spherical bolts [84] as shown in Figure 16. This causes the bearing area to lose its strength due to the damage that occurred in the material. Dimples can be observed in the surfaces, which were also like bubble structures.

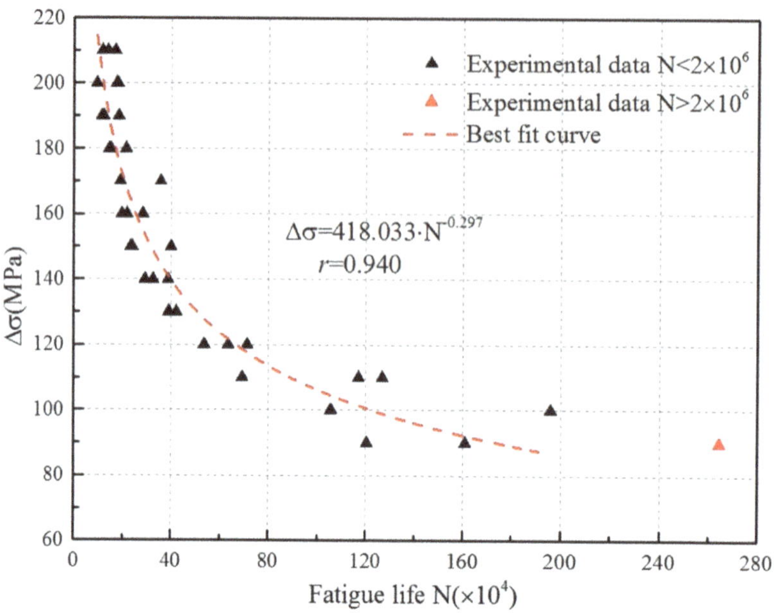

Figure 15. Fatigue life under low and higher cycles with numerical simulation. Reprinted with permission from Ref. [84]. Copyright 2024 Elsevier.

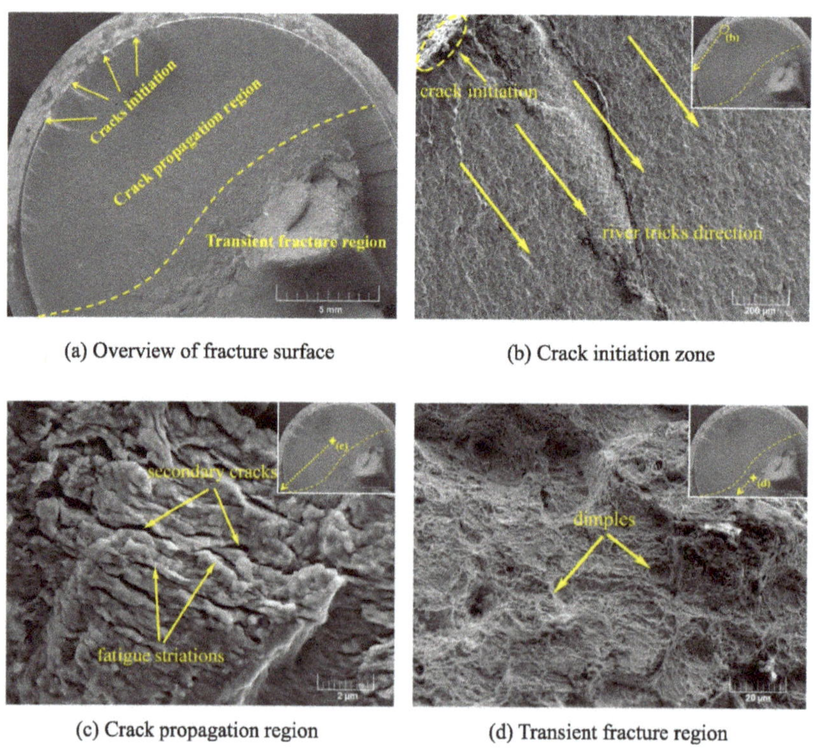

Figure 16. Fatigue crack behaviour of 5th stress loading specimen. Reprinted with permission from Ref. [84]. Copyright 2024 Elsevier.

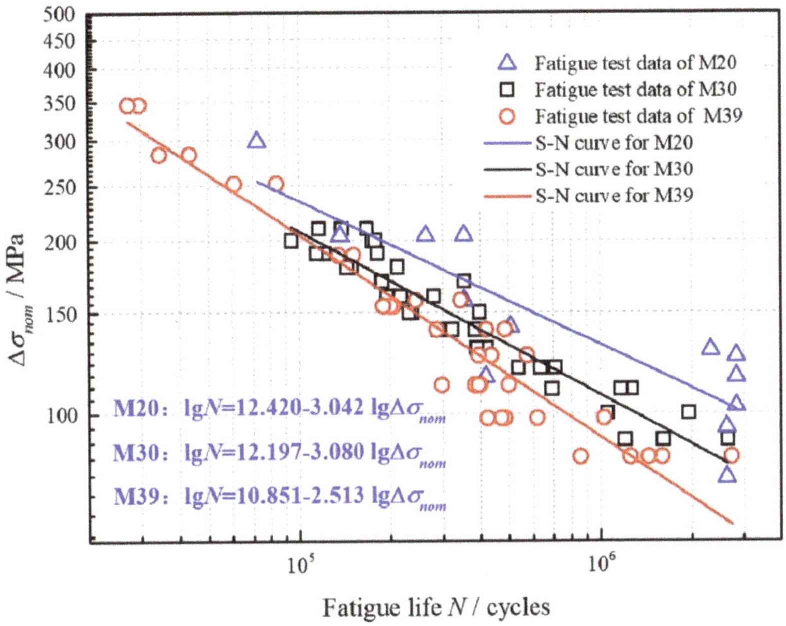

Figure 17. Fatigue life BSJs under different bolts: M20, 30 and 39. Reprinted with permission from Ref. [83]. Copyright 2024 Elsevier.

Another investigation on BSJs, as investigated by Zhou et al. [85] on M60 high-strength bolts, expresses similar findings. Qiu et al. (2020) reported the role of different bolts on BSJs as shown in Figure 17. From the fatigue testing, they reported that all the specimens failed at the thread of the bolts. The increased stress concentration led to fatigue crack initiating in the transient fracture zone which led to fatigue crack. The fatigue life of the M60 bolts in BSJs was said to be less compared to the M30 bolt since one had a bigger diameter than the other, which indicates that the stress concentration was higher. Thus, lower diameter produced the highest fatigue life, which supports the statement mentioned earlier about the effect of diameters on the fatigue of BSJs. Overall, these investigations explained how different types of parameters can lead to improving the fatigue performance of bolted joints.

2.3. Clinched Joints

Clinched joints have been used in many applications such as door frames, air-conditioning, ventilation systems and the aerospace industry due to their lightweight structure and increased interlocking strength [86]. Clinched joints are formed by compressing two or more thin sheets of metal together using a punch and a die as shown schematically in Figure 18. This prevents the top layer from separating from the lower layer since it has already formed the interlock into one component. The clinching process has four steps, clamping, punching, locking and die return, as shown in Figure 18b.

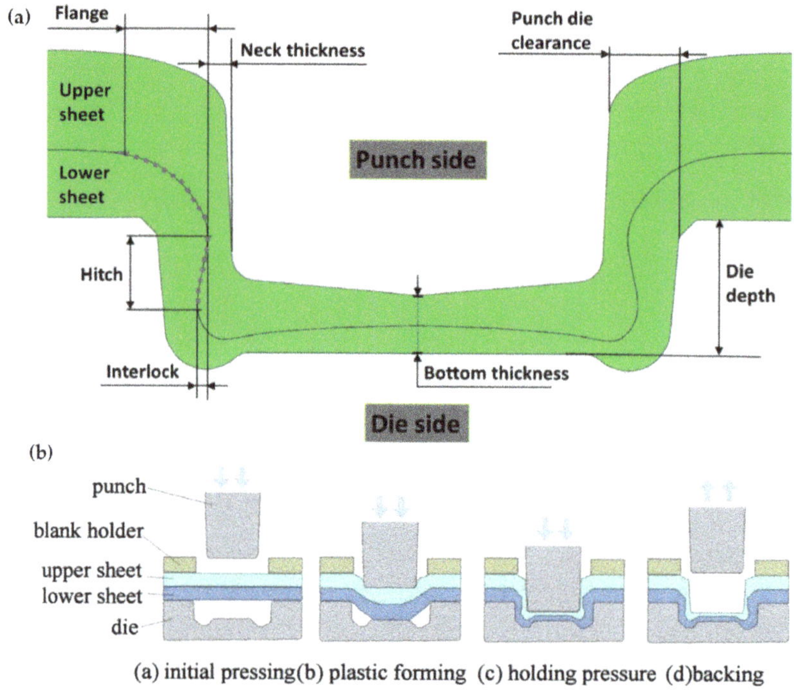

Figure 18. (**a**) Schematic drawing of clinched joint. Reprinted with permission from Ref. [86]. Copyright 2024 Elsevier. And (**b**) different steps to form the clinched joint. Reprinted with permission from Ref. [87]. Copyright 2024 Elsevier.

Most clinched joints are formed using metallic materials since they can undergo large deformation while sustaining their mechanical properties. LeBozec et al. [88] examined different joining methods including clinched joints to investigate how these different joining techniques affect the simultaneous corrosion–fatigue performance. For the unpainted clinched joint, they found that the specimen had corrosion occurring in the overlap area of the steel sheet. However, it was stated that the strength of the joints was reduced compared to the earlier investigation conducted by Harzheim et al. [89]. Harzheim et al. [88] reported detailed studies on the role of pre-corroded clinched specimens on the fatigue life of the joint. According to their findings, the pre-corroded joints exhibited improved fatigue life as the gap in the joint filled up with zinc hydroxide. This zinc hydroxide is the corrosion product originated from the metal sheet (HCT590X) surface. The tight presence of these corrosion products between two metal sheets acts as a cushion under fluctuating loads. As a result, the stress magnitude becomes lower under a given load. Somewhat contradictory results were reported by LeBozec et al. [89] on the role of corrosion/corrosion products towards the performance of the fatigue life of clinched joints. Accordingly, it was only beneficial when the fatigue loading frequency was lowered from 0.5 to 0.25 Hz. Under relatively lower loading frequency, the corrosion products do not have the opportunity to be thrown out from the joint and thus can perform their cushioning effect [90]. They also reported that the unpainted clinched joints had a higher number of cycles to failure compared to the hybrid clinched/adhesive-bonded joints. However, the clinched/adhesive joint showed more corrosion–fatigue failures compared to the clinched joints with the interface failure mode. In all the corrosion–fatigue testing, the welded specimens performed well, which can be due to their enhanced mechanical properties compared to the other types of joints. There are many types of corrosion, such as trans-granular corrosion and intergranular, pitting and galvanic corrosion. It seems that galvanic corrosion is the primary contributor to this fact [90]. Overall, research needs to be more focused on the fatigue behaviour of clinching joints during corrosion [91,92]. There were many unanswered gaps that need to be filled before they could be implemented as part of a strategy to reduce fatigue failures and increase resistance.

Clinched joints can be used as a replacement against bolted joints in automobiles and aircraft. Both types of joints experience corrosion; however, clinched joints will display better fatigue performance compared to bolted joints as they use corrosion as an advantage to delay fatigue crack growth and enhance the durability of the joint [89]. Not only do they provide better fatigue performance, but they will also reduce overall weight as they do not require additional components to support the joint and are cost-effective. Other than the enhanced fatigue properties of sandwich-clinched joints, they could be useful in construction buildings and provide thermal insulation due to their interlayer effectiveness. The applications that have been discussed also apply to SPR and EMR joints as they possess similar fatigue abilities to clinched joints.

2.4. Riveted Joints

Like bolted joints, riveted joints also use the same concept by joining two components together using rivets instead of bolts. Even though they have the same method of joining, the installation of rivets prevents the laminates from moving, unlike bolted joints where there is still some motion between the adherents. The head of the rivet is deformed as shown in Figure 19a, whereas Figure 19b shows the schematic of multiple rivets. Some applications of rivets include cars, aircraft bodies, submarines and many more. Riveted joints are said to be more cost-effective than bolted joints since there are rivet machines that can install multiple rivets at the same time in a short duration compared to bolted joints [44,93]. Some types of riveted joints include butt and lap joints, where there can be multiple rivets installed on the laminates as can be seen in Figure 19b. Overall, riveted joints pose slightly better fatigue characteristics compared to bolted joints due to their enhanced clamping pressure. Self-piercing and electromagnetic riveted joints are two variations that have been well implemented. These joints have been implemented in some engineering applications but have not been taken seriously when it comes to fatigue testing [94].

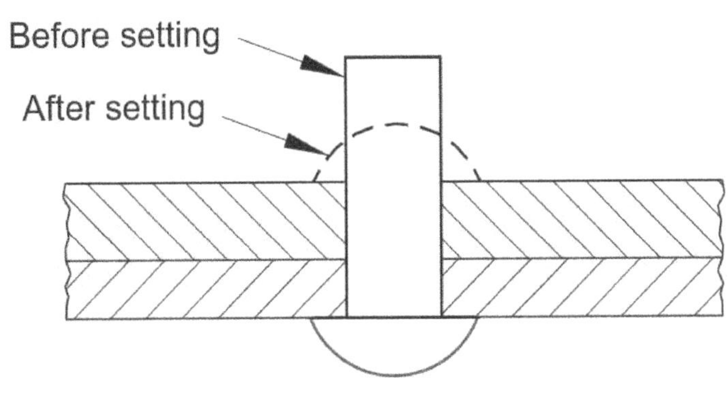

(a)

Figure 19. *Cont.*

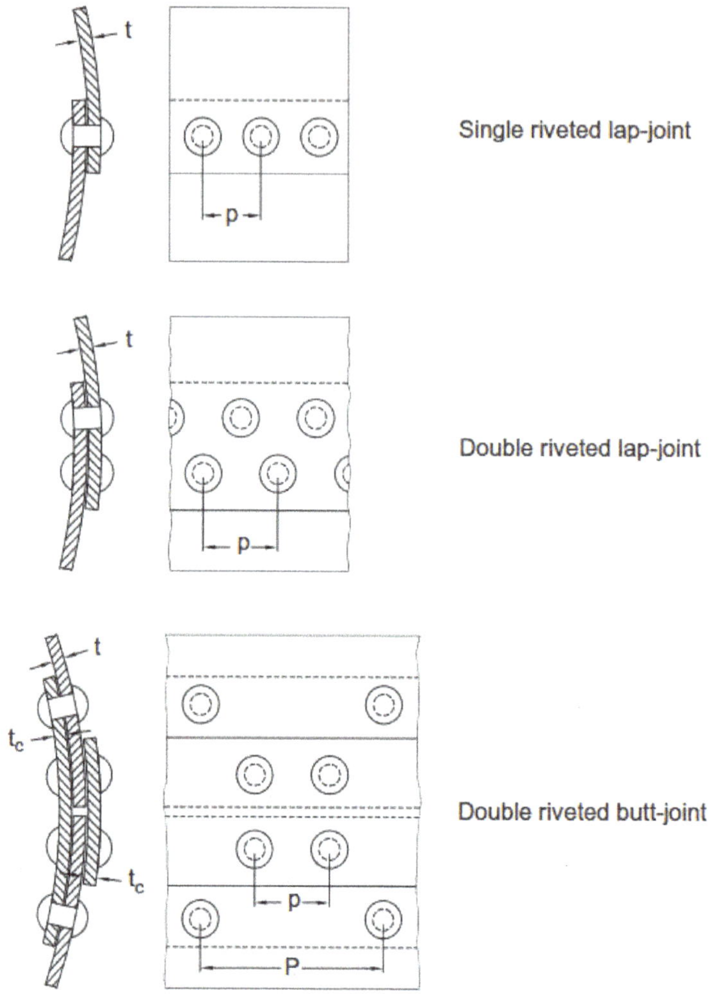

(b)

Figure 19. (**a**) Example of a rivet used between two laminates before the rivet deformation and after deformation and (**b**) an example of multiple riveted joints. Reprinted with permission from Ref. [93]. Copyright 2024 Elsevier.

2.4.1. Self-Pierce Riveting (SPR)

Self-pierce riveting (SPR) [95,96] is also known as a hybrid joint between the clinching and riveting process as shown schematically in Figure 20 and mainly used in lightweight structures. A study conducted by Gay et al. [97] examined the fatigue effects of GFRP and aluminium alloy sheets with SPRs using different rivet shapes and temperature conditions. It was found that the fatigue cracks grew rapidly during low-cyclic conditions and one of the specimens broke when the number of cycles was recorded to be 30,000. They also found that as the temperature increased, the fatigue strength decreased for the SPR joints. Thus, the mechanical properties play a vital role in determining the fatigue strength when tested against different temperature conditions, as well as the geometry of the rivets used for dissimilar sheets.

Zhang et al. [64,87] investigated SPR joints made from 1.5 mm thick steel (DP590) and aluminium alloy (AA5754) and the corresponding S-N curves are shown in Figure 21. Compared to Gay et al.'s [97] experimental results, Zhang et al.'s [87] SPR joints showed higher fatigue load limit, as the metals had better mechanical properties than GFRP, which was expected despite the effects of temperature and rivet shapes used in Gay et al.'s [97] study. In terms of the fatigue crack results, at 40 and 50% of the load level, the crack was experienced within the lower sheets of the aluminium alloy, while at 60%, the rivet experienced fatigue fracture [87]. Fatigue fracture was also observed between the lower and upper sheets, which was due to the change in load at lower cyclic conditions. As the number of cycles increased, it was found that the crack propagated along the width and thickness of the sheet and gradually propagated closer to the sheet edge.

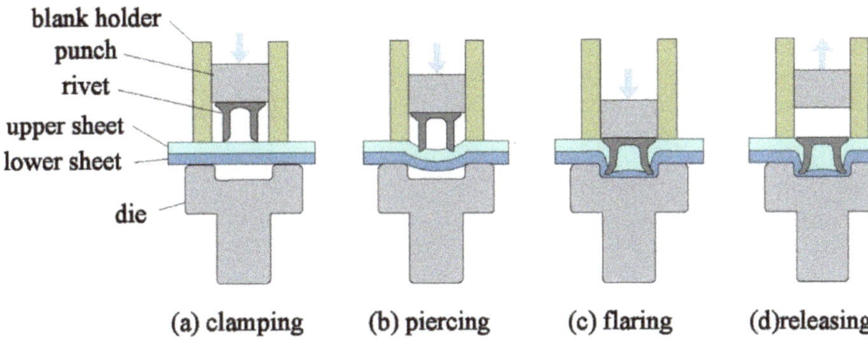

Figure 20. Schematic of the SPR process. Reprinted with permission from Ref. [87]. Copyright 2024 Elsevier.

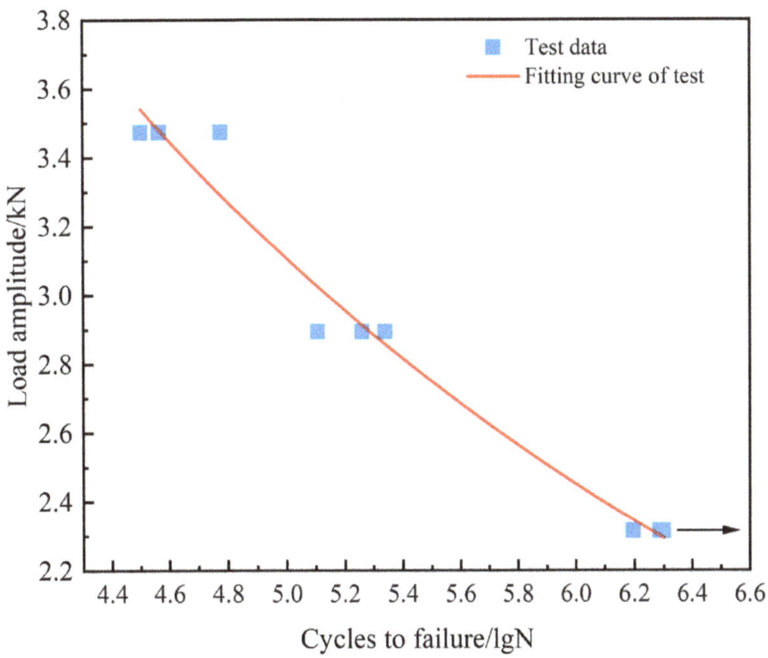

Figure 21. F-N curve of SPR joint at 2 million cycles. Reprinted with permission from Ref. [87]. Copyright 2024 Elsevier.

An investigation conducted by Zhao et al. [6] examined the effects of different sheet thicknesses of aluminium alloy sheets (AA5052). According to their findings, the fatigue life increases when the sheet size was enlarged. It was found that this delayed the crack initiation of the SPR joints when the sheet thickness was increased due to its increased resistance to high loads. However, when the sheet thickness was increased from 2 to 2.5 mm, it can be said that the fatigue loads decreased at low cyclic loads, while they increased at high cyclic loads. It still shows that the fatigue life for 2.5 mm thickness was better than the 2 mm; however, the sheet thickness was affected at low fatigue loads more than at high fatigue loads. The fatigue crack behaviour showed that all the specimens had fatigue fracture cracks around the rivet heads of the joints [87]. For the 1.5 mm thickness, the SPR joints experienced fatigue crack initiation at the rivet head, whereas as the sheet thickness increased, the crack initiated at the bottom layer of the rivet. This was due to the larger sheet thickness having larger grain sizes in the sheet, which caused the fatigue crack away from the rivet head for thinner sheets. Hence, this delays the fatigue crack growth for increased sheet thicknesses. This may be useful for dissimilar sheet materials of SPR joints if the sheet thickness is increased and may improve the joints; however, further research is needed to validate this statement.

2.4.2. Electromagnetic Riveting (EMR)

Electromagnetic riveting (EMR) has been recently developed to enhance the overall strength of riveted joints, as shown schematically in Figure 22a. The rivet die compresses the sheets together with the riveting energy provided by the equipment. Figure 22b shows that the punch deforms the rivet to form a driven head, which causes an interlock between the two sheets. The driven head influences the mechanical property of the overall joint.

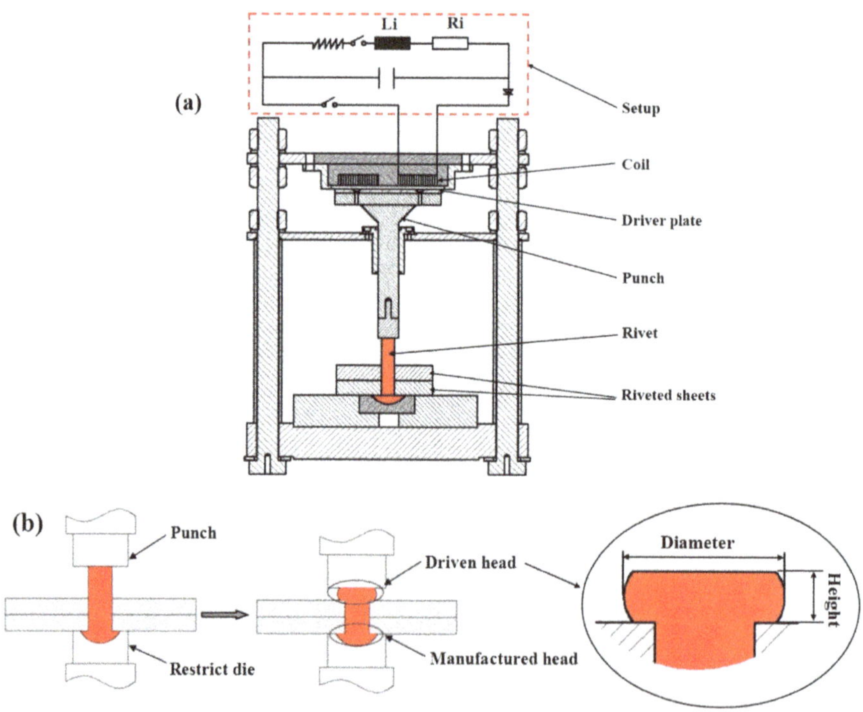

Figure 22. Schematic diagram of EMR process: (**a**) overall process and (**b**) different steps. Reprinted with permission from Ref. [98]. Copyright 2024 Elsevier.

It has better fatigue performance compared to normal riveted joints as reported by Li et al. [99] and Liao et al. [100]. A study conducted by Jiang et al. [98] reported the fatigue behaviour of riveted lap joints using EMR with CFRP and aluminium sheets with thicknesses of 2.5 and 1.8 mm, respectively. Jiang et al. [98] reported that, as the number of cycles increased, the hole of the riveted joint expanded, which eventually led to failure due to the loss of tightening strength. Jiang et al. [98] also found that the EMR specimens showed no signs of disruption during 90% of the cycles endured by the joint until the last few cycles, which were caused mainly by the fatigue crack initiation. This indicates that EMR joints displayed better performance than regular riveted and SPR joints as fatigue cracks started to initiate in the earlier stages. Overall, the EMR processes show better fatigue performance compared to regular riveted joints.

3. Discussion

This section of the manuscript aims to provide a comprehensive analysis of the fatigue characteristics of the joints reviewed in previous sections and how some of the factors considered have changed the behaviour of these joints under fatigue. The use of S-N curves has been implemented in most reports to understand the fatigue behaviour of mechanical joints. In support of that, the Paris law and Weibull distribution techniques were used to identify the crack growths and propagation, as well as the security level, which identifies how the level can be suitable for different fatigue conditions the joint experiences.

3.1. Key Findings and Trends

It has been validated that increasing the bond length can enhance the fatigue resistance of adhesive joints. Moreover, the type of adhesive used also depicts the fatigue resistance despite the bond length as some types of epoxies have different mechanical properties that could enhance fatigue performance for shorter bond lengths. However, some adhesive joints display reduced or no change in their fatigue strength despite enhancing the length of the overlap area as can be seen in Moreira et al.'s [47] investigation. It is assumed that after a certain increase in bond length, it cannot further reduce the stress concentration, and this can increase the bending, which decreases fatigue life. Moreover, increasing the laminate or adhesive thickness does not significantly increase the fatigue life, but it increases the fatigue life when the surface roughness of the adhesive is increased, as reported by Boutar et al. [55]. On the other hand, increasing the thickness of the laminate results in enhanced fatigue performance but may defer when the material used for the adherend is changed.

In bolted joints, the use of cold expansion and extrusion have shown promising results in delaying fatigue crack growth in bolted joints. Pre-tension force in bolts also plays a vital part in enhancing the fatigue strength of the joint despite some cold-extrusion bolted joints. The impact of interference fit in bolted joints also contributes to the reduction of fatigue failure due to being better than the cold-expansion and -extrusion processes. SEM statistics show that increased interference fit levels can reduce stress around the hole more effectively than cold-extrusion and -expansion methods, though it can be stated that increasing the pre-tension force further can lead to fretting within the joint despite the level of cold extrusion or expansion. Fretting is more common in bolted joints compared to the rest due to the high stress concentration factor the holes experience. The investigations reported that increased preloads led to increased fatigue strength at a certain pre-tension but showed better fatigue resistance when the spacer size was increased, which will move the stress away from the hole, delaying fatigue. The impact of surface treatment such as shot peening and lubrication also find benefits for improving fatigue performance. BSJs are a new manufacturing technique that shows better fatigue performance than standard code specifications. Increasing the thread root of BSJs reduces stress concentration, as well as decreases the diameter of bolts. These factors that enhance the fatigue durability of bolted joints must be cautiously assessed and meet the requirements for applications.

Moving on, clinched joints have better resistance to fatigue failure under corrosion as the corroded area of the joint delays the punched area from experiencing fatigue crack during the initial stages due to the stress being evenly distributed in the joint. Coated clinched joints, including hybrid clinched joints combined with adhesive, decrease fatigue life, while unpainted clinched joints show an increased fatigue life since painted clinched joints prevent the corrosion build-up and increase stress concentration near the punch area of the joint. Other than corrosion fatigue, clinched joints can also reduce fatigue failures by ensuring different connection methods are employed before the clinching process, such as cross-tension, coach-peel or tensile-shear clinched joints. Tensile shear has the best fatigue properties compared to the rest since the laminates are more aligned with each other, which allows the load to be distributed more evenly and reduces stress concentration. However, implementing an additional material between the laminates in cross-tension clinched joints can further increase fatigue life more than tensile shear, which is results in sandwich clinched joints. In terms of the manufacturing methods, different punch geometries have a significant influence on the fatigue performance since they affect the neck thickness of the overlap area. It was also stated that either annealing or quenching can reduce the service life of clinched joints and may be due to the defects in the crystal lattice microstructure.

Lastly, SPR and EMR joints are types of advanced riveted joints that display excellent fatigue mechanisms compared to normal riveted joints. In the relevant studies examined, applying different rivet shapes will alter the stress distribution in the joint and increase the fatigue strength despite being at low temperatures. However, it was also reported that high temperatures with rivets of the same shape had lower fatigue strengths, which could be due to their strength characteristic. Increasing the sheet thickness of SPR joints can enhance fatigue performance since the extra sheet layers can withstand high loads and increase their durability. EMR, on the other hand, uses electromagnetic current and relies on speed when the rivet is punched into the sheets. This provides slightly better mechanical properties compared to SPR joints. EMR joints tend to have better fatigue resistance than SPR joints since rivet holes are stretched during the later stages of the fatigue process. The setup of rivets in EMR can also affect the fatigue performance as horizontal rivet placement poses higher fatigue resistance since it is not near the edge of the joint, allowing the stress to be evenly distributed across the joint.

3.2. Knowledge Gap and Inconsistencies in the Literature

The above-mentioned literature review confirms the effect of different types of mechanical joints on their respective fatigue properties. Having said that, there are still knowledge gaps and inconsistencies in the literature that need to be critically addressed. While increasing the overlap length provides some improvements to adhesive joints, more attention needs to be given to why certain lengths prevent resistance of fatigue with a further increase in overlap length. Not much research has been focused on different laminate materials at various bond lengths, which needs to be investigated. In addition, fatigue of adhesive joints from exposure to changing environmental conditions is not critically assessed in the literature. As claimed by Meneghetti et al. [45], certain angles can evenly distribute stress along the adhesive joint, which can resist crack initiation for longer periods due to the increase in the overall structural integrity of the joint. However, this hypothesis has not been examined and needs proper attention.

Even though the cold-expansion and -extrusion techniques display improved fatigue in bolted joints, further research needs to investigate stress relaxation and different environmental effects (e.g., corrosion, temperature, UV radiation). These aspects are generally needed for fatigue testing in all mechanical joints other than bolted connections since most applications undergo various environmental conditions. Despite displaying satisfactory results when BSJs undergo fatigue testing, this new prototype is needed for further research when it comes to inspecting the defects of the threads in the bolts used in BSJs for varying loading conditions. Understanding the underlying concept can allow engineers to start implementing more BSJs in complex engineering appliances.

Moreover, one of the biggest concerns when it comes to clinched joints is the analysis of corrosion–fatigue. There have been many research studies to examine the overall strength of the fatigue joints but less analysis in terms of corrosion–fatigue. Further research is needed to explore why some clinched joints under fatigue experience corrosion to confirm their hypothesis under controlled experiments. Kim et al. [101] claimed that coach-peel clinched joints had fewer fatigue properties compared to either cross-tension or tensile-shear clinched joints with minimal evidence and this requires further validation. This also includes the lack of knowledge when it comes to investigating how sandwich clinched joints can potentially improve fatigue performance in many engineering applications. This area of investigation needs to be further studied, especially in varying loading conditions and the different interlayers used to analyse the effects they have on the joint. In terms of riveted joints, it has been mentioned that SPR joints can increase their fatigue properties when the number of layers used is more than one. However, there is a gap between different sheet materials and the increased number of sheets used, which needs further clarification.

4. Conclusions and Future Perspective

Based on the above-mentioned information, it could be concluded that though there is a large volume of information available in the literature, inconsistency arises in the given data. This is mostly due to the use of different experimental set-ups by various research groups and lack of experimental repeatability.

Bolted joints can decrease crack initiation through cold-extrusion or -expansion methods. Bolted spherical joints display enhanced fatigue performance better than single- or double-lap bolted joints. Moreover, clinched joints are said to have good fatigue resistance when they are corroded compared to the other joints. They can also withstand fatigue failure for longer periods when they are connected in tensile-shear connections compared to cross-tension or coach-peel clinched joints. Factors such as manufacturing methods such as either additional interlayer materials or heat treatment of sheets can also delay fatigue failure during the early stages.

Recent advances in riveted joint forming techniques such as self-pierce and electromagnetic riveting have caught the attention of scientists to investigate how they are different to riveted joints since riveted joints possess almost similar fatigue properties. SPRs are also known as hybrid clinched-riveted joints and have shown incredible fatigue capabilities when tested against different sheet thicknesses and rivet shapes at various temperatures. EMRs have slightly better fatigue characteristics compared to SPRs due to their machine process when tested against different materials and rivet arrangements. It is said that the crack behaviour for adhesive and clinched joints always starts at the interlayer surface and propagates slowly towards the adherents, while riveted and bolted joints have fatigue cracks initiating at the holes due to higher stress concentration.

Throughout this analysis, some limitations in the work can be found when exploring these types of joints. For instance, further research is needed to understand how certain clinched joints have a higher fatigue life when they are corroded. Scientists should focus more on the stress distributions through finite element analysis (FEA) and testing with more different types of material sheets to identify whether the theory still holds even if different fatigue parameters are tested. The research in this area is mainly focused on smaller joint dimensions, which makes it difficult to make certified validations based on one piece of evidence. One of the main research areas that needs to be focused on is how these non-thermal joints can be as useful as thermal joints. Most appliances today use welded joints more than non-thermal joints due to their enhanced structural integrity and strength, which serves as a higher advantage. For instance, some studies have compared clinched joints with spot welded joints and noticed the large fatigue behaviour difference they have between each other [102]. Hence, this would allow scientists to research more on thermal joints and utilise these joints more than non-thermal joints even though thermal joints generally are more expensive since they need skilled workers and specialised equipment for the manufacturing process. Research should be more directed towards non-thermal hybrid joints as they are the closest to possessing similar fatigue abilities to thermal joints due to their structural strength, and costs need to be reduced to manufacture such joints for future designs.

Thus, the above conclusions fulfill the objectives of the present review. Nonetheless, further exploration in this is foreseen to cover the fatigue life/behaviour of mechanical joints based on 3D-printed components, which is out of the scope of the present review.

Author Contributions: Conceptualization, A.P.; methodology, A.K.B.; software, D.S.B.; validation, A.K.B. and D.S.B.; formal analysis, D.S.B.; investigation, D.S.B.; resources, A.P.; data curation, A.K.B. and A.P.; writing—original draft preparation, D.S.B.; writing—review and editing, A.K.B. and A.P.; visualization, D.S.B.; supervision, A.K.B. and A.P.; project administration, A.P. All authors have read and agreed to the published version of the manuscript.

Funding: This research received no external funding.

Data Availability Statement: No new data were created or analyzed in this study.

Conflicts of Interest: The authors declare that they have no known competing financial interests or personal relationships whatsoever that could have influenced the work reported in this paper. The authors declare no conflict of interest.

References

1. Speck, J.A. *Mechanical Fastening, Joining, and Assembly*; CRC Press: Boca Raton, FL, USA, 2018.
2. Yokozeki, K.; Hisazumi, K.; Vallée, T.; Evers, T.; Ummenhofer, T.; Boretzki, J.; Albiez, M. Hybrid joints consisting of pre-tensioned bolts and a bonded connection, Part II: Large-scale experiments. *Int. J. Adhes. Adhes.* **2024**, *128*, 103523. [CrossRef]
3. Denkert, C.; Gerke, T.; Glienke, R.; Doerre, M.; Henkel, M.K.; Fricke, H.; Myslicki, S.; Kaufmann, M.; Voss, M.; Vallee, T. Experimental investigations on pre-tensioned hybrid joints for structural steel applications. *J. Adhes.* **2023**, *99*, 117–152. [CrossRef]
4. Boretzki, J.; Albiez, M. Static strength and load bearing behaviour of hybrid bonded bolted joints: Experimental and numerical investigations. *J. Adhes.* **2023**, *99*, 606–631. [CrossRef]
5. Abdullah, M.; Abdullah, A.; Samad, Z. Structural integrity assessment of a composite joint: A review. In *Hole-Making and Drilling Technology for Composites*; Woodhead Publishing: Sawston, UK, 2019; pp. 31–46.
6. Zhao, L.; He, X.; Xing, B.; Lu, Y.; Gu, F.; Ball, A. Influence of sheet thickness on fatigue behavior and fretting of self-piercing riveted joints in aluminum alloy 5052. *Mater. Des.* **2015**, *87*, 1010–1017. [CrossRef]
7. Adamovic, D.; Zivic, F. Hardness and Non-Destructive Testing (NDT) of Ceramic Matrix Composites (CMCs). In *Encyclopedia of Materials: Composites*; Elsevier: Amsterdam, The Netherlands, 2021; pp. 183–201. [CrossRef]
8. Russo, P.; Papa, I.; Lopresto, V. Mechanical Properties and Non-Destructive Evaluations of Joints Based on Polymer Composites. In *Encyclopedia of Materials: Composites*; Elsevier: Amsterdam, The Netherlands, 2021; pp. 334–351.
9. Okorn, I.; Nagode, M.; Klemenc, J.; Oman, S. Examining the Effects on a Fatigue Life of Preloaded Bolts in Flange Joints: An Overview. *Metals* **2024**, *14*, 883. [CrossRef]
10. Zhou, S.; Yang, B.; Huang, Y.; Wu, X. Fatigue Analysis of Composite Bolted Joints under Random and Constant Amplitude Fatigue Loadings. *Materials* **2024**, *17*, 2740. [CrossRef] [PubMed]
11. Adasooriya, N.; Pavlou, D.; Hemmingsen, T. Fatigue strength degradation of corroded structural details: A formula for S-N curve. *Fatigue Fract. Eng. Mater. Struct.* **2020**, *43*, 721–733. [CrossRef]

12. Croccolo, D.; De Agostinis, M.; Fini, S.; Khan, M.Y.; Mele, M.; Olmi, G. Optimization of Bolted Joints: A Literature Review. *Metals* **2023**, *13*, 1708. [CrossRef]
13. Pramanik, A.; Dixit, A.; Chattopadhyaya, S.; Uddin, M.; Dong, Y.; Basak, A.; Littlefair, G. Fatigue life of machined components. *Adv. Manuf.* **2017**, *5*, 59–76. [CrossRef]
14. Pollock, L.; Abdelwahab, A.K.; Murray, J.; Wild, G. The Need for Aerospace Structural Health Monitoring: A review of aircraft fatigue accidents. *Int. J. Progn. Health Manag.* **2021**, *12*, 1–16.
15. Younis, H.B.; Kamal, K.; Sheikh, M.F.; Hamza, A. Prediction of fatigue crack growth rate in aircraft aluminum alloys using optimized neural networks. *Theor. Appl. Fract. Mech.* **2022**, *117*, 103196. [CrossRef]
16. Narayanan, G. Probabilistic fatigue model for cast alloys of aero engine applications. *Int. J. Struct. Integr.* **2020**, *12*, 454–469. [CrossRef]
17. Sarkis-Onofre, R.; Catalá-López, F.; Aromataris, E.; Lockwood, C. How to properly use the PRISMA Statement. *Syst. Rev.* **2021**, *10*, 117. [CrossRef]
18. Page, M.J.; McKenzie, J.E.; Bossuyt, P.M.; Boutron, I.; Hoffmann, T.C.; Mulrow, C.D.; Shamseer, L.; Tetzlaff, J.M.; Moher, D. Updating guidance for reporting systematic reviews: Development of the PRISMA 2020 statement. *J. Clin. Epidemiol.* **2021**, *134*, 103–112. [CrossRef] [PubMed]
19. Moore, P.; Booth, G. 2—Structures under load. In *The Welding Engineer's Guide to Fracture and Fatigue*; Moore, P., Booth, G., Eds.; Woodhead Publishing: Oxford, UK, 2015; pp. 11–21. [CrossRef]
20. Sun, C.T.; Jin, Z.H. Chapter 1—Introduction. In *Fracture Mechanics*; Sun, C.T., Jin, Z.H., Eds.; Academic Press: Boston, MA, USA, 2012; pp. 1–10. [CrossRef]
21. Yan, C.; Hao, L.; Yang, L.; Hussein, A.Y.; Young, P.G.; Li, Z.; Li, Y. Chapter 8—Application and prospects. In *Triply Periodic Minimal Surface Lattices Additively Manufactured by Selective Laser Melting*; Yan, C., Hao, L., Yang, L., Hussein, A.Y., Young, P.G., Li, Z., Li, Y., Eds.; Academic Press: Boston, MA, USA, 2021; pp. 283–315. [CrossRef]
22. Mlikota, M.; Schmauder, S.; Dogahe, K.; Božić, Ž. Influence of local residual stresses on fatigue crack initiation. *Procedia Struct. Integr.* **2021**, *31*, 3–7. [CrossRef]
23. Wang, Y.; Yu, T.; Wang, Z.; Li, L.; Yuan, G.; Zhang, Z. Low cycle fatigue behaviors of CrCoNi medium-entropy alloy with different grain sizes. *Int. J. Fatigue* **2024**, *180*, 108105. [CrossRef]
24. Mlikota, M.; Dogahe, K.; Schmauder, S.; Božić, Ž. Influence of the grain size on the fatigue initiation life curve. *Int. J. Fatigue* **2022**, *158*, 106562. [CrossRef]
25. Smith, C.B.; Mishra, R.S. *Friction Stir Processing for Enhanced Low Temperature Formability: A Volume in the Friction Stir Welding and Processing Book Series*; Butterworth-Heinemann: Oxford, UK, 2014.
26. Xin, Q. Durability and reliability in diesel engine system design. In *Diesel Engine System Design*; Woodhead Publishing: Oxford, UK, 2013; pp. 113–202.
27. Smith, C.B.; Mishra, R.S. Chapter 4—Case Study of Aluminum 5083-H116 Alloy. In *Friction Stir Processing for Enhanced Low Temperature Formability*; Smith, C.B., Mishra, R.S., Eds.; Butterworth-Heinemann: Boston, MA, USA, 2014; pp. 19–124. [CrossRef]
28. Tanaka, K. 4.04—Fatigue Crack Propagation. In *Comprehensive Structural Integrity*; Milne, I., Ritchie, R.O., Karihaloo, B., Eds.; Pergamon: Oxford, UK, 2003; pp. 95–127. [CrossRef]
29. Zerbst, U.; Ainsworth, R.A.; Beier, H.T.; Pisarski, H.; Zhang, Z.L.; Nikbin, K.; Nitschke-Pagel, T.; Münstermann, S.; Kucharczyk, P.; Klingbeil, D. Review on fracture and crack propagation in weldments—A fracture mechanics perspective. *Eng. Fract. Mech.* **2014**, *132*, 200–276. [CrossRef]
30. Bolotin, V.V. *Mechanics of Fatigue*; CRC Press: Boca Raton, FL, USA, 2020.
31. Murakami, Y.; Endo, M. Prediction model of SN curve without fatigue test or with a minimum number of fatigue tests. *Eng. Fail. Anal.* **2023**, *154*, 107647. [CrossRef]
32. Murakami, Y.; Takagi, T.; Wada, K.; Matsunaga, H. Essential structure of SN curve: Prediction of fatigue life and fatigue limit of defective materials and nature of scatter. *Int. J. Fatigue* **2021**, *146*, 106138. [CrossRef]
33. Chandran, K.R. The finding of the reciprocal relationship between fatigue (SN) behavior and fatigue crack growth behavior enabling interconversion of data in structural materials. *Materialia* **2022**, *25*, 101541. [CrossRef]
34. Lee, Y.-L.; Barkey, M.E. Chapter 4—Stress-Based Uniaxial Fatigue Analysis. In *Metal Fatigue Analysis Handbook*; Lee, Y.-L., Barkey, M.E., Kang, H.-T., Eds.; Butterworth-Heinemann: Boston, MA, USA, 2012; pp. 115–160. [CrossRef]
35. Antunes, R.A.; de Oliveira, M.C.L. 9—Effect of surface treatments on the fatigue life of magnesium and its alloys for biomedical applications. In *Surface Modification of Magnesium and Its Alloys for Biomedical Applications*; Narayanan, T.S.N.S., Park, I.-S., Lee, M.-H., Eds.; Woodhead Publishing: Oxford, UK, 2015; pp. 283–310. [CrossRef]
36. Chang, K.-H. Chapter 4—Fatigue and Fracture Analysis. In *Product Performance Evaluation with CAD/CAE*; Chang, K.-H., Ed.; Academic Press: Boston, MA, USA, 2013; pp. 205–273. [CrossRef]
37. Paul, S.K. Correlation between endurance limit and cyclic yield stress determined from low cycle fatigue test. *Materialia* **2020**, *11*, 100695. [CrossRef]

38. Pang, J.-C.; Li, S.X.; Wang, Z.G.; Zhang, Z. General relation between tensile strength and fatigue strength of metallic materials. *Mater. Sci. Eng. A* **2013**, *564*, 331–341. [CrossRef]
39. Srinivasan, D.V.; Idapalapati, S. Review of debonding techniques in adhesively bonded composite structures for sustainability. *Sustain. Mater. Technol.* **2021**, *30*, e00345. [CrossRef]
40. Zuo, P.; Vassilopoulos, A.P. Review of fatigue of bulk structural adhesives and thick adhesive joints. *Int. Mater. Rev.* **2021**, *66*, 313–338. [CrossRef]
41. Ahn, J.; Stapleton, S.; Waas, A.M. 14—Advanced modeling of the behavior of bonded composite joints in aerospace applications. In *Composite Joints and Connections*; Camanho, P., Tong, L., Eds.; Woodhead Publishing: Oxford, UK, 2011; pp. 423–434. [CrossRef]
42. Delzendehrooy, F.; Akhavan-Safar, A.; Barbosa, A.Q.; Beygi, R.; Cardoso, D.; Carbas, R.J.C.; Marques, E.A.S.; da Silva, L.F.M. A comprehensive review on structural joining techniques in the marine industry. *Compos. Struct.* **2022**, *289*, 115490. [CrossRef]
43. Pramanik, A.; Basak, A.; Dong, Y.; Sarker, P.; Uddin, M.; Littlefair, G.; Dixit, A.; Chattopadhyaya, S. Joining of carbon fibre reinforced polymer (CFRP) composites and aluminium alloys–A review. *Compos. Part A Appl. Sci. Manuf.* **2017**, *101*, 1–29. [CrossRef]
44. Childs, P.R.N. Chapter 16—Fastening and Power Screws. In *Mechanical Design Engineering Handbook*; Childs, P.R.N., Ed.; Butterworth-Heinemann: Oxford, UK, 2014; pp. 677–719. [CrossRef]
45. Meneghetti, G.; Quaresimin, M.; Ricotta, M. Influence of the interface ply orientation on the fatigue behaviour of bonded joints in composite materials. *Int. J. Fatigue* **2010**, *32*, 82–93. [CrossRef]
46. Mora, R.P.; Gonzalez, C.E.C.; Gutierrez, S.D.S.; Tijerina, J.J.T.; Arista, B.V.; Pingarron, A.B. Fatigue strength evaluation and fracture behavior of joined dual phase steel/AA6061-T6 aluminum alloy. *Frat. Ed Integrità Strutt.* **2019**, *13*, 530–544. [CrossRef]
47. Moreira, R.D.F.; de Moura, M.F.S.F.; Figueiredo, M.A.V.; Fernandes, R.L.; Gonçalves, J.P.M. Characterisation of composite bonded single-strap repairs under fatigue loading. *Int. J. Mech. Sci.* **2015**, *103*, 22–29. [CrossRef]
48. de Moura, M.F.S.F.; Campilho, R.D.S.G.; Gonçalves, J.P.M. Mixed-Mode Cohesive Damage Model Applied to the Simulation of the Mechanical Behaviour of Laminated Composite Adhesive Joints. *J. Adhes. Sci. Technol.* **2009**, *23*, 1477–1491. [CrossRef]
49. Shrivastava, A. 6—Plastics Part Design and Application. In *Introduction to Plastics Engineering*; Shrivastava, A., Ed.; William Andrew Publishing: Norwich, NY, USA, 2018; pp. 179–205. [CrossRef]
50. Tang, J.H.; Sridhar, I.; Srikanth, N. Static and fatigue failure analysis of adhesively bonded thick composite single lap joints. *Compos. Sci. Technol.* **2013**, *86*, 18–25. [CrossRef]
51. Pugno, N.; Ciavarella, M.; Cornetti, P.; Carpinteri, A. A generalized Paris' law for fatigue crack growth. *J. Mech. Phys. Solids* **2006**, *54*, 1333–1349. [CrossRef]
52. Banea, M.D.; da Silva, L.F.M. Static and fatigue behaviour of room temperature vulcanising silicone adhesives for high temperature aerospace applications. *Mater. Und Werkst.* **2010**, *41*, 325–335. [CrossRef]
53. Gleich, D.; Van Tooren, M.; Beukers, A. Analysis and evaluation of bondline thickness effects on failure load in adhesively bonded structures. *J. Adhes. Sci. Technol.* **2001**, *15*, 1091–1101. [CrossRef]
54. Hart-Smith, L. Adhesive bonding of composite structures—Progress to date and some remaining challenges. *Compos. Technol. Res.* **2002**, *24*, 133–151.
55. Boutar, Y.; Naïmi, S.; Mezlini, S.; Carbas, R.J.C.; da Silva, L.F.M.; Ben Sik Ali, M. Fatigue resistance of an aluminium one-component polyurethane adhesive joint for the automotive industry: Effect of surface roughness and adhesive thickness. *Int. J. Adhes. Adhes.* **2018**, *83*, 143–152. [CrossRef]
56. Ebnesajjad, S.; Landrock, A.H. *Adhesives Technology Handbook*; William Andrew: Norwich, NY, USA, 2014.
57. Dillard, D.A. *Advances in Structural Adhesive Bonding*; Elsevier: Amsterdam, The Netherlands, 2023.
58. Akpinar, S.; Sahin, R. The fracture load analysis of different material thickness in adhesively bonded joints subjected to fully reversed bending fatigue load. *Theor. Appl. Fract. Mech.* **2021**, *114*, 102984. [CrossRef]
59. Malekinejad, H.; Carbas, R.J.; Akhavan-Safar, A.; Marques, E.A.; Castro Sousa, F.; da Silva, L.F. Enhancing fatigue life and strength of adhesively bonded composite joints: A comprehensive review. *Materials* **2023**, *16*, 6468. [CrossRef] [PubMed]
60. Moroni, F.; Musiari, F.; Favi, C. Effect of the surface morphology over the fatigue performance of metallic single lap-shear joints. *Int. J. Adhes. Adhes.* **2020**, *97*, 102484. [CrossRef]
61. Shahani, A.R.; Pourhosseini, S.M. The effect of adherent thickness on fatigue life of adhesively bonded joints. *Fatigue Fract. Eng. Mater. Struct.* **2019**, *42*, 561–571. [CrossRef]
62. Da Costa, J.; Akhavan-Safar, A.; Marques, E.; Carbas, R.; Da Silva, L. The influence of cyclic ageing on the fatigue performance of bonded joints. *Int. J. Fatigue* **2022**, *161*, 106939. [CrossRef]
63. Naat, N.; Boutar, Y.; Naïmi, S.; Mezlini, S.; Da Silva, L.F.M. Effect of surface texture on the mechanical performance of bonded joints: A review. *J. Adhes.* **2023**, *99*, 166–258. [CrossRef]
64. Zhang, X.; Ju, Y.; Zhu, A.; Zou, T. Fatigue behavior of single-lap adhesive joints with similar and dissimilar adherends under cyclic loading: A combined experimental and simulation study. *Mater. Today Commun.* **2023**, *37*, 107215. [CrossRef]

65. Sousa, F.C.; Akhavan-Safar, A.; Carbas, R.J.; Marques, E.A.; da Silva, L.F. Experimental behaviour and fatigue life prediction of bonded joints subjected to variable amplitude fatigue. *Int. J. Fatigue* **2025**, *190*, 108583. [CrossRef]
66. Sousa, F.C.; Akhavan-Safar, A.; Carbas, R.J.; Marques, E.A.; da Silva, L.F. A unified strain energy-based fatigue life prediction methodology for composite and metal adhesive joints: Effects of adhesive, geometry and, environment. *Compos. Part B Eng.* **2024**, *292*, 112022. [CrossRef]
67. Kumar, V.; Sharma, S.; Singh, A. Impact of joint variables on shear and fatigue behavior of composite-metal single-lap adhesive joint. *Eng. Fail. Anal.* **2024**, *168*, 109079. [CrossRef]
68. Machado, J.; Marques, E.; Barbosa, A.; da Silva, L. Fatigue performance of single lap joints with CFRP and aluminium substrates prior and after hygrothermal aging. *Fatigue Fract. Eng. Mater. Struct.* **2019**, *42*, 2325–2339. [CrossRef]
69. Mariam, M.; Afendi, M.; Majid, M.A.; Ridzuan, M.; Gibson, A. Tensile and fatigue properties of single lap joints of aluminium alloy/glass fibre reinforced composites fabricated with different joining methods. *Compos. Struct.* **2018**, *200*, 647–658. [CrossRef]
70. Sousa, F.C.; Akhavan-Safar, A.; Carbas, R.; Marques, E.; Goyal, R.; Jennings, J.; da Silva, L. Investigation of geometric and material effects on the fatigue performance of composite and steel adhesive joints. *Compos. Struct.* **2024**, *344*, 118313. [CrossRef]
71. Sahin, R.; Akpinar, S. The effects of adherend thickness on the fatigue strength of adhesively bonded single-lap joints. *Int. J. Adhes. Adhes.* **2021**, *107*, 102845. [CrossRef]
72. Mariam, M.; Afendi, M.; Abdul Majid, M.S.; Ridzuan, M.J.M.; Azmi, A.I.; Sultan, M.T.H. Influence of hydrothermal ageing on the mechanical properties of an adhesively bonded joint with different adherends. *Compos. Part B Eng.* **2019**, *165*, 572–585. [CrossRef]
73. Feng, N.L.; Malingam, S.D.; Irulappasamy, S. 4—Bolted joint behavior of hybrid composites. In *Failure Analysis in Biocomposites, Fibre-Reinforced Composites and Hybrid Composites*; Jawaid, M., Thariq, M., Saba, N., Eds.; Woodhead Publishing: Oxford, UK, 2019; pp. 79–95. [CrossRef]
74. Li, X.; Yang, J. Fatigue Behavior of Bolted Joints: Comparative Analysis for the Effect of Fretting and Stress Concentration. *Mater. Perform. Charact.* **2020**, *9*, 309–323. [CrossRef]
75. Zheng, M.; Chen, W.; Yan, X.; Liu, Z.; Wahab, M.A. Finite element analysis of bolted joints under torsional loads. *Tribol. Int.* **2025**, *201*, 110188. [CrossRef]
76. Navaneetha, E.; Lakshmi, A.A. Cold extrusion on bulk materials: A review. *Mater. Today Proc.* 2023; *in press*. [CrossRef]
77. Vaara, J.; Kunnari, A.; Frondelius, T. Literature review of fatigue assessment methods in residual stressed state. *Eng. Fail. Anal.* **2020**, *110*, 104379. [CrossRef]
78. Gerendt, C.; Dean, A.; Mahrholz, T.; Englisch, N.; Krause, S.; Rolfes, R. On the progressive fatigue failure of mechanical composite joints: Numerical simulation and experimental validation. *Compos. Struct.* **2020**, *248*, 112488. [CrossRef]
79. Wan, N.; He, Q.; Guo, Y.; Hu, F.; Li, G. Experimental and numerical investigations on fatigue behavior of 42CrMo4 high-strength steel single lap double-bolted joints. *J. Constr. Steel Res.* **2024**, *212*, 108322. [CrossRef]
80. Chakherlou, T.N.; Taghizadeh, H.; Aghdam, A.B. Experimental and numerical comparison of cold expansion and interference fit methods in improving fatigue life of holed plate in double shear lap joints. *Aerosp. Sci. Technol.* **2013**, *29*, 351–362. [CrossRef]
81. Chakherlou, T.N.; Shakouri, M.; Akbari, A.; Aghdam, A.B. Effect of cold expansion and bolt clamping on fretting fatigue behavior of Al 2024-T3 in double shear lap joints. *Eng. Fail. Anal.* **2012**, *25*, 29–41. [CrossRef]
82. Sun, G.; Li, B.; Wu, J.; Ma, J. Experimental study on mechanical properties of conical head for bolted spherical joints. *Structures* **2024**, *63*, 106293. [CrossRef]
83. Qiu, B.; Lan, T.; Ji, X.; Lei, H.; Zhang, S. Optimizing fatigue life prediction of high strength bolts in bolted spherical joints. *J. Constr. Steel Res.* **2024**, *217*, 108635. [CrossRef]
84. Qiu, B.; Yang, X.; Zhou, Z.; Lei, H. Experimental study on fatigue performance of M30 high-strength bolts in bolted spherical joints of grid structures. *Eng. Struct.* **2020**, *205*, 110123. [CrossRef]
85. Zhou, Z.; Zhang, S.; Lei, H.; Qiu, B.; Zhang, L.; Wang, G. Experimental Study on Variable-Amplitude Fatigue Performance of M60 High-Strength Bolts of Grid Structures with Bolted Spherical Joints. *Materials* **2022**, *15*, 8939. [CrossRef] [PubMed]
86. Nourani, S.A.; Stilwell, G.; Pons, D.J. Shear testing of clinch joints at different temperatures: Explanation of the failure sequence. *J. Adv. Join. Process.* **2023**, *7*, 100140. [CrossRef]
87. Zhang, Y.; Xu, C.; Peng, R.; Lei, B.; Jiang, J. A comparative study of self pierceing riveting and mechanical clinch of DP590-Al5754 high-strength steel and aluminum alloys. *Eng. Fail. Anal.* **2024**, *166*, 108835. [CrossRef]
88. LeBozec, N.; Thierry, D. A new device for simultaneous corrosion fatigue testing of joined materials in accelerated corrosion tests. *Mater. Corros.* **2015**, *66*, 893–898. [CrossRef]
89. Harzheim, S.; Ewenz, L.; Zimmermann, M.; Wallmersperger, T. Corrosion phenomena and fatigue behavior of clinched joints: Numerical and experimental investigations. *J. Adv. Join. Process.* **2022**, *6*, 100130. [CrossRef]
90. Harzheim, S.; Chen, C.; Hofmann, M.; Wallmersperger, T. Coupled chemo-electro-mechanical model for galvanic corrosion in clinched components. *PAMM* **2024**, e202400028. [CrossRef]
91. Basak, A.; Matteazzi, P.; Vardavoulias, M.; Celis, J.-P. Corrosion–wear behaviour of thermal sprayed nanostructured FeCu/WC–Co coatings. *Wear* **2006**, *261*, 1042–1050. [CrossRef]

92. Basak, A.K.; Celis, J.-P.; Vardavoulias, M.; Matteazzi, P. Effect of nanostructuring and Al alloying on friction and wear behaviour of thermal sprayed WC–Co coatings. *Surf. Coat. Technol.* **2012**, *206*, 3508–3516. [CrossRef]
93. Childs, P.R.N. 16—Fastening and power screws. In *Mechanical Design Engineering Handbook*, 2nd ed.; Childs, P.R.N., Ed.; Butterworth-Heinemann: Oxford, UK, 2019; pp. 773–832. [CrossRef]
94. Nejad, R.M.; Berto, F.; Wheatley, G.; Tohidi, M.; Ma, W. On fatigue life prediction of Al-alloy 2024 plates in riveted joints. *Structures* **2021**, *33*, 1715–1720. [CrossRef]
95. Zhao, H.; Han, L.; Liu, Y.; Liu, X. Analysis of joint formation mechanisms for self-piercing riveting (SPR) process with varying joining parameters. *J. Manuf. Process.* **2022**, *73*, 668–685. [CrossRef]
96. Wu, G.; Li, D.; Lai, W.-J.; Shi, Y.; Kang, H.; Peng, Y.; Su, X. Fatigue behaviors and mechanism-based life evaluation on SPR-bonded aluminum joint. *Int. J. Fatigue* **2021**, *142*, 105948. [CrossRef]
97. Gay, A.; Lefebvre, F.; Bergamo, S.; Valiorgue, F.; Chalandon, P.; Michel, P.; Bertrand, P. Fatigue of Aluminum/Glass Fiber Reinforced Polymer Composite Assembly Joined by Self-piercing Riveting. *Procedia Eng.* **2015**, *133*, 501–507. [CrossRef]
98. Jiang, H.; Luo, T.; Li, G.; Zhang, X.; Cui, J. Fatigue life assessment of electromagnetic riveted carbon fiber reinforce plastic/aluminum alloy lap joints using Weibull distribution. *Int. J. Fatigue* **2017**, *105*, 180–189. [CrossRef]
99. Li, G.; Jiang, H.; Zhang, X.; Cui, J. Mechanical properties and fatigue behavior of electromagnetic riveted lap joints influenced by shear loading. *J. Manuf. Process.* **2017**, *26*, 226–239. [CrossRef]
100. Liao, Y.; Sun, H.; Wu, G.; Jin, C.; Cui, J.; Li, G.; Jiang, H. Effect of rivet arrangement on fatigue performance of electromagnetic riveted joint with Φ10 mm diameter rivet. *Int. J. Fatigue* **2023**, *176*, 107892. [CrossRef]
101. Kim, J.B.; Kim, H.K. Fatigue behaviour of clinched joints in a steel sheet. *Fatigue Fract. Eng. Mater. Struct.* **2015**, *38*, 661–672. [CrossRef]
102. Mori, K.; Abe, Y.; Kato, T. Mechanism of superiority of fatigue strength for aluminium alloy sheets joined by mechanical clinching and self-pierce riveting. *J. Mater. Process. Technol.* **2012**, *212*, 1900–1905. [CrossRef]

MDPI AG
Grosspeteranlage 5
4052 Basel
Switzerland
Tel.: +41 61 683 77 34

Metals Editorial Office
E-mail: metals@mdpi.com
www.mdpi.com/journal/metals

www.ingramcontent.com/pod-product-compliance
Lightning Source LLC
Chambersburg PA
CBHW040826250926
51850CB00007BA/1271
* 9 7 8 3 7 2 5 8 6 4 0 2 7 *